Vehicular Electric Power Systems

POWER ENGINEERING

Series Editor

H. Lee Willis
ABB Inc.
Raleigh, North Carolina

Advisory Editor

Muhammad H. Rashid
University of West Florida
Pensacola, Florida

1. Power Distribution Planning Reference Book, *H. Lee Willis*
2. Transmission Network Protection: Theory and Practice, *Y. G. Paithankar*
3. Electrical Insulation in Power Systems, *N. H. Malik, A. A. Al-Arainy, and M. I. Qureshi*
4. Electrical Power Equipment Maintenance and Testing, *Paul Gill*
5. Protective Relaying: Principles and Applications, Second Edition, *J. Lewis Blackburn*
6. Understanding Electric Utilities and De-Regulation, *Lorrin Philipson and H. Lee Willis*
7. Electrical Power Cable Engineering, *William A. Thue*
8. Electric Systems, Dynamics, and Stability with Artificial Intelligence Applications, *James A. Momoh and Mohamed E. El-Hawary*
9. Insulation Coordination for Power Systems, *Andrew R. Hileman*
10. Distributed Power Generation: Planning and Evaluation, *H. Lee Willis and Walter G. Scott*
11. Electric Power System Applications of Optimization, *James A. Momoh*
12. Aging Power Delivery Infrastructures, *H. Lee Willis, Gregory V. Welch, and Randall R. Schrieber*
13. Restructured Electrical Power Systems: Operation, Trading, and Volatility, *Mohammad Shahidehpour and Muwaffaq Alomoush*
14. Electric Power Distribution Reliability, *Richard E. Brown*
15. Computer-Aided Power System Analysis, *Ramasamy Natarajan*
16. Power System Analysis: Short-Circuit Load Flow and Harmonics, *J. C. Das*
17. Power Transformers: Principles and Applications, *John J. Winders, Jr.*
18. Spatial Electric Load Forecasting: Second Edition, Revised and Expanded, *H. Lee Willis*
19. Dielectrics in Electric Fields, *Gorur G. Raju*

20. Protection Devices and Systems for High-Voltage Applications, *Vladimir Gurevich*
21. Electrical Power Cable Engineering: Second Edition, Revised and Expanded, *William A. Thue*
22. Vehicular Electric Power Systems: Land, Sea, Air, and Space Vehicles, *Ali Emadi, Mehrdad Ehsani, and John M. Miller*

ADDITIONAL VOLUMES IN PREPARATION

Power Distribution Planning Reference Book: Second Edition, Revised and Expanded, *H. Lee Willis*

Power System State Estimation: Theory and Implementation, *Ali Abur and Antonio Exposito*

Vehicular Electric Power Systems
Land, Sea, Air, and Space Vehicles

Ali Emadi
Illinois Institute of Technology
Chicago, Illinois, U.S.A.

Mehrdad Ehsani
Texas A&M University
College Station, Texas, U.S.A.

John M. Miller
J-N-J Miller Design Services, P.L.C.
Cedar, Michigan, U.S.A.

MARCEL DEKKER, INC.   NEW YORK · BASEL

Although great care has been taken to provide accurate and current information, neither the author(s) nor the publisher, nor anyone else associated with this publication, shall be liable for any loss, damage, or liability directly or indirectly caused or alleged to be caused by this book. The material contained herein is not intended to provide specific advice or recommendations for any specific situation.

Trademark notice: Product or corporate names may be trademarks or registered trademarks and are used only for identification and explanation without intent to infringe.

Library of Congress Cataloging-in-Publication Data
A catalog record for this book is available from the Library of Congress.

ISBN: 0-8247-4751-8

This book is printed on acid-free paper.

Headquarters
Marcel Dekker, Inc., 270 Madison Avenue, New York, NY 10016, U.S.A.
tel: 212-696-9000; fax: 212-685-4540

Distribution and Customer Service
Marcel Dekker, Inc., Cimarron Road, Monticello, New York 12701, U.S.A.
tel: 800-228-1160; fax: 845-796-1772

Eastern Hemisphere Distribution
Marcel Dekker AG, Hutgasse 4, Postfach 812, CH-4001 Basel, Switzerland
tel: 41-61-260-6300; fax: 41-61-260-6333

World Wide Web
http://www.dekker.com

The publisher offers discounts on this book when ordered in bulk quantities. For more information, write to Special Sales/Professional Marketing at the headquarters address above.

Copyright © 2004 by Marcel Dekker, Inc. All Rights Reserved.

Neither this book nor any part may be reproduced or transmitted in any form or by any means, electronic or mechanical, including photocopying, microfilming, and recording, or by any information storage and retrieval system, without permission in writing from the publisher.

Current printing (last digit):

10 9 8 7 6 5 4 3 2 1

PRINTED IN THE UNITED STATES OF AMERICA

To the memory of my sister, Annahita
Ali Emadi

to my wife, Zohreh
Mehrdad Ehsani

for Doreen and for Mike
John M. Miller

Series Introduction

Power systems are the most fundamental aspect of electrical engineering, because such systems create and control the energy that enables—literally powers—all electric and electronic capabilities. Power engineering is by far the oldest and most traditional of the various areas within electrical engineering. While initially restricted to only stationary electric system applications (in the early 20th century), power systems—engineered combinations of generation, distribution, and control—gradually worked their way into all manner of marine, automotive, and aerospace applications. Today, no significant vehicle—whether robotic or manned, whether military or civilian, whether designed to move on, under, or above land or sea, or in outer space—is designed without its power system being a core part of its overall design.

There are significant differences between vehicular and stationary (utility, industrial, building) power systems engineering. But there are also great similarities. Ultimately, all share the same overall mission, are subject to the same physical principals and limitations, and, perhaps in different measure, use the same basic technical approaches and rules. Most important, however, there is tremendous transfer of technology between the two main branches of power systems engineering. Early marine and airborne power systems borrowed much from existing electric utility power system technology. Today, vehicular technologies such as fuel cells and advanced power electronics controls are making their way into electric utility and industrial power applications.

Vehicular Electric Power Systems: Land, Sea, Air, and Space Vehicles provides a very thorough and complete discussion of theory and application of power systems engineering to anything that moves. The book's greatest strength is its combination of a thorough exploration of the needs and limitations of each type of vehicular power application (e.g., automotive), with very detailed discussions of the specific technologies available for power systems, such as fuel cells and multiconverter systems, and their control.

Like all the books in Marcel Dekker's Power Engineering series, *Vehicular Electric Power Systems: Land, Sea, Air, and Space Vehicles* puts modern technology in the context of practical application; it is useful as a reference book as well as for self-study and advanced classroom use. The series includes books covering the entire field of power systems engineering, in all of its specialties and subgenres, all aimed at providing practicing electrical and design engineers with the knowledge and techniques they need to meet our society's energy and engineering challenges in the 21st century.

H. Lee Willis

Preface

Mechanical, electrical, hydraulic, and pneumatic systems are conventional power transfer systems in different land, sea, air, and space vehicles. In order to improve vehicle fuel economy, emissions, performance, and reliability, the more electric vehicle (MEV) concept emphasizes the utilization of electrical power systems instead of non-electrical power transfer systems. In addition, the need for improvement in comfort, convenience, entertainment, safety, communications, maintainability, supportability, survivability, and operating costs necessitates more electric vehicular systems. Therefore, electric power distribution systems with larger capacities and more complex configurations are required to facilitate increasing electrical demands in advanced vehicles.

In MEVs, solid-state switching power converters are extensively used for generating, distributing, and utilizing electrical energy throughout the system. Different converters such as DC/DC choppers, DC/AC inverters, AC/AC converters, and AC/DC rectifiers are used in source, load, and distribution subsystems to provide power at different voltage levels and in both DC and AC forms. Most of the loads are also in the form of power electronic converters and electric motor drives. Therefore, in these vehicles, different converters are integrated together to form complex and extensively interconnected multi-converter systems. The number of power electronic converters in these systems varies from a few converters in a conventional car to tens of converters in the advanced aircraft and spacecraft power systems, to hundreds of converters in the international space station. Recent advancements in the areas of power electronics, electric motor drives, fault tolerant electrical power distribution systems, control electronics, digital signal processors (DSPs), and microprocessors are already providing the impetus towards MEVs.

These unconventional power systems have unique system architectures, characteristics, dynamics, and stability problems that are not similar to those of conventional electrical power systems. The purpose of this book is to present a conceptual definition and a comprehensive description of these systems. In addition, an inclusive explanation of the conventional and advanced architectures, role of power electronics, and present trends is given. Furthermore, this book addresses the fundamental issues faced in these systems, both before and after their implementation.

This book consists of thirteen chapters. It starts with an introduction to electrical power systems, basics of electric circuits, and principles of control systems in Chapter 1. Chapters 2 and 3 are also introductory chapters about fundamentals of power electronics and electric machines, respectively. Conventional and advanced power electronic AC/DC, DC/DC, DC/AC, and AC/AC converters are presented in Chapter 2. Chapter 3 deals with the conventional DC, AC induction, and AC synchronous machines and their associated power electronic drivers. Advancements in the areas of power electronics and motor drives facilitate electrification of vehicular systems and enable the introduction of more electric vehicles with improved performance, efficiency, volume, and weight.

Chapter 4 presents a comprehensive description of automotive power systems including conventional automobiles and more electric cars. At present, most automobiles use a 14V DC electrical system. However, demands for higher fuel economy, performance, and reliability as well as reduced emissions push the automotive industry to seek electrification of ancillaries and engine augmentations. In advanced cars, throttle actuation, power steering, anti-lock braking, rear-wheel steering, air conditioning, ride-height adjustment, active suspension, and the electrically heated catalyst will all benefit from the electrical power system. Therefore, a higher system voltage, such as the proposed 42V PowerNet, is necessary to handle these newly introduced loads.

Chapter 5 deals with electric and hybrid electric vehicles. Principles of hybrid electric drivetrains, system configurations, electrical distribution system architectures, control strategies, hybridization effects, low-voltage traction systems, and design methodologies are presented. In addition, electrical systems of heavy duty vehicles and electric dragsters are explained. Modeling and simulation of automotive power systems are also described in Chapter 5.

Chapter 6 concentrates on air vehicles. Conventional aircraft power systems, electrical loads, power generation systems, AC and DC distribution systems, and the concept of more electric aircraft are presented in Chapter 6. In addition, space power systems including spacecraft and the international space station are described in Chapter 7. Chapter 7 also explains modeling, real-time state estimation, and stability assessment of aerospace power systems.

In Chapter 8, sea and undersea vehicles are comprehensively studied. Propulsion and non-propulsion electric loads, more electric ships, integrated power systems, and pulsed power technology, as well as advanced sea and

Preface

undersea vehicles are introduced. Chapter 9 concentrates on the applications of fuel cells in different land, sea, air, and space vehicles. It explains structures and operations of fuel cells as well as their utilization. In Chapter 10, modeling techniques for energy storage devices including batteries, fuel cells, photovoltaic cells, and ultracapacitors are presented in detail.

Advanced motor drives for vehicular applications are reviewed in Chapter 11. Brushless DC (BLDC) and switched reluctance motor (SRM) drives are comprehensively presented as advanced motor drive technologies for different vehicles. Furthermore, motoring and generating modes of operation as well as sensorless techniques are explained.

Chapter 12 introduces multi-converter vehicular dynamics. In Chapter 12, electrical loads of advanced vehicles are categorized to two groups. One group is constant voltage loads, which require constant voltage for their operation. The other group is constant power loads that sink constant power from the source bus, which is a destabilizing effect for the system and known as negative impedance instability. Effects of such loads on the dynamic behavior of different vehicles are comprehensively studied.

The purpose of Chapter 13 is to present an assessment of the effects of constant power loads in AC vehicular distribution systems. Furthermore, recommendations for the design of AC vehicular systems to avoid negative impedance instability are provided. Guidelines for designing proper distribution architectures are also established.

The material in this book is recommended for a graduate or senior-level undergraduate course. Depending on the background of the students in different disciplines such as electrical and mechanical engineering, course instructors have the flexibility to choose the material or skip the introductory sections/chapters from the book for their lectures. This text has been taught at Illinois Institute of Technology as a graduate level course titled Vehicular Power Systems. An earlier version of this text has been revised based on the comments and feedback received from the students in this course. We are grateful to the students for their help.

This book is also an in-depth source for engineers, researchers, and managers who are working in vehicular and related electrical, electronic, electromechanical, and electrochemical industries.

We would like to acknowledge gratefully the contributions of many graduate and undergraduate students at Illinois Institute of Technology in different sections/chapters of this book. They are Mr. Ranjit Jayabalan contributing in Chapter 1 and Section 4.5, Mr. Ritesh Oza contributing in Sections 2.1 and 2.2, Mr. Sheldon S. Williamson contributing in Sections 2.3 and 7.1-7.3, Mr. Basem Fahmy contributing in Chapter 3, Mr. Erwin Uy contributing in Section 4.6, Mr. Fernando Rodriguez contributing in Section 4.6, Mr. Arjun Shrinath contributing in Section 4.8, Mr. Srdjan M. Lukic contributing in Sections 5.5 and 5.10, Ms. Valliy Dawood contributing in Section 5.8, Mr. Rajat Bijur contributing in Section 5.9, Mr. Sachin A. Borse contributing in Chapter 8, and Mr. Yogesh P. Patel contributing in Section 11.2. In addition, Chapters 9 and 10 draw heavily from the graduate research work of Mr. Sheldon S. Williamson, which is gratefully acknowledged.

We would also like to acknowledge the efforts and assistance of the staff of Marcel Dekker, Inc.

Ali Emadi
Mehrdad Ehsani
John M. Miller

Contents

Preface .... *vii*

1 Introduction to Electrical Power Systems 1
   1.1   Fundamentals of Electric Circuits 1
   1.2   Control Systems 8
   1.3   Electrical Systems 11
   1.4   References 13

2 Fundamentals of Power Electronics 15
   2.1   AC/DC Rectifiers 16
   2.2   DC/DC Converters 25
   2.3   DC/AC Inverters 39
   2.4   Selected Readings 47

3 Electric Machines 49
   3.1   Electro-mechanical Power Transfer Systems 49
   3.2   Fundamentals of Electromagnetism 53
   3.3   DC Machines 55
   3.4   Induction Machines 59
   3.5   Synchronous Machines 63
   3.6   Selected Readings 65

4 Automotive Power Systems 67
   4.1   Conventional 14V Electrical System Architecture 70
   4.2   Advanced Electrical Loads 72
   4.3   Increasing the System Voltage to 42V 73

	4.4	Advanced Distribution Systems	77
	4.5	Starter, Alternator, and Integrated Starter/Alternator	78
	4.6	Automobile Steering Systems	120
	4.7	Semiconductors for Automotive Applications	132
	4.8	Automotive Communication Networks and Wireless Techniques	139
	4.9	References	183
5	Electric and Hybrid Electric Vehicles	189
	5.1	Principles of Hybrid Electric Drivetrains	190
	5.2	Architectures of Hybrid Electric Drivetrains	194
	5.3	Electrical Distribution System Architectures	195
	5.4	More Electric Hybrid Vehicles	197
	5.5	Hybrid Control Strategies	198
	5.6	Hybridization Effects	206
	5.7	42V System for Traction Applications	208
	5.8	Heavy Duty Vehicles	211
	5.9	Electric Dragsters	219
	5.10	Modeling and Simulation of Automotive Power Systems	224
	5.11	References	229
6	Aircraft Power Systems	231
	6.1	Conventional Electrical Systems	231
	6.2	Power Generation Systems	234
	6.3	Aircraft Electrical Distribution Systems	235
	6.4	Stability Analysis	237
	6.5	References	238
7	Space Power Systems	241
	7.1	Introduction	242
	7.2	International Space Station	242
	7.3	Spacecraft Power Systems	251
	7.4	Modeling and Analysis	257
	7.5	Real-Time State Estimation	272
	7.6	Stability Assessment	282
	7.7	References	290
8	Sea and Undersea Vehicles	295
	8.1	Power System Configurations in Sea and Undersea Vehicles	295
	8.2	Power Electronics Building Blocks (PEBBs)	298
	8.3	Controller Architecture for Power Electronic Circuits	300
	8.4	Power Management Center (PMC)	303
	8.5	Electrical Distribution System in Sea and Undersea Vehicles	304
	8.6	Advanced Electrical Loads in Sea and Undersea Vehicles	307

Contents

	8.7	Advanced Electric Drives in Sea and Undersea Vehicles	309
	8.8	References	330
9	Fuel Cell Based Vehicles		335
	9.1	Structures, Operations, and Properties of Fuel Cells	335
	9.2	Important Properties of Fuel Cells for Vehicles	348
	9.3	Light-Duty Vehicles	353
	9.4	Heavy-Duty Vehicles	356
	9.5	Current Status and Future Trends in Fuel Cell Vehicles	360
	9.6	Aerospace Applications	362
	9.7	Other Applications of Fuel Cells	364
	9.8	Conclusion	380
	9.9	References	381
10	Electrical Modeling Techniques for Energy Storage Devices		385
	10.1	Battery Modeling	386
	10.2	Modeling of Fuel Cells	388
	10.3	Modeling of Photovoltaic (PV) Cells	390
	10.4	Modeling of Ultracapacitors	393
	10.5	Conclusion	397
	10.6	References	397
11	Advanced Motor Drives for Vehicular Applications		399
	11.1	Brushless DC Motor Drives	399
	11.2	Switched Reluctance Motor Drives	413
	11.3	References	420
12	Multi-Converter Vehicular Dynamics and Control		425
	12.1	Multi-Converter Vehicular Power Electronic Systems	426
	12.2	Constant Power Loads and Their Characteristics	428
	12.3	Concept of Negative Impedance Instability	430
	12.4	Negative Impedance Instability in the Single PWM DC/DC Converters	435
	12.5	Stability of PWM DC/DC Converters Driving Several Loads	445
	12.6	Stability Condition in a DC Vehicular Distribution System	450
	12.7	Negative Impedance Stabilizing Control for PWM DC/DC Converters with Constant Power and Resistive Loads	453
	12.8	Conclusion	461
	12.9	References	462
13	Effects of Constant Power Loads in AC Vehicular Systems		465
	13.1	Vehicular AC Distribution Systems	465
	13.2	Modeling of AC Constant Power Loads	468
	13.3	Negative Impedance Instability Conditions	476

13.4 Hybrid (DC and AC) Vehicular Systems with Constant Power Loads.. 483
13.5 Conclusion 490
13.6 References.. 491

Index *493*

Vehicular Electric Power Systems

1

Introduction to Electrical Power Systems

1.1 Fundamentals of Electric Circuits

1.1.1 Ohm's Law

Ohm's law states that the potential difference across any element is directly proportional to the current carried by it, if the physical conditions (temperature, dimensions, etc.) remain unchanged.

$$V \alpha R \tag{1.1}$$

$$V = RI \tag{1.2}$$

R is a constant of proportionality and is called the resistance of an element. R may be represented in terms of element length (l), area of the cross-section (A), and resistivity of the element (ρ).

$$R = (\rho l) / A \tag{1.3}$$

Resistance of an element is its opposition to the flow of current. Not all materials follow Ohm's law like superconductors, which have zero resistance. Such materials are called non-ohmic materials, while all others that follow the Ohm's law are called ohmic materials.

1.1.2 Kirchhoff's Law

Any electric circuit is composed of elements such as resistors, capacitors, and inductors, which carry current and have a voltage that exists across them. As a result, these elements give rise to a relationship with current and voltage referred to as Kirchhoff's current and voltage laws.

Kirchhoff's current law states that the current at any node of an electric circuit is zero. Here, a node in an electric circuit is defined as the point of intersection of two or more electrical components. Kirchhoff's current law is shown in Figure 1.1.

$$\sum i = i_1 + i_2 + i_3 - i_4 = 0 \tag{1.4}$$

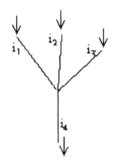

Figure 1.1 Representation of Kirchhoff's current law.

Kirchhoff's voltage law states that the sum of the products of the voltage and current across each element is equal to the sum of the potential sources in the closed circuit. Figure 1.2 shows the representation of Kirchhoff's voltage law.

$$\sum V = 0 \tag{1.5}$$

$$V = iR_1 + iR_2 \tag{1.6}$$

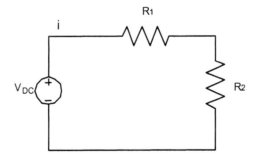

Figure 1.2 Representation of Kirchhoff's voltage law.

Introduction to Electrical Power Systems

1.1.3 Network Elements Voltage Current Relations

The common network elements in an electric circuit are resistors, inductors, and capacitors. The relation of these elements with the current they carry and the voltage that exists across their terminals are given below with equations.

Voltage current relation for resistor:

$$V = RI \qquad (1.7)$$

V = Voltage across the element

I = Current flowing in the element

Voltage current relation for inductor:

$$V = L\,(di/dt) \qquad (1.8)$$

L = Inductance of the element

Voltage current relation of capacitor:

$$I = C\,(dv/dt) \qquad (1.9)$$

C = Capacitance of the element

1.1.4 Transient Circuit Analysis for RC, RL and RLC Circuits

In general cases, the voltage and current measured in a circuit are during the steady state condition, that is, when the source provides a constant DC or AC signal. However, when a circuit is switched on or off, it tends to change from one steady state to another. This transition period between the two steady states is called transient period and its analysis is referred to as transient analysis. The transient behavior of a system primarily exists due to the presence of energy storage devices like inductances and capacitances that have a high inertia to a sudden change in current and voltage.

In the transient analysis of RL circuits, consider the switch initially open. At this point, the current and voltage in the inductor is zero. When the switch is closed, the source is across the resistor and inductor. There is an instantaneous change in the voltage across the inductor, but the current in it cannot change instantaneously. The current before and after the switching is almost the same. Once the switched is closed and the circuit is in steady state, the inductor behaves as a short circuit. The RL circuit and the change in voltage and current with time on switching are shown in the Figure 1.3.

In the transient analysis of RC circuits, when the switch is closed, the capacitor acts as an open circuit. The capacitor current becomes zero; but, the voltage is that of the source. The change in capacitor voltage is not instantaneous and the voltage before and after the switching is almost the same.

A first order differential equation of the circuit gives the solution. Figure 1.4 shows the RC circuit and its transient behavior.

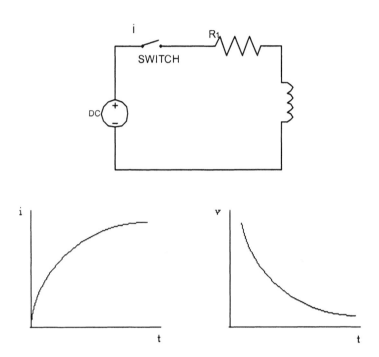

Figure 1.3 The RL circuit and the change in current and voltage during transient period.

The transient analysis of RLC circuits is similar to that for RL and RC circuits. In an RLC circuit, the components may be arranged in parallel or series combination. The only difference in this analysis is that a second order differential equation will be used to give the solution of the RLC circuit. An RLC circuit is shown in Figure 1.5.

1.1.5 Introduction to Laplace Transforms

The Laplace transform is a powerful tool in solving a wide range of initial value problems. It transforms ordinary and partial differential equations to simple algebraic problems where solution can be easily obtained. Applying inverse Laplace transform on the algebraic problem gives the solution to the ordinary and partial differential equations.

A function $f(t)$ has a Laplace transform $F(s)$ when defined over the interval of $0 < t < \infty$,

Introduction to Electrical Power Systems

$$L[f(t)] = F(s) = \int_0^\infty e^{-st} f(t) dt \tag{1.10}$$

where s is real and L is called the Laplace transform.

Some of the conditions for the existence of the Laplace transform are that it should be piecewise continuous on $0 < t < \infty$ and $f(t)$ should be of exponential order as t reaches infinity. Although these two conditions are sufficient, they do not necessitate the existence of $F(s)$.

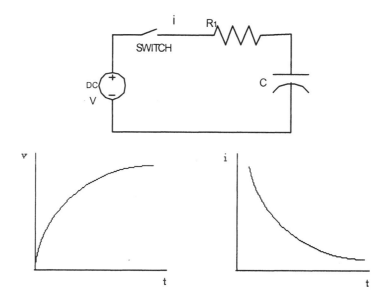

Figure 1.4 RC circuit and its transient behavior.

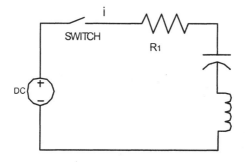

Figure 1.5 RLC circuit.

1.1.6 Sinusoidal Excitation and Phasors

A sinusoidal waveform is an alternating current (AC) [as opposed to direct current (DC)] that flows first in one direction and then in the opposite direction, as shown in Figure 1.6. The ampere current is a function of time and is not constant like in DC. Such a waveform may be represented as

$V = V_m \ Sin \ (\omega t)$ and $I = I_m \ Sin \ (\omega t)$ (1.11)

where V_m and I_m are peak values. A system excited by such a source will give a linear or non-linear output depending on whether the system is a linear or non-linear.

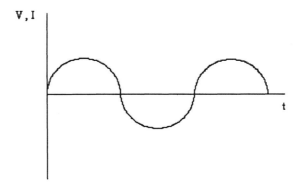

Figure 1.6 A sinusoidal signal.

Phasors are the complex numbers that we multiply with e^{jwt} in the expression for $i(t)$ and $v(t)$.

V is a voltage phasor,

$V(t) = V \ e^{jwt} = |V| \ e^{j} \ e^{jwt} = |V| \ e^{(jwt + j)}$ (1.12)

Phasor laws:

Ohms law, $V=RI$

Inductance, $V = jwLI$

Capacitor, $V = [1/(jwC)]I$

Impedance, $(V/I) = Z$

Impedance in a generalized circuit plays the role of resistance in a resistive circuit. It is a complex number (not a phasor) and has a real (resistive) component and an imaginary (reactive) component. An important note that needs to be made when dealing with phasors is that they are useful for finding the forced response of a system.

Introduction to Electrical Power Systems

1.1.7 Fourier Series

Fourier series are expansions of periodic functions $f(x)$ in terms of an infinite sum of sine and cosine functions such as

$$F(x) = a_n Cos(nx) + b_n Sin(nx) \tag{1.13}$$

The values of the coefficients a_n and b_n are determined from the orthogonality of the sine and cosine functions. Fourier series are computed using the following integrals.

$$\int_{-\pi}^{\pi} Sin(mx) Sin(nx) dx = \pi \delta_{mn}, \quad m, n \neq 0 \tag{1.14}$$

$$\int_{-\pi}^{\pi} Cos(mx) Cos(nx) dx = \pi \delta_{mn}, \quad m, n \neq 0 \tag{1.15}$$

$$\int_{-\pi}^{\pi} Sin(mx) dx = 0 \tag{1.16}$$

$$\int_{-\pi}^{\pi} Cos(mx) dx = 0 \tag{1.17}$$

δmn is Kronekar delta, which is a discrete version of delta function defined by $\delta ij = 0$ for $i \neq j$ and 1 for $i = j$.

1.1.8 Digital Systems

Primarily, a system may be classified as an analog or digital system based on the nature of the signal used by them. An analog system uses signals (voltages and currents) which are continuous and may be sinusoidal or even constant DC. Digital systems, on the other hand, have a discontinuous signal, whose waveform is composed of pulse with a non-zero voltage level and a zero voltage level.

A digital system may use a unipolar signal having two voltage levels – zero and a positive or negative voltage level – or may be a bipolar signal, consisting of a zero, positive, and negative voltage level. Digital systems are generally related to low signal low power applications such as calculators; however, they can be used in high power applications as well, such as in power converter circuits.

Digital systems are primarily composed of discrete elements such as transistors, resistors, capacitors, and logic gates or may be integrated together to form integrated circuits (IC). Microprocessors and memory chips are examples of digital circuits packaged together to form a single digital component that can be used in larger digital circuits. Digital components such as analog to digital and digital to analog converters help in interfacing analog and digital circuits to provide a more comprehensive system design.

1.2 Control Systems

1.2.1 Signal Flow Graphs and Block Diagrams

Signal flow graphs and block diagrams serve as an invaluable means of system representation in control system analysis. They help in defining the basic control concept and simplifying the representation of complex system. Both can represent almost all possible systems. The signal flow graphs and block diagrams for a simple control system are shown in Figure 1.7.

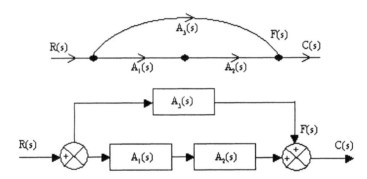

Figure 1.7 Signal flow graph and block diagram representation of a simple control system.

The input/output relationship is given as

$A(s) = C(s)/R(s)$ (1.18)

where

$A(s)$ = Transfer function of the system

$R(s)$ = Input or reference variable

$C(s)$ = Output or controlled variable

$F(s)$ = Error variable

$F(s)$ may be a feedback or feedforward variable depending on whether the connector is negative or positive, respectively. These variables may represent any physical parameter such as velocity and temperature. The upper parallel

Introduction to Electrical Power Systems

path in Figure 1.7 is the forward loop. In the case of the signal flow graph, the sign of the variable in the parallel branch indicates whether it is a feedback loop or not.

As stated above, the signal flow graph is a simplified graphical representation of linear systems only that makes use of linear algebraic equations to establish an input/output relationship. Additionally, the representation of signal flow graph is more mathematically constrained compared to a block diagram.

1.2.2 Feedback Control Techniques

Feedback control techniques refer to a specific concept or an idea that is involved with the control of a physical system. The control here is taken in the context of the control of an entire system as an entity and not the control of a segment of a system. The concept of feedback control system and techniques have largely evolved from linear system theory, which forms the foundation for much of the advanced technologies that exists today. A generalized representation of feedback control system is given in Figure 1.8.

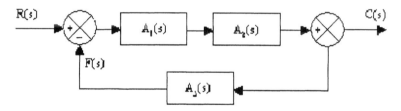

Figure 1.8 A generalized form of feedback control system.

There are numerous feedback control techniques and they are classified based on the approach of the design. However, the common ones are those like linear and non-linear control. Almost all systems are non-linear in nature; but, linear feedback control techniques are idealized models for test purposes. They function as linear systems as long as the applied signals are within the linear operating range. Non-linear control techniques are used sometimes to make the control more effective and precise and to achieve minimum time to carry out a desired control.

In time invariant systems, the control parameters are stationary with time; but, in time variant systems, the control parameters tend to vary. A linear time variant system design and analysis is however more complex to solve compared to a linear time invariant system. In the case of continuous data control system, the control signal is a function of time at various parts of the system. If the signal has a carrier, it is an AC system; otherwise, it is a DC system. However, a DC system may have a few signals that are of AC nature.

A sampled data control system involves the use of pulse signals at various parts of the system, while a digital control system makes use of a binary array of numbers for the control. Both control techniques are collectively referred to as discrete control systems. The inherent benefit of such a control is component sharing for multiple functions, downsizing control layout, and an increased degree of flexibility.

1.2.3 Stability and Routh-Hurwitz Criterion

Stability of a system refers to its ability to come back to its stable operating condition, which is also often referred to as a steady state in which the input and output variables or functions are related through numerical constants. Mathematically, the system may be stated as stable if it does not have any positive real poles, which, in other words, sharply defines the system as stable and unstable otherwise. In practical systems, the possibility to reach instability exists, but many times due to system component saturation, the system remains in stable state. Systems with negative feedbacks ensure that the system stability is maintained and are used when system components natural limitation are insufficient in maintaining system stability.

The Routh-Hurwitz criterion is one of the methods of determining system stability without solving for the roots of the characteristics equation of a system. Instead, it involves computation of a triangular array that is a function of the coefficients of the characteristic equation. The necessary and sufficient condition for stability is that the element in the first column be positive. If these are all positive, then all the zeros are on the left hand half of the s-plane. However, if all the roots are not positive, then the number of zeros lying on the right hand half plane equals the number of sign changes in the first column.

1.2.4 Time Domain and Frequency Domain Analysis

In most control systems, time is taken as an independent variable with respect to which the output or the system state is referred to. In time domain analysis, a sample or reference input signal is applied to the system and the output obtained is used for evaluating the system. The system behavior with time or time response $C(t)$ has two segments – transient response $C_t(t)$ and steady state response $C_s(t)$.

$$C(t) = C_t(t) + C_s(t) \qquad (1.19)$$

In control systems, the transient response is that part of the response that goes to zero at infinity.

$$Limit_{t \to \infty} C_t(t) = 0 \qquad (1.20)$$

The steady state response, on the other hand, is that portion of the response at which the system response becomes constant at infinity or changes uniformly

Introduction to Electrical Power Systems

with time. For designing a new control system, transient and steady state responses are given as the system specifications.

Although time domain is more realistic for control system analysis, the frequency domain analysis is preferred for certain systems like those used in communications. This is primarily due to the fact that high order systems are difficult to handle. There is no specific approach to design systems with specifications like rise time and delay time in the time domain analysis approach. However, the frequency domain analysis has graphical tools as a simplified approach for solving linear control systems. Also, the relationship that exists between time domain and frequency domain makes it possible to interpret the time domain system parameters from the frequency domain parameters. The frequency domain analysis takes the approach of using the system transfer function.

1.2.5 State Space Description

The state space representation provides a holistic and complete representation of a system. This representation can be used to describe a large family of systems like single input single output, multi-input single output, multi-input multi-output, time variant and time invariant systems. The state space formulation is given as

$$v(n+1) = Av(n) + Bu(n) \tag{1.21}$$

$$y(n) = Cv(n) + Du(n) \tag{1.22}$$

This formulation describes a discrete time system, but can be easily extended to describe continuous time systems. Here, $v(n+1)$ is a vector of states at strategically placed nodes in the system at time $n+1$. The matrix A here is termed a state transition matrix and models the dynamic behavior of the system. The state space vector is especially powerful for multi-input multi-output linear systems and also for time varying systems.

1.3 Electrical Systems

1.3.1 Power Network Models

A power network model is a simplified representation of a power network (single or three phase) by a single line diagram. It involves the use of standard symbols to denote different components of the system. A simple power system network comprises an alternator, transmission line, and load unit. The power network diagram is shown in Figure 1.9.

1.3.2 Per-Unit Quantity

In a power network, all the components like alternators, transformers, and loads may work at different voltages, current, and power levels. It will be convenient for the analysis of such power systems if the voltage, current, power, and impedances are expressed with reference to a common base value. This base value is any arbitrary value chosen for the simplification of analysis only and does not have any physical meaning. All the voltages, currents, powers, and impedances of the components are then expressed as a percent or per unit of the base value.

The base values may thus be defined as the ratio of the actual value to the base value expressed in decimals or in percent.

Per unit value = Actual Value/Base Value (1.23)

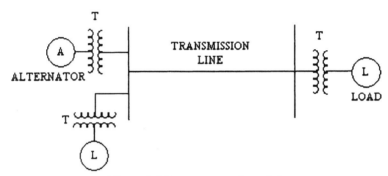

Figure 1.9 Power network model.

1.3.3 Gauss-Seidel Load Flow

Load flow studies involve the solution of an electric power system under steady state condition. On the basis of certain inequality constraints imposed on the node voltages and reactive power of the generators, the solution is obtained. The load flow studies relieve information about magnitude and phase angle of the voltages at each bus and the real and reactive power flowing in each segment of the system. It also gives the initial system conditions when transient behaviors are to be studied. The load flow analysis is essential from the point of operating the power network in the most optimal condition, to expand an existing network system, and also while designing new power network.

The Gauss-Seidel method is an iterative procedure for solving a set of non-linear load flow equations. The Gauss-Seidel load flow equation is

$V_p = [1/Y_{pp}] * [(P_p - jQ_p)/V_p - \sum Y_{pq} V_p]$ (1.24)

where

$P = 1, 2, 3, \ldots$

V_1, V_2, V_3 – Bus voltages

V^0_1, V^0_2, V^0_3 – Initial bus voltages

Introduction to Electrical Power Systems

P and Q are real and reactive power, respectively. The iterative process of computing the bus voltages is repeated until the bus voltage converges to the desired accuracy. The choice of initial values significantly affects rate of convergence and the knowledge of selecting these initial values is obtained by experience.

1.3.4 DC Power Systems

The most commonly used is the AC power system, where power is generated, transmitted, and used in the AC form. In the case of DC power systems, power could be generated as DC and then transmitted in the same DC form. Another option would be to generate power conventionally as AC and rectify it to convert to DC and then have it transmitted in the same DC form. The latter is preferred as rectification is a simple process involving conversion of AC to DC (rectifier) compared to conversion of DC to DC (chopper) when power is generated in the DC form. Also, a rectifier is more economical and has higher efficiency compared to choppers. In vehicular electrical power systems, both in land and space vehicles, power is essentially generated in the AC form and then rectified to DC to be used for storage in the battery or for use by the loads. At the transmitted end of the DC power system if required, DC is converted back to AC using inverters for using AC loads. Figure 1.10 depicts a typical block diagram layout of a DC power system.

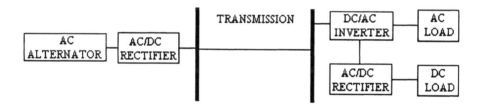

Figure 1.10 Block diagram of a DC power system.

1.4 References

[1] C. R. Paul, S. A. Nasar, and C. E. Unnewehr, *Introduction to Electrical Engineering*, McGraw Hill Co., New York, US, 1986.
[2] A. N. Kani, *Power System Analysis*, 1st Edition, RBA Publication, India, 1999.
[3] B. C. Kuo, *Automatic Control System*, 5th Edition, Prentice-Hall Inc., New Jersey, US, 1987.
[4] G. F. Franklin, J. D. Powell, and A. E. Naeini, *Feedback Control of Dynamic System*, 4th Edition, Prentice-Hall Inc., New Jersey, US, 2002.

[5] H. F. Davis, *Fourier Series and Orthogonal Functions*, Dover, New York, US, 1963.
[6] T. W. Korner, *Fourier Analysis*, Cambridge University Press, Cambridge, England, 1988.
[7] S. J. Chapman, *Electrical Machines and Power System Fundamentals*, 1st Edition, McGraw Hill, New York, US, 2002.

2
Fundamentals of Power Electronics

Power electronics is the technology for conversion of one type or level of an electric waveform to another. Power electronic converters are increasingly utilized in different vehicular applications. These converters include AC/DC rectifiers, DC/DC choppers, DC/AC inverters, and AC/AC voltage controllers. The impetus towards this expansion of power electronics has been provided by recent advancements in the areas of semiconductor switching devices, control electronics, and advanced microcontrollers and digital signal processors (DSPs). In fact, these advancements enable the introduction of power electronic converters with reduced cost, highest performance, maximum efficiency, and minimum volume and weight.

Power diodes, thyristors, transistors, MOSFETS, and isolated gate bipolar transistors (IGBTs) are the main power electronic switches. Power diodes are the simplest, uncontrollable power electronic switches. Power diodes are forward biased (ON) when their current is positive and reverse biased (OFF) when their voltage is negative. Thyristors are controllable three-terminal devices. If a current pulse applies to its gate, a thyristor can be turned on and conduct current from its anode to its cathode provided there is a positive anode-to-cathode voltage. However, in order to turn a thyristor on, gate current must be above a minimum value called I_{GT}.

Power transistors have the characteristics of conventional transistors. However, they have the capability of conducting higher collector current. They also have higher breakdown voltage (V_{CEO}). Power transistors are designed for high current, high voltage, and high power applications. They are usually operated either in the fully on or fully off state.

Power MOSFETs are voltage-controlled devices. They are usually N-channel and of the enhancement type. Most power MOSFETs are off when

$V_{GS} < 2v$ and are on when $V_{GS} > 4v$. When a power MOSFET is on, there is a small resistance, i.e., less than $1\,\Omega$, between drain and source and when it is off, there is a large resistance (almost open circuit) between drain and source.

IGBTs are equivalent to power transistors whose bases are driven by MOSFETs. Similar to a MOSFET, an IGBT has a high impedance gate, which requires only a small amount of energy to switch the device. Like a power transistor, an IGBT has a small on-state voltage.

2.1 AC/DC Rectifiers

An AC/DC rectifier is a power electronic circuit, which converts its input AC voltage into a DC output voltage. In this section, single-phase types of AC/DC converters and their operating modes in different loading conditions are presented.

2.1.1 Single-Phase, Half-Wave, Uncontrolled Rectifiers

The diagram shown in Figure 2.1 is the power stage of a single-phase, half-wave, uncontrolled rectifier. The input to the converter is single-phase supply and the switch used in this converter is diode, which is uncontrollable. If we look at the output voltage, the waveform is half of the input voltage. Thus, this converter is known as a single-phase, half-wave, uncontrolled rectifier.

The operation of this converter is very simple. During the positive half cycle of the input supply voltage, positive Vs will appear across the diode, which is forward biased (Figure 2.2 (a)). Thus, the output voltage in a positive half cycle will be the input or supply voltage Vs. The current will be Vs/R. When supply voltage enters into the negative half cycle, the diode is reverse biased. Thus, no voltage appears at the output and there is no current.

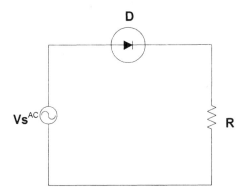

Figure 2.1 Single-phase, half-wave, uncontrolled rectifier with resistive load.

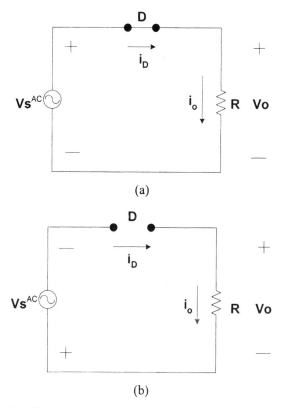

Figure 2.2 Circuit configuration: (a) diode is on (b) diode is off.

Power stage is the same for an inductive load. This converter includes only one diode. Again, during the positive half cycle of the input voltage, diode is forward biased and Vs appears across the resistor and inductor. When supply voltage goes in negative half cycle, due to the stored energy into the inductor, the output current will not be zero as soon as the input voltage becomes zero and then negative. Therefore, the diode will still conduct. The duration of this period is decided by the size of the inductor. Once the energy stored in the inductor is dissipated completely into the resistor, the output current will go to zero and diode will be reverse biased. Output voltage will be zero at this instant.

For loads with an internal battery or back-emf in case of DC motors, there is one source on each side of the diode. If amplitude of the battery is greater than the input or source voltage, the diode will be reverse biased and it will not conduct. Thus, output voltage of the rectifier will be equal to that of battery. As soon as input (AC source) voltage gets higher than the battery voltage, the diode will be forward biased and it will start conducting. Output voltage will be source voltage following input voltage until the current through the inductor dies to zero. Once energy stored in the inductor is completely depleted, the current through it will be zero. Output voltage will be equal to the battery voltage.

2.1.2 Single-Phase, Half-Wave, Controlled Rectifiers

As is shown in Figure 2.3, these rectifiers have a controlled switch (fully controlled or semi-controlled) instead of a simple diode (which is an uncontrolled switch). Input to the rectifier is single-phase supply and the output waveform is half of the input voltage; thus, these are single-phase, half-wave, controlled rectifiers. Similar to the uncontrolled rectifiers, operation of controlled rectifiers is explained below in different loading conditions and different operating modes.

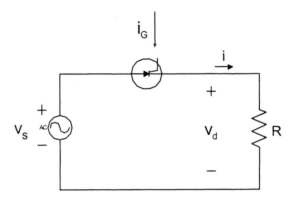

Figure 2.3 Single-phase, half-wave, controlled rectifier with resistive load.

Here the load is purely resistive and the switch is a silicon controlled rectifier (SCR) or thyristor. A thyristor conducts only when its gate is given signal and it is forward biased. Once the gate signal is received, it starts conducting. It stops conducting only if either current through it dies to zero or it gets reverse biased. Figure 2.4 shows the power stage and operating mode circuit diagrams of this rectifier with a purely resistive load. In the first half cycle of the input voltage, even though the thyristor is forward biased, it will not conduct unless gate signal is applied. Thus, at this time output current is zero. Output voltage is also zero. If the gate signal is applied at angle α, thyristor starts conducting. Now, the output voltage is equal to the supply voltage. The output current is given by output voltage divided by the load resistance. This will continue until the half period. Then, supply voltage will be negative and, thus, the thyristor will be reverse biased. As mentioned earlier, as soon as thyristor gets reverse biased, it will turn off and there will not be any output current and, thus, no output voltage. Thyristor will conduct again only when next gate signal is applied and the sequence will continue. Figure 2.5 shows the input and output waveforms.

Fundamentals of Power Electronics

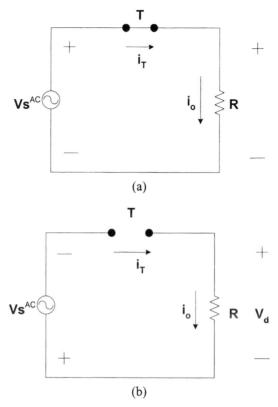

Figure 2.4 Circuit configuration: (a) thyristor is on (b) thyristor is off.

For an inductive load, during the first half cycle of the input voltage, the thyristor will be forward biased and ready to conduct. It will start conducting as soon as the gate signal is applied. Once the thyristor is triggered, input voltage will appear across the load. Now, current starts flowing through switch and load. The inductor will store energy during this time. At the end of the half period, current through the inductor is not zero and, thus, it has to pass through the switch in the circuit. Until current through the inductor becomes zero, the source voltage will continue appearing at the output of the rectifier. When the current reaches zero, the thyristor will be reverse biased. Output voltage will remain zero until SCR is fired again.

For a load with an internal battery, when thyristor is not triggered, there is no output current and the voltage is equal to the battery voltage. Here one important point is that if the thyristor is triggered before input voltage becomes greater than the battery voltage, it will not turn on because it is reverse biased. Once input or source voltage is greater than the battery voltage, the switch is ready to turn on. When the thyristor is triggered, current starts flowing through the load. Energy will be stored in the inductor. Once load current becomes zero, the switch will turn off and output voltage will no more be the source voltage. But, as soon as current becomes zero, voltage will have a jump from source voltage to the battery voltage. Output voltage will stay at this level until the switch is triggered again. Once the switch is triggered, the whole sequence will be repeated.

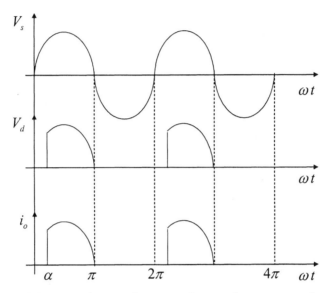

Figure 2.5 Input voltage and output voltage and current waveforms.

2.1.3 Single-Phase, Full-Wave, Uncontrolled Rectifiers

These rectifiers are supplied from single-phase AC source. Output voltage waveform has both halves of the source voltage. Diodes, which are uncontrollable power switches, are used. Figure 2.6 depicts a single-phase, full-wave, uncontrolled rectifier. Looking at the power stage diagram of these rectifiers, they resemble bridge topology and, thus, they are also called diode bridge rectifiers.

Circuit state diagrams, in different switching conditions, are shown in Figure 2.7. In the first half cycle of source voltage, diodes D1 and D2 are forward biased. Thus, source voltage will appear at the load and current starts flowing through the resistive load. After the half period, voltage goes in negative region; therefore, diodes D3 and D4 are forward biased. Due to this, negative voltage will appear across diodes D1 and D2 and they will be reverse biased. Current through output will be still there in the same direction and negative of the source voltage will appear at the output. Output voltage and output current will always be positive and in the same direction.

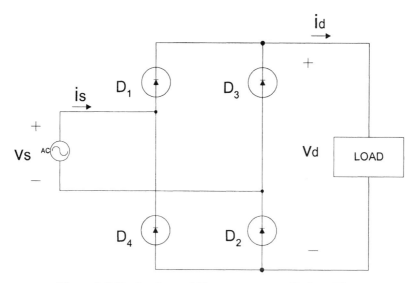

Figure 2.6 Single-phase, full-wave, uncontrolled rectifier.

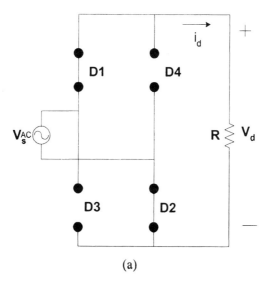

(a)

Figure 2.7 Circuit configuration: (a) D1, D2 are conducting (b) D3, D4 are conducting.

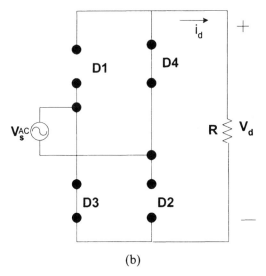

(b)

Figure 2.7 (Continued)

2.1.4 Single-Phase, Full-Wave, Controlled Rectifiers

Figure 2.8 depicts a single-phase, full-wave, controlled rectifier with resistive load. All the switches are thyristors, which are controlled devices.

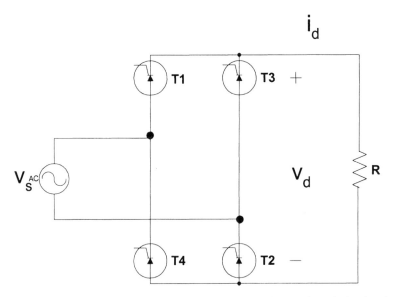

Figure 2.8 Single-phase, full-wave, controlled rectifier with resistive load.

Fundamentals of Power Electronics

Below is the operating stage diagram when the load is resistive. We can see that, at any time, two switches are closed. But, a point to be kept in mind is that at no instant are two switches from one leg closed. This is simply because if it happens the source gets short-circuited. This phenomenon is also known as shoot-through fault.

Thyristors will only conduct when they are forward biased and a gate signal is applied to its gate terminal. Therefore, in a positive half cycle, even though thyristors T1 and T2 are forward biased, they will not conduct unless they are fired. As soon as a gate signal is applied, T1 and T2 start conducting. At the same time, T3 and T4 are reverse biased and they will not conduct. In the operating stage diagram, T1 and T2 are shown as closed switches and T3 and T4 are open switches. Current starts flowing through the load. In other words, load is being supplied and source voltage will appear across it. The load current will be the source voltage divided by the load resistance. When source voltage becomes zero and, thus, load current also reaches zero, both thyristors T1 and T2 are reverse biased and stop conducting. At the same time, thyristors T3 and T4 are forward biased and ready to conduct, but not conducting. When gate signal is applied to them, they start conducting and load current starts flowing through them. Or in other words load is being supplied and a negative source voltage will appear across the load.

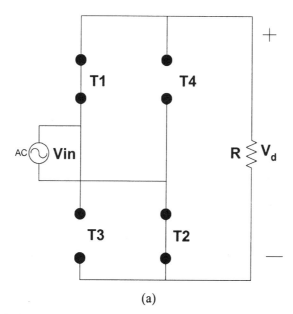

(a)

Figure 2.9 Circuit configuration: (a) T1, T2 are conducting (b) T3, T4 are conducting.

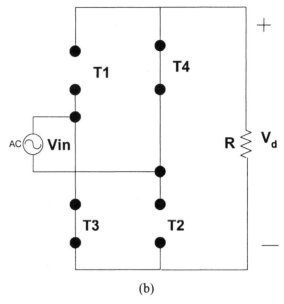

(b)

Figure 2.9 (Continued)

Figure 2.10 depicts input and output waveforms for a single-phase, full-wave, controlled rectifier when operated with purely resistive load. Here, α is the firing angle. This is given in degrees. The firing angle can be defined as the angle or instant at which the gate signal is applied to the thyristor. As the load is purely resistive and there is not any internal source, the operation of this converter is solely defined by the source voltage and firing angle. As the source voltage always encounters zero magnitude twice in each complete cycle, this configuration always operates in a discontinuous mode.

For an inductive load, we have one energy storage device. Therefore, circuit operation will also depend on the state of the inductor. When current though this converter is zero during the converter operation, the converter is in discontinuous conduction mode (DCM) of operation. If during the operation of the converter, inductor current never reaches zero, converter is in continuous conduction mode (CCM) of operation.

In CCM operation, during the positive half cycle of the source voltage, thyristors T1 and T2 are forward biased and ready to conduct. As soon as the gate signal is applied, they start conducting and source voltage appears across the load. When the source voltage becomes zero, thyristors are about to be reversed biased, but the load is inductive. Current through the inductor cannot change instantaneously and, thus, current is still positive, not zero. Therefore, after π, unless the load current becomes zero, source voltage will appear across the load. Gate signals are applied to thyristors T3 and T4 at $\alpha + \pi$. As soon as

Thyristors T3 and T4 are fired, they start conducting and T1 and T2 are forced to be off. Now, negative of the source voltage will appear across the load.

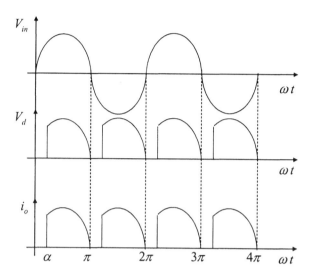

Figure 2.10 Input and output waveforms.

In DCM operation, current through the inductor reaches zero after π and before $\alpha + \pi$. When inductor current reaches zero, output voltage will be zero.

Output voltage for this rectifier in continuous conduction mode of operation is as follows:

$$v_s = \sqrt{2}V_s Sin(\omega t)$$

$$V_d = \frac{1}{\pi}\int_{\alpha}^{\pi+\alpha}\sqrt{2}V_s Sin(\omega t)d(\omega t) = \frac{2\sqrt{2}}{\pi}V_s Cos(\alpha) \tag{2.1}$$

If the firing angle is greater than 90 degrees, the output voltage is negative. Current is still in the same direction. Thus, this rectifier is a two-quadrant AC/DC converter. In fact, the full bridge controlled converter has the ability to operate as an inverter.

2.2 DC/DC Converters

Power electronic converters which change the level of DC source to a different level of DC, keeping regulation in consideration, are known as DC/DC converters. They are also popularly known as choppers. These converters can be regarded as DC transformers. But they are much more efficient with less volume, cost, and size.

2.2.1 Buck Converters

A DC/DC Buck converter steps down the input voltage source. Figure 2.11 depicts the power circuit of this converter. As discussed for rectifiers, here also DC/DC converter operation is discussed in both continuous and discontinuous conduction modes of operation.

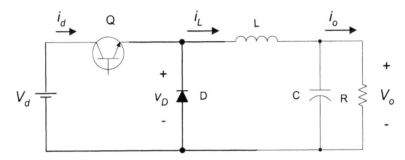

Figure 2.11 DC/DC Buck converter.

From the diagram shown above, it can be seen that there are two variables which should be monitored to have a complete idea about the performance of this converter. They are inductor current and capacitor voltage (output voltage). Opening and closing of the semiconductor switch changes the stage of the power circuit. Therefore, circuit performance is discussed in relation to the semiconductor switch. The switch can be a transistor, MOSFET, IGBT, or any other full-controlled power electronic switch.

In CCM operation, when switch is on ($0 < t < DT$), the circuit diagram for the converter is as shown in Figure 2.12. As soon as the switch is turned on, the current starts flowing through it. As there is an inductor in the path, the current cannot have a step change, but it starts increasing linearly. It is also seen that negative of input voltage appears across the diode and, thus, it will get reverse biased. Looking to the flow of the current, one can say that during this period, output capacitor is being charged by source and load is supplied from the source. The other variable, the inductor current, also increases linearly and energy is stored into it. In this interval, diode current is zero, as it is reverse biased. As there are two voltages (input and output) on either end of the inductor, the difference of them will appear across it. If we look into the state equations, the following observations can be made.

$0 < t < DT$

$$V_d = v_L + V_o$$
$$\frac{di_L}{dt} = \frac{V_d - V_o}{L}$$

(2.2a)

Fundamentals of Power Electronics

$$i_L(t) = \frac{V_d - V_o}{L} t + I_{L,\min}$$

$$i_L(t = DT) = I_{L,\max}$$ (2.2b)

$$\Delta I_L = I_{L,\max} - I_{L,\min} = \frac{V_d - V_o}{L} DT$$

In the above equations, V_d is source voltage, V_o is output voltage, i_L is inductor current, and D is the duty.

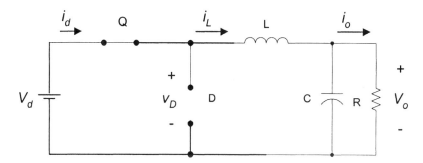

Figure 2.12 DC/DC Buck converter when switch is on.

When switch is off ($DT < t < T$), the circuit diagram for the converter is as shown in Figure 2.13. As can be seen from this circuit diagram, the path to the load current is provided by the diode. The diode is forward biased. The energy stored in the inductor is released and will supply the load. The job of the capacitor in this interval is to keep the output voltage within the ripple limit set at the time of design. Thus, the amount of load not supplied from the inductor is maintained by the output capacitor. But, as there is not any source and energy stored in the inductor is depleting, the current through the inductor decreases linearly in this duration. The capacitor is discharged in this interval. The only voltage now appearing across the inductor is a negative output voltage. Below are the mathematical equations for this interval.

$$DT < t < T$$

$$v_L = -V_o$$

$$\frac{di_L}{dt} = \frac{-V_o}{L}$$ (2.3a)

$$i_L(t) = \frac{-V_o}{L}(t - DT) + I_{L,\max}$$

$$i_L(t=T) = I_{L,\min}.$$

$$\Delta I_L = I_{L,\max.} - I_{L,\min.} = \frac{V_o}{L}(1-D)T \qquad (2.3b)$$

From (2.2) and (2.3), we have

$$\frac{V_d - V_o}{L} DT = \frac{V_o}{L}(1-D)T \qquad (2.4)$$

$$V_o = DV_d, \quad D = \frac{t_{on}}{T} \qquad (2.5)$$

Duty ratio D varies between 0 and 1. Thus, one can easily say that the output voltage will always be less than the input voltage. Figure 2.14 depicts the inductor voltage and current waveforms in CCM operation. We can see that output voltage is not plotted below. The reason for that is we assume here that output voltage is essentially kept constant by good design.

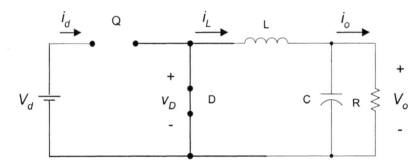

Figure 2.13 DC/DC Buck converter when switch is off.

Linear increment and decrement in the inductor can be observed from the waveform above, which justifies our description. It is also seen that the inductor current does not reach zero and, thus, the converter operates in CCM operation. One justifiable reason for the CCM operation is the large value of the output inductor. Obviously, load will always play its role in deciding the mode of operation. Therefore, one can say that if operating load range, input voltage, required output voltage, and switching frequency are given, it is possible to design the value of both duty ratio and inductor, to operate the converter in CCM operation. It is also possible to design the output capacitor, if it is given how much ripple voltage in output voltage is allowed.

The critically discontinuous conduction mode (CDCM) of operation is the boundary between continuous and discontinuous conduction modes of

Fundamentals of Power Electronics

operation. Figure 2.15 shows the inductor voltage and current waveforms for CDCM operation.

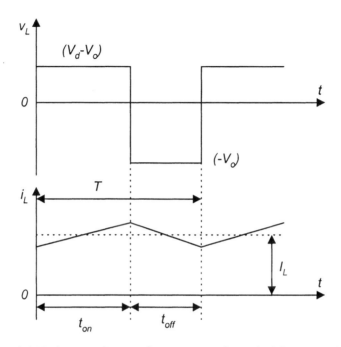

Figure 2.14 Inductor voltage and current waveforms in CCM operation.

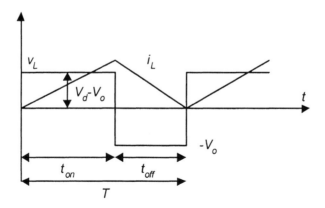

Figure 2.15 DC/DC Buck converter waveforms in CDCM operation.

In CDCM operation, the inductor current reaches zero, but as soon as it reaches zero, switch is immediately turned on and current again becomes non-zero. Thus, it is said that the converter is operating in boundary condition.

Figure 2.16 depicts the DC/DC Buck converter waveforms in DCM operation. It is clear that the duration for which the inductor current stays at zero is comparatively much larger than the CDCM. In fact, in CDCM, the inductor current just hits the zero line and increases. But, in this case, the current stays at zero. Here also minimum instantaneous inductor current is zero: it reaches zero before T, while in the case of CDCM, it reaches zero exactly at T.

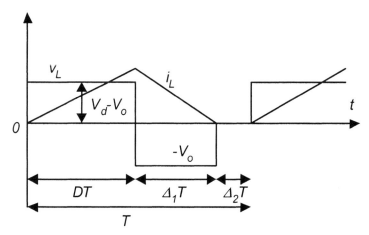

Figure 2.16 DC/DC Buck converter waveforms in DCM operation.

2.2.2 Boost Converters

A DC/DC Boost converter steps up the input voltage source. Figure 2.17 depicts the power circuit of this converter. This converter increases the level of the DC voltage source and can operate in both continuous and discontinuous conduction modes of operation, per the designed system parameters.

In this converter, there is a switch in parallel with the source, instead of series as it was in Buck converter. Therefore, drive is not floating and common ground is available to the semiconductor switch. One other difference, which can be seen from the power stage, is that there is an inductor immediately after the source. This inductor can be output of a rectifier with DC link. This makes it possible to have small size of EMI filter. The problem is that natural short circuit protection is not available, as switch is in parallel.

In CCM operation, when switch is on ($0 < t < DT$), the circuit diagram for the converter is as shown in Figure 2.18.

Fundamentals of Power Electronics 31

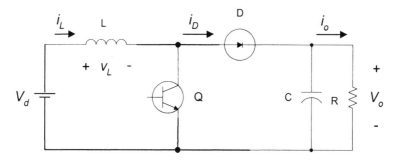

Figure 2.17 DC/DC Boost converter.

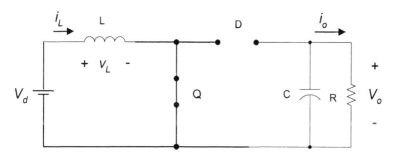

Figure 2.18 DC/DC Boost converter when switch is on.

As can be seen in Figure 2.18, due to closure of the switch, the negative voltage will appear across the diode and it will reverse biased. During this time, the switch is conducting. Inductor current is equal to the source current and the same current flows through the switch. Source voltage appears across the inductor. It is obvious that diode current is zero and, thus, the load current comes from the capacitor only. Energy is being stored in the inductor. Capacitor is being discharged through the load. Below are the state variable equations during this period.

$0 < t < DT$

$V_d = v_L$

$\dfrac{di_L}{dt} = \dfrac{V_d}{L}$ (2.6a)

$i_L(t) = \dfrac{V_d}{L} t + I_{L,\min}$

$i_L(t = DT) = I_{L,\max}$

$$\Delta I_L = I_{L,\max} - I_{L,\min} = \frac{V_d}{L} DT \qquad (2.6b)$$

When the switch is off ($DT < t < T$), the circuit diagram for the converter is as shown in Figure 2.19.

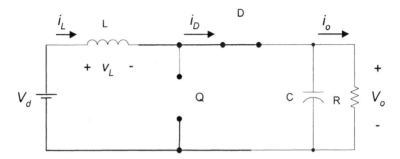

Figure 2.19 DC/DC Boost converter when switch is off.

As the switch is off, there is not any negative voltage across the diode. In fact, the diode will be forward biased and ready to conduct. The only path for the inductor current is through the diode to the load. Therefore, in this interval, switch current becomes zero; the inductor current is still the input current and the same current flows through the diode. The inductor is depleting the energy in charging the capacitor, which was discharged during the previous interval, and supplying the load. Voltage across the inductor will be a negative difference of two voltage sources on either side of it. As energy is depleting from the inductor, the current decreases linearly. State equations during this time interval are as follows:

$$DT < t < T$$
$$v_L = V_d - V_o$$
$$\frac{di_L}{dt} = \frac{-(V_o - V_d)}{L}$$
$$i_L(t) = \frac{-(V_o - V_d)}{L}(t - DT) + I_{L,\max} \qquad (2.7)$$
$$i_L(t = T) = I_{L,\min}$$
$$\Delta I_L = I_{L,\max} - I_{L,\min} = \frac{(V_o - V_d)}{L}(1 - D)T$$

As current was increasing until the switch was turned off, the initial condition for this period is maximum inductor current and final condition will be minimum of inductor current. As converter is in CCM operation, before the

Fundamentals of Power Electronics

current reaches zero, the switch is turned on and the cycle repeats. From (2.6) and (2.7), we have

$$\frac{V_d}{L} DT = \frac{(V_o - V_d)}{L}(1-D)T \qquad (2.8)$$

$$V_o = \frac{1}{1-D} V_d \qquad (2.9)$$

Figure 2.20 depicts the inductor voltage and current waveforms in CCM operation.

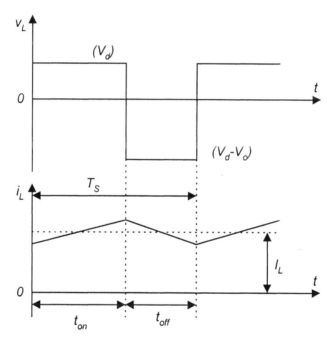

Figure 2.20 Inductor voltage and current waveforms in CCM operation.

As explained before, during the time when the switch is on, the current increases linearly. When the switch is off, the current starts decreasing linearly. The inductor current never reaches zero throughout the operating range. Average of this inductor current is equal to the load current. Voltage across the inductor is also input voltage. Output voltage is not shown in Figure 2.20 and it is assumed to be kept constant at the required voltage level higher than the input voltage.

Different loading condition can bring the converter from continuous to discontinuous conduction mode. But if the input voltage, required output voltage, and operating load range are given, it is possible to design the converter in such a way that it will never go into discontinuity. In that case, only the value of the inductor is the parameter, which is responsible for continuous or discontinuous conduction mode of operation. Value of this parameter can be decided based on studying when the converter will go into a discontinuous conduction mode. In the boundary condition, when the converter is not in completely continuous conduction and not completely in discontinuous conduction mode, inductor voltage and current waveforms are as depicted in Figure 2.21. It is mentioned previously that boundary condition refers to CDCM. Converter performance in the first interval of the switching period in this operating condition is the same as in CCM. The only difference is initial condition. Here, initially inductor current is zero. As soon as the inductor current hits zero, the switch is turned on again.

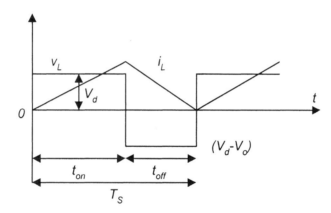

Figure 2.21 DC/DC Boost converter waveforms in CDCM operation.

Figure 2.22 depicts the DC/DC Boost converter waveforms in DCM operation. It is clear that the duration for which the inductor current stays at zero is comparatively much larger than the CDCM. In fact, in CDCM, the inductor current just hits the zero line and increases. But, in this case, the current stays at zero. Here also the minimum instantaneous inductor current is zero, but it reaches zero before T. In the case of CDCM, on the other hand, it reaches zero exactly at T.

2.2.3 Buck-Boost Converters

In Buck converters, output voltage is always less than the input voltage. On the other hand, in Boost converters, output voltage is always greater than the input voltage. However, a Buck-Boost converter can have its output voltage both higher and lower than the input voltage, depending on the duty ratio. Figure 2.23 depicts the power circuit of this converter.

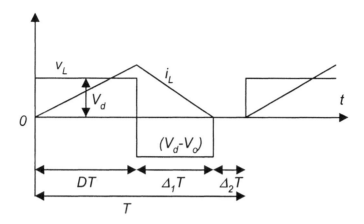

Figure 2.22 DC/DC Boost converter waveforms in DCM operation.

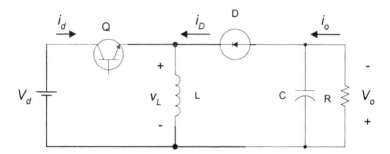

Figure 2.23 DC/DC Buck-Boost converter.

From the circuit diagram shown in Figure 2.23, the first difference between other topologies studied and this topology is that the output voltage here is of opposite polarity to the input voltage. Therefore, the common ground cannot be employed for this topology. Depending on the component values of the circuit, this converter can be operated in continuous or discontinuous conduction modes of operation.

In CCM operation, when the switch is on ($0 < t < DT$), the circuit diagram for the converter is as shown in Figure 2.24.

As shown in Figure 2.24, when switch is closed, input voltage appears across the inductor. On the other side, due to the negative voltage across the diode, it will reverse biased and there will not be any diode current. Load is supplied by the capacitor only as the source is disconnected from it. Thus, when the switch is closed, inductor current rises linearly. The inductor stores energy during this switching interval. The capacitor discharges through the load.

Mathematical formulation of different parameters and variables of the converter during the on interval are as follows:

$$0 < t < DT$$
$$V_d = v_L$$
$$\frac{di_L}{dt} = \frac{V_d}{L}$$
$$i_L(t) = \frac{V_d}{L}t + I_{L,\min}$$
$$i_L(t = DT) = I_{L,\max}$$
$$\Delta I_L = I_{L,\max} - I_{L,\min} = \frac{V_d}{L}DT$$

(2.10)

When the switch is off ($DT < t < T$), the circuit diagram for the converter is as shown in Figure 2.25.

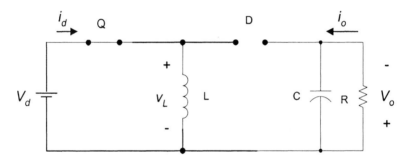

Figure 2.24 DC/DC Buck-Boost converter when switch is on.

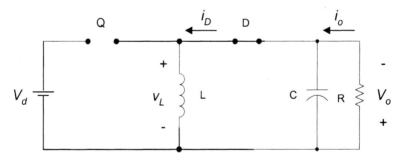

Figure 2.25 DC/DC Buck-Boost converter when switch is off.

From the circuit diagram, it is clear that the diode provides the path for the inductor current. Because the switch is open, there is not any input current from the source and the same with the switch current. On the contrary, the diode is conducting and, thus, it has the current, which is equal to the inductor current now. Practically, the inductor supplies both capacitor and load. During this switching interval, the inductor depletes its energy and the capacitor is charged. Decrease in the inductor current is linear and in the same direction. During this off interval, circuit state equations are as follows:

$$DT < t < T$$
$$v_L = -V_o$$
$$\frac{di_L}{dt} = \frac{-V_o}{L}$$
$$i_L(t) = \frac{-V_o}{L}(t - DT) + I_{L,\max}$$
$$i_L(t = T) = I_{L,\min}$$
$$\Delta I_L = I_{L,\max} - I_{L,\min} = \frac{V_o}{L}(1 - D)T$$

(2.11)

From (2.10) and (2.11), we have

$$\frac{V_d}{L}DT = \frac{V_o}{L}(1 - D)T \tag{2.12}$$

$$V_o = \frac{D}{1 - D}V_d \tag{2.13}$$

Figure 2.26 depicts the inductor voltage and current waveforms in CCM operation.

As explained earlier, inductor current increases during the on time and decreases during the off time. When the switch is on, input voltage appears across the inductor and when, the switch is off, output voltage appears across the load. In fact, an advantage of this converter topology is that the input and output currents are independent. They do not interact with each other via the inductor. This feature makes it possible to design a much more robust controller.

As explained for the previous two topologies, CDCM operation is important for design considerations. For a given input voltage range, load range, and required output voltage, by studying CDCM operation, duty ratio and inductance value can be determined. If voltage ripple is specified, then it is possible to design output capacitor as well. When the converter operates in boundary condition, the inductor current and voltage waveforms are as shown in Figure 2.27.

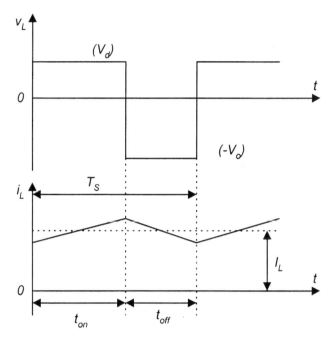

Figure 2.26 Inductor voltage and current waveforms in CCM operation.

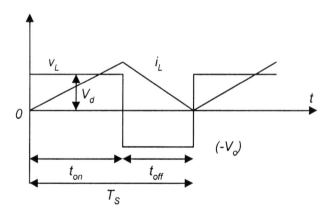

Figure 2.27 DC/DC Buck-Boost converter waveforms in CDCM operation.

Figure 2.28 depicts the DC/DC Buck-Boost converter waveforms in DCM operation. We can see that the duration of zero inductor current is much longer than that of the CDCM operation.

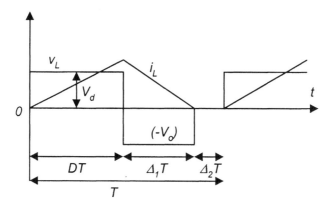

Figure 2.28 DC/DC Buck-Boost converter waveforms in DCM operation.

2.3 DC/AC Inverters

2.3.1 Principles of Operation

Generally, the single-phase, full bridge DC/AC inverters are popularly known as "H-Bridge" inverters. These DC/AC inverters are basically either voltage source/fed inverters (VSI) or current source/fed inverters (CSI). In the case of a VSI, the input voltage is considered to remain constant, whereas in a CSI, the input current is assumed to be constant. Furthermore, the single-phase DC/AC inverter could also be termed as "variable DC linked inverter," if the input voltage is controllable. A simple representation of a single-phase DC/AC inverter is as shown in Figure 2.29.

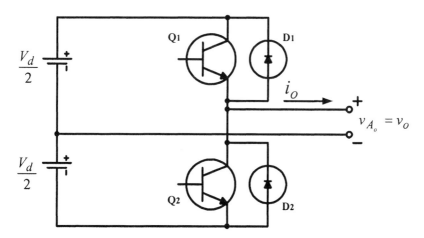

Figure 2.29 Basic circuit diagram of a single-phase DC/AC inverter.

In Figure 2.29, when switch Q1 is turned on for a time period of T/2, the instantaneous voltage across the load is Vd/2. If, on the other hand, switch Q2 is turned on for a time period of T/2, the instantaneous voltage across the load is -Vd/2. The control circuit is designed in such a way that Q1 and Q2 are not on at the same time. The typical output voltage waveform for the circuit arrangement of Figure 2.29 is as depicted in Figure 2.30.

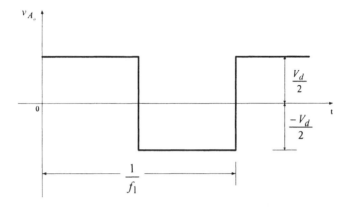

Figure 2.30 Basic square wave output voltage waveform.

The single-phase DC/AC inverter of Figure 2.29 is a half-bridge inverter and its RMS output voltage is calculated from Figure 5.30 as follows:

$$v_o = \sqrt{\frac{2}{T} \int_0^{T/2} \frac{V_s^2}{4} dt} = \frac{V_d}{2} \tag{2.14}$$

Furthermore, the instantaneous output voltage can be expressed in a Fourier series as:

$$v_o = \sum_{1,3,5,\ldots}^{\infty} \frac{2V_s}{n\pi} Sin(n\omega t) \tag{2.15}$$

Here, $\omega = 2\pi f_0$ is the frequency of the output voltage in rad/s. The output voltage of practical inverters is not purely sinusoidal, and contains harmonics, which distort the shape of the waveform. Hence, the quality of power achieved from practical inverters is lower and needs to be improved by the usage of suitable output filters. Two important parameters generally defined with regards to the output power quality are total harmonic distortion (THD) and distortion factor (DF). THD is basically a measure of how close the shape of the attained output waveform is to its fundamental component. It is expressed as follows:

$$THD = \frac{1}{V_1} \sqrt{\sum_{n=2,3,...}^{\infty} V_n^2} \qquad (2.16)$$

Here, V_1 is the fundamental component of the output voltage waveform, whereas V_n is the value of the n number of harmonic components in the output voltage. The output voltage THD is generally expressed as a percentage of its fundamental component.

Another important factor to be considered while evaluating the output power quality is the distortion factor (DF). This is necessary to be calculated since THD does not indicate the level of each harmonic component. The DF is, thus, a measure of the effectiveness in reducing the unwanted harmonics, without having to specify the values of the second-order load filter. It can be defined as follows:

$$DF = \frac{1}{V_1} \sqrt{\sum_{n=2,3,...}^{\infty} \left(\frac{V_n}{n^2}\right)^2} \qquad (2.17)$$

Furthermore, the DF for an individual harmonic component can be defined as follows:

$$DF_n = \frac{V_n}{V_1 n^2} \qquad (2.18)$$

Thus, having studied the basic operation of a 1-phase DC/AC inverter and the important factors affecting its power quality, it is now imperative to understand the various topology switching schemes for the design. In the sections to follow, these switching schemes are explained in brief.

2.3.2 Square-Wave, Full-Bridge Inverters

The basic layout of a single-phase, full-bridge DC/AC inverter is as shown in Figure 2.31.

The full-bridge inverter is made up of two 1-leg inverters, similar to the one discussed in the previous section. When power demands are higher, these inverters are generally preferred over half-bridge inverters. The maximum output voltage for this arrangement is twice that of a half-bridge inverter, for the same value of DC input voltage. Therefore, at higher power levels of operation, this is a major advantage.

As shown in Figure 2.31, the full-bridge DC/AC inverter consists of 4 high-frequency switches, *TA+, TA-, TB+,* and *TB-*. When switches *TA+* and *TB-* are switched on, the input voltage *Vd* appears across the load. If switches *TA-* and *TB+* are switched on at the same time, the voltage across the load is reversed, and is equal to *-Vd*. Typical waveforms for the circuit of Figure 2.31 is as shown in Figure 2.32.

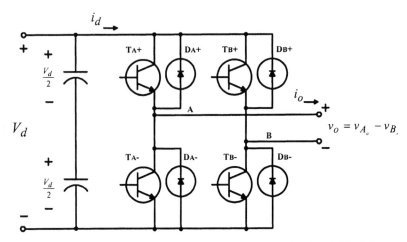

Figure 2.31 Basic circuit diagram of a single-phase, full-bridge DC/AC inverter.

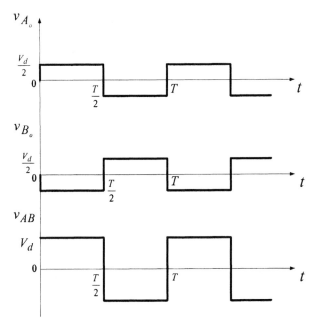

Figure 2.32 Output phase and line voltage waveforms for a full-bridge DC/AC inverter.

From Figure 2.32, the RMS of the output voltage for the single-phase DC/AC inverter can be expressed as follows:

Fundamentals of Power Electronics

$$V_o = \sqrt{\frac{2}{T}\int_0^{T/2} V_d^2 dt} = V_d \qquad (2.19)$$

Furthermore, the output voltage can also be expressed as a Fourier series expansion, as depicted below:

$$v_o = \sum_{n=1,3,5,\ldots}^{\infty} \frac{4V_d}{n\pi} Sin(n\omega t) \qquad (2.20)$$

Hence, if $n = 1$, we have the following expression for the RMS value of the fundamental component:

$$V_1 = \frac{4V_d}{\sqrt{2}\pi} \qquad (2.21)$$

Each of the switches *TA+*, *TA-*, *TB+*, and *TB-* are paralleled with diodes *DA+*, *DA-*, *DB+*, and *DB-*, respectively. These diodes are known as freewheeling diodes and, during their conduction period, energy is fed back to the input DC source.

2.3.3 Pulse-Width Modulated (PWM) Inverters

For many applications, it is required to control the output voltage of the DC/AC inverter. For this purpose, many popular switching schemes have been developed. The output voltage of the inverter is basically controlled in order to cope with the variations in input DC voltage, for voltage regulation of the inverter, and for constant voltage/frequency control. The most popular technique that was developed for this purpose was pulse-width modulation (PWM). Within the PWM technique, various variations are possible, which give finer and better control of the inverter switching for different applications.

Primarily, the DC/AC inverter output voltage needs to be sinusoidal, with its magnitude and frequency controllable. In order to do so, in the case of the PWM technique, a sinusoidal reference signal, oscillating at the desired frequency, is compared with a high-frequency triangular carrier waveform, as shown in Figure 2.33. This PWM technique is basically known as the bipolar switching scheme. The frequency of the triangular carrier waveform determines the inverter switching frequency and is generally kept constant at amplitude of $\hat{V}_{tri}$.

It is mandatory that a few terms are made clear before moving on to the actual discussion with regards to the PWM behavior. The triangular waveform, v_{tri}, in Figure 2.33, basically oscillates at the switching frequency, f_s. It is this frequency that decides the frequency at which the inverter switches are turned on and off. On the other hand, the control signal, $v_{control}$, is used to modulate the switch duty ratio, and oscillates at a frequency of f_1. This is basically the desired frequency, at which the fundamental component of the output voltage oscillates. Furthermore, the output voltage, as mentioned earlier, is not a pure sinusoidal waveform and consists of voltage components oscillating at harmonic frequencies of f_1.

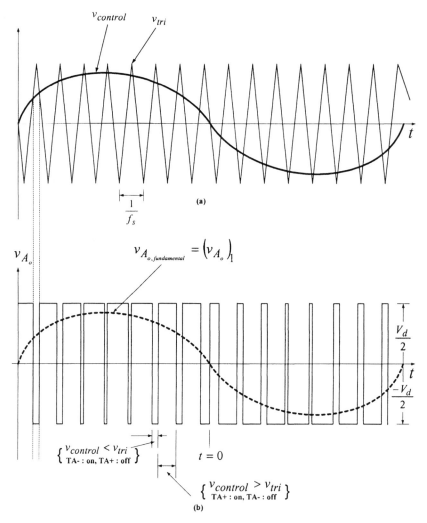

Figure 2.33 Diagrammatic representation of bipolar PWM switching technique.

Fundamentals of Power Electronics

From Figure 2.33, the important terms of amplitude and frequency modulation ratio can be defined. The amplitude modulation ratio, m_a is defined as:

$$m_a = \frac{\hat{V}_{control}}{\hat{V}_{tri}} \qquad (2.22)$$

Here, $\hat{V}_{control}$ is the peak amplitude of the control signal, whereas $\hat{V}_{tri}$ is the peak amplitude of the triangular carrier signal, and is generally maintained constant. The value of the ratio m_a is generally < 1, and in that case, the PWM scheme is called "undermodulation," whereas, when $m_a > 1$, the scheme is called "overmodulation".

Referring back to Figure 2.33, the frequency modulation ratio (m_f) can be defined as follows:

$$m_f = \frac{f_s}{f_1} \qquad (2.23)$$

In the full-bridge DC/AC inverter of Figure 2.31, the switches T_{A+} and T_{A-} are controlled by comparing the magnitudes of $v_{control}$ and v_{tri}. The output voltage across the load may vary accordingly, and is as depicted below:

$$v_{control} > v_{tri}, \qquad T_{A+} \text{ is ON,} \qquad v_{A_o} = \frac{V_d}{2} \qquad (2.24)$$

$$v_{control} < v_{tri}, \qquad T_{A-} \text{ is ON,} \qquad v_{A_o} = \frac{-V_d}{2} \qquad (2.25)$$

Thus, since the 2 switches are never off simultaneously, the output voltage v_{A_o} fluctuates between $V_d/2$ and $-V_d/2$. Another popular variation of the PWM technique, as mentioned before, is the unipolar switching scheme. The typical waveforms for this scheme are as shown in Figure 2.34, and are generated from simulations.

As is clear from Figure 2.34, in the case of a unipolar switching scheme for PWM, the output voltage is either switched from high to zero or from low to zero, unlike in the bipolar switching scheme, where the switching takes place directly between high and low. Referring back to Figure 2.31, the full-bridge DC/AC inverter may have the following switch controls under a unipolar PWM switching scheme:

$v_{control} > v_{tri}$, $\quad T_{A+}$ is on $\quad\quad$ (2.26)

$-v_{control} < v_{tri}$, $\quad T_{B-}$ is on $\quad\quad$ (2.27)

$-v_{control} > v_{tri}$, $\quad T_{B+}$ is on $\quad\quad$ (2.28)

$v_{control} < v_{tri}$, $\quad T_{A-}$ is on $\quad\quad$ (2.29)

Again, as aforementioned, the switch pairs (T_{A+}/T_{A-}) and (T_{B+}/T_{B-}) are complementary to each other: that is, when one of the switch pairs are open, the other switch pair must be closed.

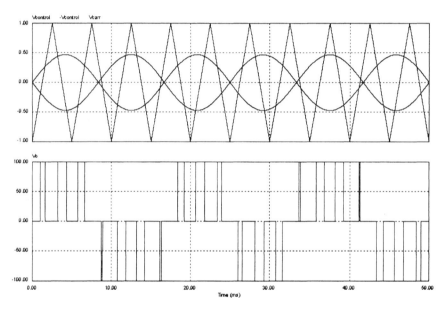

Figure 2.34 Diagrammatic representation of unipolar PWM switching technique.

In some cases, the PWM technique used may be modified in accordance with the application demanded. It is noticeable that the pulse widths in the output voltage waveform do not change significantly near the peak of the sinusoidal modulating signal. This is due to the inherent characteristics of the sine wave itself. To avoid this minor drawback, some kind of modified PWM switching technique may be used, involving the introduction of the carrier signal only during the beginning and end of the half cycle. Such a modified PWM technique increases the fundamental component in the output voltage, hence improving its harmonic characteristics. Furthermore, such modified PWM techniques help in reducing the number of power components, as well as in

Fundamentals of Power Electronics

reducing switching losses. Apart from the PWM technique of output voltage control of a single-phase DC/AC inverter, there are various other advanced techniques developed, depending on the type of application to be satisfied.

2.4 Selected Readings

[1] N. Mohan, T. M. Undeland, and W. P. Robbins, *Power Electronics: Converters, Applications, and Design*, John Wiley & Sons, 1995.
[2] P. T. Krein, *Elements of Power Electronics*, Oxford University Press, 1998.
[3] D. H. Hart, *Introduction to Power Electronics*, Prentice-Hall, 1997.
[4] R. W. Erickson and D. Maksimovic, *Fundamentals of Power Electronics*, 2nd Edition, Kluwer Academic Publishers, 2001.
[5] T. L. Skvarenina, *The Power Electronics Handbook*, CRC Press, 2002.
[6] M. H. Rashid, *Power Electronics*, 2nd Edition, Prentice Hall, 1993.
[7] J. G. Kassakian, M. F. Schlecht, and G. C. Verghese, *Principles of Power Electronics*, Addison Wesley, 1991.
[8] A. M. Trzynadlowski, *Introduction to Modern Power Electronics*, John Wiley and Sons, 1998.
[9] N. Mohan, *Power Electronics: Computer Simulation, Analysis and Education using PSpice*, MNPERE, Minneapolis, 2000.
[10] B. K. Bose, *Problems Manual for Modern Power Electronics and AC Drives*, Prentice Hall, 2001.

3
Electric Machines

In this chapter, fundamentals of electro-mechanical systems, electromagnetism, electric machines, and utilization of electric machines to drive mechanical loads are presented.

3.1 Electro-Mechanical Power Transfer Systems

Electric machines can be utilized in either the motoring mode or the generating mode of operation. In the motoring mode, these machines use electricity to drive mechanical loads, while, in the generating mode, they are used to generate electricity from mechanical prime movers. Figures 3.1 and 3.2 show typical graphical representations of the operating modes of electric machines. Figures 3.1 and 3.2 are depicted for a three-phase supply source, but it is also possible to have a single-phase or DC supply source. In the next sections, we will examine the basic operating principles, characteristics, and driving equations for these machines. With the use of advanced power electronic tools, it is possible to achieve desired characteristics out of simple electric machines.

As mentioned earlier, a mechanical system can either be a mechanical load or a prime mover to drive the electric machine. In mechanical systems, basic parameters of the interest are distance/position, speed, acceleration, moment of inertia, friction, and torque. It is also important to know the relation between these parameters. The following equations define the relation between all these parameters.

- Acceleration, $a = d\omega/dt$
- Speed, $\omega(t) = \omega(0) + \int a(\tau)\, d\tau$
- Position, $\theta = \theta(0) + \int \omega(\tau)\, d\tau$

- Electromagnetic Power, $P_{em} = T_{em} \cdot \omega$
- Load Power, $P_L = T_L \cdot \omega$
- Kinetic Energy, $K = \tfrac{1}{2} J \omega^2$

Speed varies with respect to both power and torque. It is possible to keep one of them constant and let the other vary with speed, i.e., power can be constant and speed can vary with respect to torque or torque can be kept constant and speed can vary with respect to power. The relationship between torque and speed can also be derived from the equation below.

$$T = T_r \left(\frac{n}{n_r} \right)^k \tag{3.1}$$

where T, T_r, n, n_r, and k are torque, rated torque, speed, rated speed, and exponential coefficient, respectively. Different torque-speed characteristics are possible with different values of k. The dynamic mechanical equation for the system is as follows:

$$T_{em} = T_L + J \frac{d\omega}{dt} + B\omega \tag{3.2}$$

where T_{em}, T_L, J, and B are developed electromagnetic torque, load torque, moment of inertia, and bearing and friction coefficient, respectively.

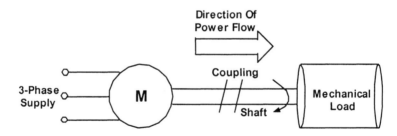

Figure 3.1 A typical electric machine in motoring mode.

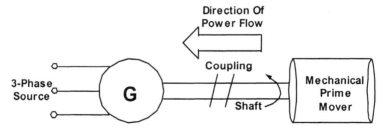

Figure 3.2 A typical electric machine in generating mode.

It is well known that, in mechanical systems, to change the speed with respect to load, a gearing mechanism is used. Below is a figure that gives an idea how gearing on the shaft can change the speed.

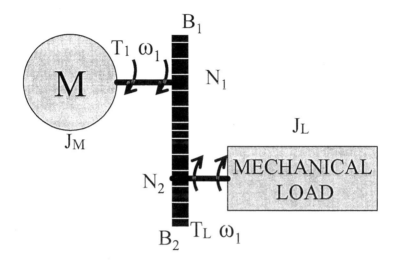

Figure 3.3 A typical mechanical coupling using gears.

For the gearing system of Figure 3.3, we have

$$T_1\omega_1 = T_L\omega_2$$
$$\frac{\omega_2}{\omega_1} = \frac{N_1}{N_2} \qquad (3.3)$$
$$T_1 = \left(\frac{N_1}{N_2}\right)T_L$$

where

N_1, N_2 = Number of teeth in gears

ω_1, ω_2 = Angular speed of the motor and load

B_1, B_2 = Bearing and friction coefficient of the motor and load

T_1, T_L = Developed electromagnetic and load torque

As we want our electric machine/system to respond to the mechanical behavior, it is necessary to have all the parameters on the mechanical system side to be considered on the motor side.

$$J_{L,reflected} = \left(\frac{N_1}{N_2}\right)^2 J_L$$

$$B_{2,reflected} = \left(\frac{N_1}{N_2}\right)^2 B_2 \qquad (3.4)$$

$$J = J_M + J_{L,reflected} = J_M + \left(\frac{N_1}{N_2}\right)^2 J_L$$

$$B = B_1 + B_{2,reflected} = B_1 + \left(\frac{N_1}{N_2}\right)^2 B_2$$

$$T_1 = T_{L,reflected} + J\frac{d\omega_1}{dt} + B\omega_1 \qquad (3.5)$$

Depending on the direction of the produced torque and speed/rotation of the shaft, an electric machine has the capability to operate in both motoring and generating modes and, correspondingly, a mechanical system can work as load and prime mover. If we combine both of them, we can have four-quadrant operation out of the combined electro-mechanical system, as is shown in Figure 3.4.

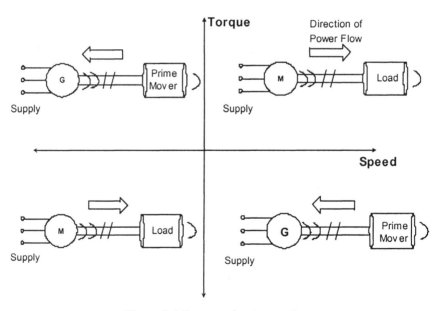

Figure 3.4 Four-quadrant operation.

3.2 Fundamentals of Electromagnetism

Ampere's law states that, in any closed path, a magnetic field is established due to the current passing through that path. In fact, at any time, the line integral of the magnetic field intensity along any closed path equals the total current enclosed by the path. The density of the flux due to the magnetic field depends on the core material as follows:

$$B = \mu H$$
$$\mu = \mu_r \mu_0$$
(3.6)

where μ is the permeability of the material, H is the magnetic field intensity, and B is the flux density in Tesla. μ_r is the relative permeability; the relative permeability of air or free space is 1; it ranges from 1000 to 80000 for different magnetic materials.

If a current I is applied to a core, it creates a magneto-motive force (mmf) that depends on the number of turns N and the current passing through that coil. $F = N \times I$ is the magneto-motive force. Flux Φ is generated due to the magneto-motive force F. This is similar to the electric circuits in which the current is established due to the applied voltage. The magneto-motive force relates to the flux by the relation $F = \Phi \times R$, where R is the reluctance of the magnetic circuit.

The relationship curve between the magnetic flux density B and the magnetic field intensity H is shown in Figure 3.5.

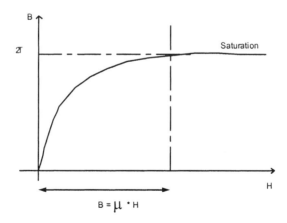

Figure 3.5 Saturation curve.

From the saturation curve, we notice that the magnetic field intensity is proportional to the magnetic flux density to a certain point where it is approximately 2 Tesla. Passing this point, the increase in the magnetic field intensity will result in a smaller increase in the magnetic flux density and soon any increase in the magnetic field will gradually result in no increase at all, meaning that the core is saturated.

Figure 3.6 Non-magnetic material; the net magnetic field is zero.

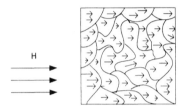

Figure 3.7 A magnetized material.

Figure 3.8 shows the hysteresis characteristic of magnetic materials. When an AC current is applied to the winding of a core with zero magnetic flux density, as the current increases, the flux density increases from point a to point b, as in Figure 3.8. When the current decreases, the flux density will decrease correspondingly and will follow the b-c-d path. As the current increases, the magnetic flux density will increase, following the path d-e-b. In fact, when a magneto-motive force is applied to the core windings and removed, the flux density will follow the a-b-c path. After removing this magneto-motive force, the flux density will not follow the same path it originated from because some magnetic field will remain even after removing the magneto-motive force; this field is known as the residual flux. To push the flux density back to zero, a force needs to be applied in the opposite direction of the flux; this force is known as the coercive force H_c, and it pushes the flux back to zero from c to a.

Hysteresis characteristic is a source of power loss in magnetic materials. Other than hysteresis loss, eddy current loss is the main power loss. Eddy currents are generated in the core causing heat inside the magnetic material. The amount of heat generated by eddy currents is related to the path size that the currents follow. In order to eliminate losses caused by eddy currents, the core is constructed from small strips of magnetic materials which are known as laminations. This helps to reduce the heat inside the core, which helps to reduce the heat without causing any effects.

Electric Machines

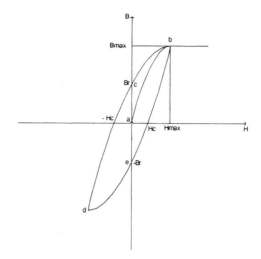

Figure 3.8 Hysteresis characteristic.

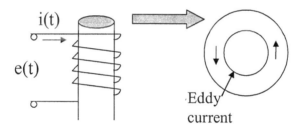

Figure 3.9 Eddy currents.

3.3 DC Machines

DC machines are the simplest machines in terms of operation and control. Therefore, power electronic drivers for DC motors are simple compared to the drivers for AC motors. However, DC machines are relatively expensive; they also require high maintenance. DC machines consist of two main parts: stator and rotor. The stator is the fixed part of a DC machine; it contains field windings that produce magnetic field. The rotor is the moving part of a DC machine; it contains windings where the back emf is produced.

3.3.1 Separately-Excited DC Machines

In separately-excited DC machines, the armature and field windings are connected to two separate voltage sources, as is shown in Figure 3.10. Machine equations used for this configuration are as follows:

$$v_a = R_a i_a + L_a \frac{di_a}{dt} + e_a$$

$$v_F = R_F i_F + L_F \frac{di_F}{dt} \qquad (3.7)$$

$$e_a = K i_F \omega$$

$$T = K i_F i_a$$

Figure 3.11 shows the armature current-speed characteristics of a separately-excited DC machine. In this machine, as we increase the armature resistance, the speed decreases. In addition, as we decrease the field current, the machine speed increases. At starting ω=0, the starting current is very high. In order to avoid a high starting current, we can use a starter to reduce the input voltage or increase the armature resistance.

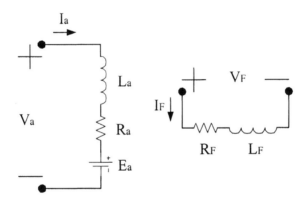

Figure 3.10 Equivalent circuit of a separately-excited DC machine.

3.3.2 Shunt DC Machines

In shunt DC machines, the armature and field windings are connected to the same voltage source, as is shown in Figure 3.12. Shunt machine equations are as follows:

$$v_a = R_a i_a + L_a \frac{di_a}{dt} + e_a$$

$$v_F = v_a = R_F i_F + L_F \frac{di_F}{dt} \qquad (3.8)$$

$$e_a = K i_F \omega$$

$$T = K i_F i_a$$

Electric Machines

If the direction of current flow is changed in either the armature or field windings, then the machine will run in the opposite direction. In order to avoid demagnetizing the exciter field, it is preferable to reverse the direction of current flowing in the armature. The speed of the machine can be controlled by controlling the field current and the field voltage. In shunt DC machines, the starting current decreases as we increase the starting resistance. Also, as we increase the input voltage the starting current increases.

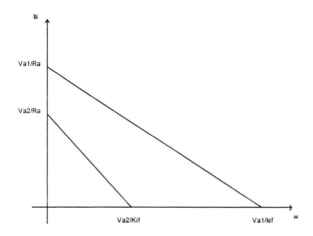

Figure 3.11 Armature current-speed characteristics of separately-excited motor.

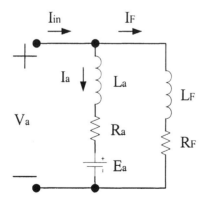

Figure 3.12 Equivalent circuit of a shunt DC machine.

3.3.3 Series DC Machines

In a series DC motor, the field and armature windings are connected in series, as is shown in Figure 3.13. The load current flows through both windings. To keep the voltage drop in the field windings to a minimum; the winding consists of a few turns of thick copper wire.

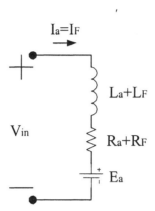

Figure 3.13 Equivalent circuit of a series DC machine.

If a series DC motor runs without a load, a small current flows in the field and armature windings. The magnetic field is weak. In spite of this weak magnetic field, in order that a sufficiently large back-emf is induced, the armature has to turn at a high speed. Thus, when off-load, a series DC motor is susceptible to "runaway". Since the magnetic field is dependent on the load current, the speed is also very dependent on loading. A series DC motor develops a high torque and is used in preference to a shunt DC machine, where heavy starting is involved. Machine equations are:

$$v_{in} = e_a + (R_a + R_F)i_a + (L_a + L_F)\frac{di_a}{dt}$$
$$e_a = Ki_a \omega \qquad (3.9)$$
$$T = Ki_a^2$$

From the torque-speed curve shown in Figure 3.14 for a series DC machine, we can conclude that with a small voltage applied, a high torque is developed. In addition, series DC machines have high starting torque and have low operating torque. As is expressed in (3.9), we can get unidirectional torque with a positive or a negative armature current, meaning that regardless of the current direction, the torque will always have one direction; therefore, we can drive a series DC machine from an AC voltage as well as DC voltage. These series DC machines are named universal motors.

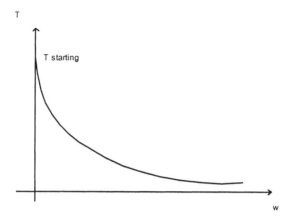

Figure 3.14 Torque-speed characteristics of series DC machines.

3.3.4 Compound DC Machines

A compound DC machine has one series and one shunt winding. The windings are connected in such a way that their effects assist each other. Compound DC machines are used where the starting torque of shunt motors are too small, i.e., hoist machines. They are also used where the fluctuation in speed of a series motor under load is to be avoided. At the instant of switching on, the series winding produces a high starting torque. When the load is increased, the shunt winding has the effect of retaining an almost constant speed, preventing the reduction in speed that occurs in a series motor. Machine equations are as follows:

$$v_a = e_a + (R_a + R_{F1})i_a + (L_a + L_{F1})\frac{di_a}{dt}$$
$$v_F = R_{F2}i_F + L_{F2}\frac{di_F}{dt}$$
$$e_a = (K_1 i_a + K_2 i_F)\omega$$
$$T = (K_1 i_a + K_2 i_F)i_a$$

(3.10)

3.4 Induction Machines

A three-phase induction motor has a stator winding with the pole-pair P in which a rotating field is generated from the three-phase mains supply. A voltage is induced in the rotor windings by the rotating field of the stator, which, in turn, produces a current in the rotor. This produces a force which sets the armature rotating. The armature accelerates and the armature current decreases. This decrease continues until the torque generated by the motor decreases to a value equal to the external mechanical loading torque. Armature speed N is less than the speed of the rotating field N_s by an amount equal to the slip speed s.

$$s = \frac{N_s - N}{N_s} \qquad (3.11)$$

Induction machines are simply nothing but a transformer with a rotating secondary winding. The primary induces voltage in the secondary; unlike the transformer, the frequency of an induction machine at the rotor side isn't the same as the frequency at the primary side.

$$f_r = sf_s \qquad (3.12)$$

where f_r and f_s are rotor current frequency and mains supply frequency, respectively. Induction machines cover a variety of applications in the industry; more than 50 percent of the electrical power usage is for electric motors; 90 percent of this power is for induction motors and more than 80 percent of them are three phase induction motors. In the industrial cases most of the motor drives are induction machines because of their low maintenance cost, high reliability, high efficiency, and high power density.

The rotor can be of two types: squirrel cage or wound rotor. The squirrel cage rotor has conductivity bars mounted on slots and they are shorted by a large shorting ring. In wound rotors, we have the accessibility to add resistors to the rotor side circuit, as is shown in Figure 3.15. This is an advantage for this kind of rotors because, by adding resistors, we can control the torque-speed characteristics of the motor. There are three rotor wires connected to the slip rings that are mounted on the rotor shaft.

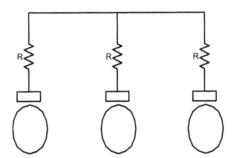

Figure 3.15 Slip rings and external resistors of a wound rotor motor.

Figure 3.16 depicts the equivalent circuit of induction machines. At high speeds where the mechanical speed is slightly smaller than the synchronous speed, we find that slip is very small; so, in the equivalent circuit for induction motor, R'_2/s is very large compared to the reactance. On the other hand, at low

Electric Machines

speeds where the mechanical speed is close to zero or it is zero, e.g., a locked-rotor test, R'_2/s is very small and the reactance is much larger.

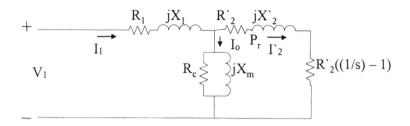

Figure 3.16 Equivalent circuit of induction machines.

3.4.1 Torque-Speed Characteristics

At starting, the speed of the motor is zero and, therefore, slip is one. When the speed reaches the synchronous speed, slip reaches zero. Torque-slip or torque-speed relationship of the machine can be expressed as

$$T = \frac{3V_1^2}{\omega_s} \frac{R'_2/s}{\left(R_1 + R'_2/s\right)^2 + \left(X_1 + X'_2\right)^2} \qquad (3.13)$$

Maximum torque can be calculated from (3.13) as follows:

$$T_{max.} = \frac{3V_1^2}{\omega_s} \frac{1}{R_1 + \sqrt{R_1^2 + \left(X_1 + X_2\right)^2}} \qquad (3.14)$$

From this maximum torque, we can conclude that whatever load value we add to the rotor side, the maximum torque does not change. In addition, at the synchronous speed where the rotor speed is the same as asynchronous speed, the developed torque is zero. We can change the developed torque by varying the terminal voltage, the frequency, the rotor resistance, or the number of poles. Figures 3.17 and 3.18 depict torque-speed characteristics when supply frequency and voltage are controlled, respectively.

If we increase the voltage, the machine might go to saturation; on the other hand, if we decrease the voltage, the machine will not operate in optimal condition. Furthermore, if we decrease frequency, the machine might go to saturation as well. By increasing frequency, we will not utilize the machine optimally as well. In order to overcome the saturation problem and optimize the operation of induction motors, we can maintain the voltage/frequency (V/f) ratio constant. Maximum torque will remain unchanged as well (Figure 3.19).

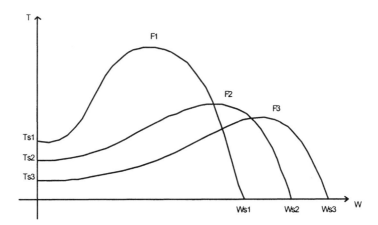

Figure 3.17 Torque-speed characteristics for different frequencies.

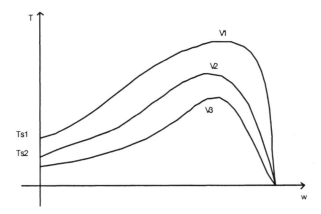

Figure 3.18 Torque-speed characteristics for different voltages.

Using a Steinmetz circuit, three-phase motors can also be operated on a single-phase mains supply with an additional capacitor. In a Steinmetz circuit, the torque is reduced to about one-third and the power is reduced to approximately 80%. If a similar starting torque is required, as with three-phase operation, an additional starting capacitor must be fitted. This capacitor is connected in parallel to the operating capacitor. After running up to speed, the capacitor is switched out of the circuit by means of a speed monitoring circuit breaker. A Steinmetz circuit is suitable for motors with a power rating of up to 2kW due to the large value of capacitance required at higher power ratings.

Electric Machines 63

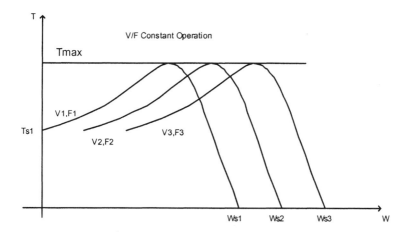

Figure 3.19 Torque-speed characteristics for V/f constant operation.

3.5 Synchronous Machines

Synchronous machines are used mainly in the generating mode. In fact, most of the power generators are synchronous machines. This is mainly due to the high quality of the output AC voltage. Therefore, there is no need for any power conversion to achieve the required AC voltage. Control of the frequency and amplitude of the output voltage is much simpler than other generators. Output frequency is proportional to the rotor speed. Therefore, rotor speed is controlled to be constant. Rotor windings are connected to a DC voltage source.

Synchronous machines are also used in the motoring mode at very high power ratings. Advanced synchronous motors with permanent magnets are emerging as one of the best candidates for low power applications. They are permanent magnet synchronous motor (PMSM) drives, which have applications in vehicular power systems. These advanced motor drives are explained in the next chapters.

In synchronous machines, at starting, the rotor vibrates and due to these vibrations, there is no starting torque. In order to have a starting torque, there are two methods that may be used. In the first method, the machine is started as an induction machine by adding more windings mounted on the rotor, resembling the cage of an induction motor. This results in an additional torque at starting; this type of winding is known as damper. In second method, a variable frequency supply is used where we can adjust the frequency starting from zero. The torque equation for a synchronous machine is as follows:

$$T = \frac{P}{\omega_s} = \frac{3}{\omega_s} \frac{V_t E_f}{X_s} Sin(\delta) \qquad (3.15)$$

where δ, ω_s, V_t, E_f, and X_s are power angle, synchronous speed, terminal voltage, back-emf, and synchronous reactance, respectively.

Figure 3.19 shows the relation between the power angle δ and the torque for a synchronous machine. From the torque-power angle curve, we notice that the maximum torque occurs at $\delta=90°$. If we exceed the maximum torque, the rotor will gradually stop.

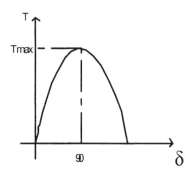

Figure 3.19 Torque-power angle characteristics of synchronous machines.

3.5.1 Torque-Speed Characteristics

Figure 3.20 depicts torque-speed characteristics of synchronous machines. We can conclude that for any synchronous machine, it can be a motor at nothing but speed Ns; it also cannot provide torque at anything but Ns. These motors have applications as constant speed drives, as are explained in Chapter 6. However, in order to have a variable speed operation, the frequency of the mains supply must be variable. Permanent magnet adjustable speed drives are explained in the next chapters.

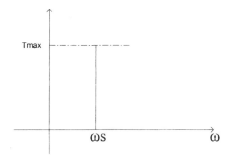

Figure 3.20 Torque-speed characteristics of synchronous machines.

3.6 Selected Readings

[1] A. E. Fitzgerald, C. Kingsley Jr., and S. D. Umans, *Electric Machinery*, 6th Edition, McGraw-Hill, 2002.
[2] R. Krishnan, *Electric Motor Drives: Modeling, Analysis, and Control*, Prentice-Hall, 2001.
[3] M. A. El-Sharkawi, *Fundamentals of Electric Drives*, PWS Pub. Co., 2000.
[4] P. C. Krause, O. Wasynczuk, and S. D. Sudhoff, *Analysis of Electric Machinery and Drive Systems*, 2nd Edition, John Wiley & Sons, 2002.
[5] N. Mohan, *Electric Drives: An Integrative Approach*, MNPERE, Minneapolis, 2001.
[6] N. Mohan, *Advanced Electric Drives*, MNPERE, Minneapolis, 2001.
[7] B. K. Bose, *Modern Power Electronics and AC Drives,* Prentice-Hall PTR, 2002.
[8] B. K. Bose, *Problems Manual for Modern Power Electronics and AC Drives,* Prentice Hall, 2001.
[9] T. Wildi, *Electrical Machines, Drives, and Power Systems*, 4th Edition, Prentice-Hall, 2000.
[10] B. S. Guru and H. R. Hiziroglu, *Electric Machinery and Transformers*, 3rd Edition, Oxford University Press, 2001.
[11] A. M. Trzynadlowski, *The Field Orientation Principle in Control of Induction Motors*, Kluwer Academic Publishers, Norwell, MA, 1994.
[12] P. Vas, *Vector Control of AC Machines*, Oxford Science Publications, New York, 1990.

4
Automotive Power Systems

The automobile electrical system, from the very first days of propulsion by internal combustion engine technology, was used only for igniting fuel in the engine's cylinders. In the more than 100 years of the automobile, the electrical system has become a major consideration in the design and performance of the entire vehicle. In the early years leading up to 1912, ignition of the fuel-air mixture in the cylinders was one of the major challenges. There were attempts at ignition using open-flame, hot-wire, and hot-tube igniters, but none were satisfactory due to the fact that the timing of the ignition event was highly variable and critical for proper function of the engine, not to mention the difficulty of implementation of the various alternatives. Creating an electric spark inside the cylinder was the most straightforward approach [1]. The earliest electrical ignition systems were actually low voltage systems in which a cam outside the cylinder drove a mechanical arm in the cylinder that made intermittent contact with a stationary rod that was inserted into the cylinder through an insulating sleeve and connected to a series of dry cells through a small inductor. When the contact in the cylinder closed a current would build up in the inductor, and when the contact opened a spark discharge occurred, igniting the gas mixture. This low voltage ignition system was used on single cylinder and multi-cylinder engines. Because control of spark timing was so difficult, the next step was to place a spark plug in the cylinder and to apply very high voltage to ionize the spark gap under the high chamber pressure existing in the cylinder at the time of ignition. These early systems used a magneto, a type of alternating current generator, that was either low tension or high tension depending on whether or not a high voltage winding was placed on the same core as the 6V primary. Power for the primary was again provided by a series of dry cells to power the system while the engine was hand cranked.

Later a system based on a Rumkoff coil located in the engine compartment was used to provide nearly continuous high voltage to the center contact of a distributor. This ignition system relied on a small wooden cased transformer module having a magnetic armature that energized a relay-like switch that would open the primary winding when sufficient magnetic charge was accumulated. Interruption of the primary winding current by opening the contact resulted in high voltage induced across the secondary winding as the stored magnetic field collapsed, sending 20kV or higher to the distributor and on to the appropriate cylinder. This electro-mechanical ignition system was very noisy both audibly and electrically since the spark source was virtually continuous.

A major development in the automobile electrical system occurred in 1908 when Charles Kettering developed the single spark, breaker type ignition system. This system was developed to overcome the many disadvantages of the magneto and "trembler" or "vibrator" coil systems. There were several other types of breaker systems produced by various suppliers but all worked at 6V and relied on a mechanical contact, or breaker points, to interrupt current to the high voltage coil. Arcing associated with interruption of primary current to the coil caused pitting and erosion of the contacts and eventual failure. To alleviate the damage caused by current interruption and subsequent arcing a capacitor was shunted across the points to absorb the current. When V8 engines were introduced by Ford in 1932 with their demand for higher compression ratio[1], it was necessary to increase the primary coil current to more than 10A. It was not uncommon for the contacts in breaker ignition systems to require maintenance or replacement every 6000 miles or less due to breaking such high currents.

Automotive batteries were at first dry cell for ignition only and, when they became depleted, it was necessary to replace them with fresh cells. The earliest wet type lead acid secondary cell was invented in 1859 by Gaston Plante by rolling thin strips of lead foil with porous insulating material sandwiched in between the films. When submerged in a sulfuric acid electrolyte, it became a relatively high energy capacity storage unit. By 1881, this "jelly roll" lead-acid battery had evolved to the flat lead grid structure commonplace today. For powering ignition only, a 60Ah lead-acid battery was very typical.

When the early oil and acetylene illumination lamps were replaced by incandescent lamps, the battery capacity had to be increased to 120Ah so that reasonable life could be realized. Power was relatively low for these early batteries, but with a cell voltage of 2.25V, it only required 3 such cells for a module. The automobile electrical system up to this time consisted of primarily the ignition system and the lights. There was still no generator to replenish the battery; thus, cars had to be periodically recharged and, if dry cells were being used, these had to be replaced. The year was 1912 when Cadillac Motor Co.

---

[1] Compression ratio, CR, was generally <6:1; but, with V8s and the demand for performance, ratios greater than 8:1 were introduced placing even more demand on the breaker ignition points due to high currents at 6V.

introduced the electric self-starter. This had a major impact on convenience and acceptability of the automobile to the general public since all automobiles up to this time required hand cranking to start. The early starter motor was actually a starter-generator because it had both sets of windings on one armature each with its own set of brushes and commutator. The impact of the electric starter on the automobile industry was so great that, within the following two years (1913 and 1914), 90% of all cars built would rely on it. In 1913, the Robert Bosch Co. in Germany produced a geared starter in which a planetary gear and an overrunning mechanism were designed in. These features became characteristic of all future starter motors.

Charles Kettering realized that compact size and low mass would be critical for an engine starter motor when he invented the self-starter. Early starter motors were bulky compared to today's units, which are required to have low mass, compact dimensions, low noise, low cost, lifetime durability of 30k to 40k engagements and be capable of delivering high torque necessary for cold starts at $-28°F$ at 80% SOC. In the future, starter motors will be required to deliver even faster cranking to reduce emissions and noise and have durability of up to 400k engagements should idle stop be employed. It is interesting to note that, today, the commonplace 1.1kW starter motor, a size that accommodates engines in the 1.4 liter to 2.8 liter displacement class, has a mass of only 2.4kg.

The issue with starter motors now, as in the early days of the automobile, is that the 200 to 320 rpm capability of these geared devices does not deliver the crankshaft speed and, hence, the piston velocity over top dead center necessary for reliable starting. It was recognized early on that direct drive starter motors, although more bulky and heavier, would however deliver crankshaft speeds in excess of the 450 rpm necessary to insure more reliable starting without need for over-fueling during cranking [2]. The integral starter motor and DC generator remained in production well into the 1920's; but, as time went on, the starter became independent of the generating function. Mechanical complexity of needing a high reduction gear ratio >20:1 for engine cranking yet a reasonable gear ratio of <3:1 for the generator so that it would not over speed during running became major hurdles in maintaining a common unit. The early DC generator was rated from 15 to 20A at 6V. With a generator, the electrical system battery capacity was reduced to some 30Ah because the generator maintained the battery SOC and provided power for ignition and lights.

Maintaining regulation of the early DC electrical system was implemented in the DC generator first by a single current sensing solenoid and later by a combination of three relays. A current sensing relay was in reality a linear solenoid in series with the generator output so that its armature moved in response to current and connected to a coil of resistance wire that was placed in series with the generator field winding to implement field weakening control of the generator. Innovations in DC generator regulation lead to use of a third brush to supply current for the generator field. Later a voltage sensing and current sensing relay were added to improve the regulation. A third relay was added in series with the generator to prevent the battery from discharging into the DC generator when the engine was not running. Temperature compensation of the generator output was implemented so that the charge acceptance characteristics of the lead-acid battery could be matched. Early systems used a bimetallic element and a set of contacts so that as temperature dropped, a resistance could be added in the field winding circuit to change the output voltage. The major improvement in the automobile electrical system during the 1920's was the introduction of the hard rubber cased battery to replace the earlier wooden cased types. Cadillac continued to use the Kettering breaker point ignition and Ford continued to use the magneto and vibrator coils until the introduction of the Model A in 1928. The years between 1912 and 1930 were a time of the most innovation in the vehicle electrical system and a time when all the fundamental concepts were established.

4.1 Conventional 14V Electrical System Architecture

Developments of the automobile electrical system remained relatively static in the years between 1930 and 1960. There were improvements in ignition to match the demands of higher compression engines and durability of components was improved but no real changes in concepts. By the 1950's, the ever present issue of reliable ignition again came to the forefront to challenge the automobile electrical system. With newer engines and higher compression ratio, the 6V ignition was exhibiting signs of becoming overwhelmed and durability was becoming a major obstacle. In 1952, during a meeting of the chief engineers of all the automobile companies in North America, the realization was present that unless something was done to improve the ignition system, there would be an epidemic of failed engines on the streets due to welded breaker points. The decision was made to increase the electrical system voltage from a 6V battery with 7V DC nominal distribution potential to a 12V battery with 14V DC distribution level. The first 12V systems were introduced by General Motors in 1955 followed by Ford and Chrysler. The same rationale was present in Europe and soon all but VW would switch to 12V systems. There were some engineering difficulties stemming from the transition to 12V, primarily in lighting, but other systems adapted quickly. The starter motor for example, already overrated for its function, was even more overrated at 12V. The battery became a six cell module instead of three cell at approximately the same energy rating. The battery's physical shape changed to a module that was narrower and longer than before. The generator armature windings were changed to twice the number of turns with small gauge wire to match 12V system needs. Beyond that, the generator, regulator, and electrical system in general were not impacted very much. The ignition system however was and it was now capable of delivering more energetic sparks, with improved reliability.

During the period from 1960 to 1980, the most significant change was the replacement of the DC generator with the Lundell alternator. The driving force behind the change was the increasing demand for electrical power. Electrical system demand had risen from the 100W of 1912 to typically 500W in 1960 and, by 1980, it increased even more to 1500W as more and more electrically power devices were installed: the car radio, introduced in the early 1920's, more electrical lighting, electric windshield wipers, and cabin heating using blower motors for air circulation. The enabling technology that resulted in the practical introduction of the 12V Lundell alternator was the availability of the semiconductor bipolar diode. In 1958, Delco-Remy introduced diode rectified generators for vehicles having high electrical loads. The Lundell alternator was invented by Chrysler Motor Co. engineers in 1933, but it was impractical for mass production because of the unavailability of efficient rectification. Early press pack bipolar diodes changed that and, in 1961, Chrysler introduced the 12V alternator in high volume passenger cars. This move was followed quickly by GM in 1963 and Ford in 1965.

The diode bridge had inherent protection against battery drain during engine off periods so the generator cutout relay could be omitted but the field relay was retained. It wasn't long before the field regulator relay was replaced by an electronic voltage regulator and, by the mid-1960's, electronic regulators were used in most cars. In fact, electrification of the basic engine functions began in earnest during the late 1960's with the introduction of electronic ignition to eliminate the mechanical breaker points with a transistor switch. In 1963, the first magnetic pulse pickup distributor was introduced by Delco-Remy and, by 1975, most vehicles produced in the U.S. relied on pulse pickup transistorized ignition. Electrification continued with the introduction of electronic engine controls. In 1978, Ford introduced a microprocessor based system to control engine spark timing and exhaust gas recirculation (EGR). Motivation for electronic engine control enabled fuel economy and reduced emissions were a direct result of the oil shortage and energy crisis of 1974-1975. Later, electronic fuel injection was included. From 1980 to the present, the capability and pervasiveness of electronic engine controls have increased to the point of taking complete control of the engine combustion process, air and fuel metering, exhaust gas treatment, alternator control, and recently transmission control. Next, we turn our attention to the electrical system architecture and how it is being impacted by power train electrification and the impending demand for a more electric vehicle.

The conventional electrical system in an automobile can be divided into the architectural elements of energy storage, generation, starting, and distribution. We have discussed energy storage, generation, and starting from a historical perspective. Electrical distribution systems provide electrical power to connected consumers including ignition, interior and exterior lighting, electric motor driven fans pumps and compressors, and instrumentation subsystems. In order for the power available at the sources to be made available at the terminals of the loads, some organized form of distribution throughout an automobile is essential. At present, most automobiles use a 14V DC electrical system. Figure

4.1 shows the conventional electrical distribution system for automobiles. This has a single voltage level, i.e., 14V DC, with the loads being controlled by manual switches and relays [3]-[10]. Because of the point-to-point wiring, the wiring harness is heavy and complex.

The present average power demand in an automobile is approximately 1kW. The voltage in a 14V system actually varies between 9V and 16V at the battery terminals, depending on the alternator output current, battery age, state of charge, and other factors. This results in overrating the loads at nominal system voltage. There are several other disadvantages, which have been addressed in [5]-[8].

Besides all the disadvantages, the present 14V system cannot handle future electrical loads to be introduced in the more electric environment of the future cars, as it would be expensive and inefficient to do so.

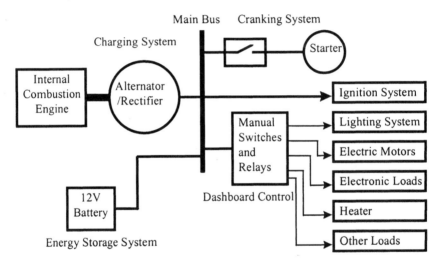

Figure 4.1 Conventional 14V DC distribution system architecture.

4.2 Advanced Electrical Loads

In more electric cars (MEC), there is a trend towards expanding electrical loads and replacement of more engine driven mechanical and hydraulic systems with electrical systems. These loads include the well-known lights, pumps, fans, and electric motors for various functions. They will also include some less well-known loads, such as electrically assisted power steering, electrically driven air conditioner compressor, electromechanical valve control, electrically controlled suspension and vehicle dynamics, and electrically heated catalytic converter. In fact, electrical subsystems may require a lower engine power with higher efficiency. Furthermore, they can be used only when needed. Therefore, the

Automotive Power Systems

MEC can have optimum fuel economy and performance. There are also other loads such as anti-lock braking, throttle actuation, ride-height adjustment, and rear-wheel steering, which will be driven electrically in the future.

Figure 4.2 shows electrical loads in MEC power systems. As is described in [7]-[10], most of the future electric loads require power electronic controls. In future automobiles, power electronics will be used to perform three different tasks. First task is simple on/off switching of loads, which is performed by mechanical switches and relays in conventional cars. Second task is the control of electric machines. Third one is not only changing the system voltage to a higher or lower level, but also converting electrical power from one form to another using DC/DC, DC/AC, and AC/DC converters.

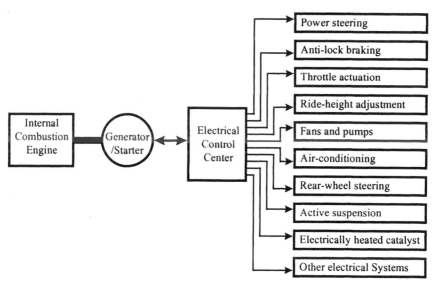

Figure 4.2 Electrical loads in the MEC power systems.

4.3 Increasing the System Voltage to 42V

Due to the increasing electrical loads, automotive systems are becoming more electric. Therefore, MEC will need highly reliable, fault-tolerant, autonomous controlled electrical power systems to deliver high quality power from the sources to the loads. The voltage level and form in which power is distributed are important. A higher voltage (such as the proposed 42V) will reduce the weight and volume of the wiring harness, among several other advantages [6], [7]. In fact, increasing the voltage of the system, which is 14V in the conventional cars, is necessary to cope with the greater loads associated with the more electric environments in future cars. The near-future average power demand is anticipated to be 3kW and higher.

Figure 4.3 shows the concept of a dual-voltage automotive power system architecture of the future MEC. Indeed, it is a transitional two-voltage system, which can be introduced until all automotive components evolve to 42V. Finally, the future MEC power system will most likely be a single voltage bus (42V DC) with a provision for hybrid (DC and AC), multi-voltage level distribution and intelligent energy and load management.

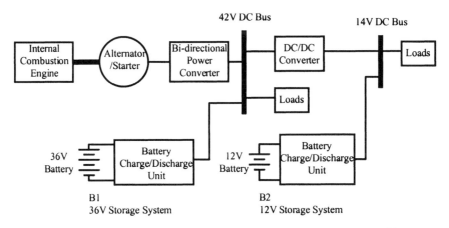

Figure 4.3 The concept of a dual voltage automotive power system architecture of the future MEC.

Transition to 42V PowerNet vehicle electrical system will mean an overall reduction in vehicle wiring harness mass and wire gauge. Figure 4.4 illustrates this impact.

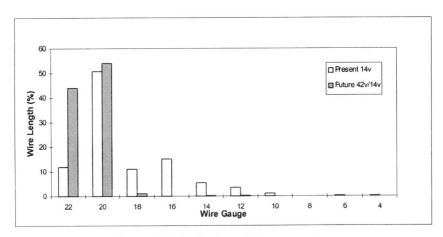

Figure 4.4 Distribution of wire gauge size.

Automotive Power Systems 75

There will be a similar impact on vehicle electronic content, as depicted in Figure 4.5. The introduction of power electronics-enabled functions will have a concomitant increase in silicon content for the average passenger car.

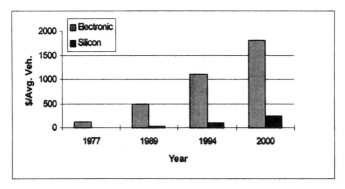

Figure 4.5 Value of electronic content and silicon content of average car.

Wire harness current carrying capacity is illustrated in Figure 4.6, where it can be seen that current capacity decreases somewhat linearly for decreasing wire gauge.

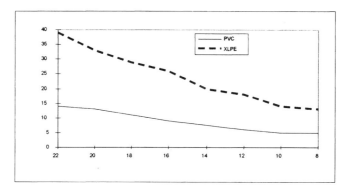

Figure 4.6 Wire current carrying capacity (A/mm^2) vs. wire gauge. Allowable current densities (A/mm^2): Δ T= 50°C for PVC, Δ T= 95°C for XLPE.

Polyvinyl chloride (PVC) is the most common insulation system for wire harness in passenger cars. The higher temperature-capable cross-linked PVC known as XLPE is used in wet locations such as the door interior harness because of its moisture resistance and insulation creep integrity, which helps maintain environmental seals. Wire gauge selection can be computed using the geometric series relation of wire gauge number to diameter, as described below.

Note from Figure 4.4 that 20 AWG is the most common wire gauge in both the present 14V electrical system and in the proposed 42V PowerNet system. We take the properties of 20 AWG wire as a baseline and from it compute the characteristics of any other wire gauge in the electrical distribution system (EDS) using the relations below, where $R_{20}(x,z,T)$ is the known wire of AWG "x", length "z" in feet and at temperature T_o in °C. The reference temperature, $T_o = 20$ °C in all wire tables.

$$R_{20}(x_o, z_0, T_0) = 10.15 \quad (m\Omega/ft) \tag{4.1}$$

$$R_{cable}(y, z, T) = R_{20}(x, z_0, T_0) * z * (1 + \gamma(T - T_0)) * \frac{1}{2^{(x-y)/3}} \quad (m\Omega) \tag{4.2}$$

For example, the battery cable in a conventional 14V automotive system is 2 AWG stranded copper wire of total length of approximately 7 feet when the battery is located underhood. For this cable, equation (4.2) predicts a cable resistance R_{cable} at the underhood temperature of 70 °C of 1.327 mΩ. Each connector due to crimping and material properties will have a resistance of 0.5 mΩ. A relay contact or switch contact is on this same order of resistance.

To further illustrate the impact of harness and connector resistance on electrical system performance, consider now that the cable discussed in relation to equation (4.2) has 8 connections (terminations): 1 at the ground wire to the engine block, 2 at the battery terminals, 2 at the starter motor solenoid contactor, 2 at the starter motor brushes, and 1 at the starter motor case ground to engine block. In this complete circuit, the battery will have an internal resistance that is a function of its temperature, SOC, and age. We can assume this to be a typical 70 Ah Pb-acid battery with internal resistance of 7 mΩ if new and at better than 80% SOC at room temperature. To calculate the maximum current to the starter motor under these conditions, we assume that the starter motor armature has a resistance matching the battery internal resistance at nominal conditions, but that it has the same temperature dependence as the cable (both use copper) and that the starter motor brushes develop a net voltage drop of 1.1V. This yields a voltage drop of 0.55V per brush, which is very typical of DC motor characteristics and brush-to-commutator properties. Since the battery is nearly fully charged, it will have an internal potential of 6*2.1V/cell = 12.6V. We calculate the maximum current delivered to the starter motor as:

$$I_{starter} = \frac{(V_{batt} - 2V_{brush}) * 10^3}{R_{cable}(y, z, T) + N_c * R_{conn} + R_{int} + R_{arm}} \quad (A) \tag{4.3}$$

For the conditions noted above and where N_c = number of interconnects, we find that starter motor current is 556A maximum. In today's automobile, the starter motor is internally geared with a planetary set having a ratio G_{gr}= 3.6:1

Automotive Power Systems

and externally geared at the pinion to ring gear at the crankshaft of G_{rg}= 14:1 or slightly higher depending on the engine. In fact, the same starter motor can be used for several different engine displacements by changing the ring gear number of teeth (gear ratio). Now using the fact that this starter motor has a torque constant, k_t = 0.011 Nm/A, we can calculate the torque delivered to the engine crankshaft as:

$$T_{crank} = N_{gr} N_{rg} k_t I_{starter} \quad (N-m) \tag{4.4}$$

For the conditions given and for the approximations made, we find that this permanent magnet, geared, starter motor capable of delivering 308 N-m of torque to the engine crankshaft. If we now assume the temperature is at -30 °C, we can recalculate the starter motor maximum current by first noting that although the cable, termination, and starter motor armature resistances will decrease in proportion to temperature, the battery internal resistance will actually increase due to slower ion transport dynamics and increased polarization. We will make the approximation that cable and armature resistance, since these are copper wire based, will have the same resistance change. Terminations on the other hand will be assumed to have negligible change. The battery internal resistance for this example is approximated according to the following expression (α = 0.0003, β =2):

$$R_{int}(T) = R_{int}(T_0) * (1 + \alpha * (T - T_0)^\beta) \quad (\Omega) \tag{4.5}$$

With this modification to equation (4.2), the starter motor current at -30 °C actually drops to 504A, yielding a cranking torque of 280 N-m. In reality, the starter motor brush voltage drop will also increase, further reducing the available torque. Typically, starter motor torque capability at cold temperatures is still far in excess of the torque necessary to breakover and crank a cold engine. Cranking speeds at cold temperatures are also slower due to higher friction in the engine so that crank speeds of 100 rpm or less are typical.

One further point to make regarding brushed DC motors and cold conditions is that if improperly designed, the cold in-rush current, assuming a fresh battery, may be higher than the room temperature design point, resulting in demagnetization of the ceramic magnets. Unlike rare earth magnets used in starter-alternator and other high performance electric machines, a starter motor and most DC motors in the car will rely on ceramic 7 or ceramic 8 magnets, which have less coercive force at cold temperatures and so could be partially demagnetized.

4.4 Advanced Distribution Systems

The conventional automotive electrical power system is a point-to-point topology in which all the electrical wiring is distributed from the main bus to different loads through relays and switches of the dashboard control. As a result, the distribution network has expensive, complicated, and heavy wiring circuits. However, in the advanced automotive electrical systems, multiplexed architectures with separate power and communication buses are used to improve the system. In a multiplexed network, loads are controlled by intelligent remote modules. Therefore, the number and length of wires in the harness are reduced. In addition, these systems have a power management system (PMS). The primary function of the PMS is time phasing of the duty-cycle of loads in order to reduce the peak power demand. Other functions of the PMS are battery management, load management, and management of the starter/generator system including the regulator. Figure 4.7 shows typical inputs and outputs of a power management center.

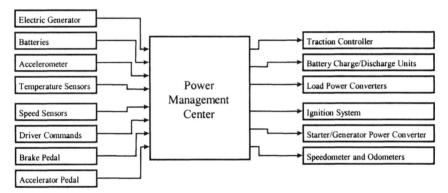

Figure 4.7 Power management system.

Figure 4.8 shows advanced multiplexed automotive power system architectures of the future with power and communication buses. The distribution control network of Figure 4.8 simplifies vehicle physical design and assembly process and offers additional benefits from the amalgamation with intelligent power management control.

4.5 Starter, Alternator, and Integrated Starter/Alternator

Article I. The increasing power demand in vehicles has resulted in a need for a higher onboard generation capacity. With the increasing generation requirement, the torque characteristics of the generator are found to closely converge with that of the starter motor. Hence, integrating the two machines and using a single machine for the two purposes would be technically viable and economically advantageous. It would result in a more compact design solution as well.

Projections for electrical load increase today are following a year-over-year growth of roughly 4.5%, on top of which we envision loads representative

Automotive Power Systems

of power train and vehicle electrification. Figure 4.9 illustrates this trend in EDS load growth as higher power functions such as electric assist power steering, electromechanical valve actuation (EVA) and electric driven brakes, water pump, and fans are incorporated.

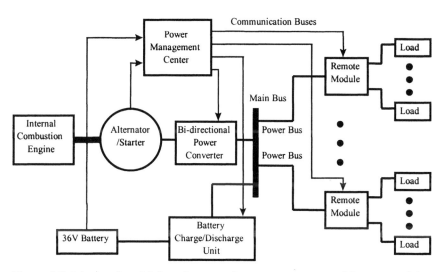

Figure 4.8 Advanced multiplexed automotive power system architectures of the future with power and communication buses.

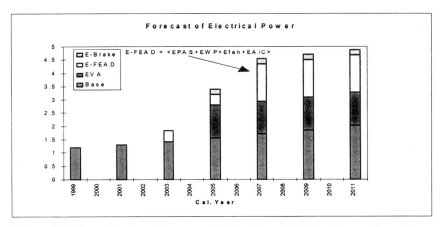

Figure 4.9 Electrification of power train functions (FEAD = front end ancillary drive): average power (kW) vs. calendar year.

4.5.1 Introduction

The automotive industry has gone through many radical changes and one of them is the present change in the automotive power system. In the first half of the past century, the 6V electrical system in automobiles served the purpose of ignition, cranking, and a few lighting loads. Since then, there has been a constant rise in the power requirement, as illustrated in Figure 4.9.

The major reason for such a high power system and, hence, a need for a high power onboard generation system is due to the increasing number of electrical loads [11]-[13]. Performance loads, such as electric steering, that were traditionally driven by mechanical, pneumatic, and hydraulic systems are now increasingly being replaced by the electrically driven systems to increase the performance and efficiency of operation. When driven electrically, power is available on demand, which enables reduced and optimal power consumption. Luxury loads have also increased over time, imposing a higher demand of electrical power. The increasing induction of safety loads such as electrically active suspension and roll-over stability have also imposed high power demands on the vehicular power system. However, a major concern with safety loads is not just the need for high power but a highly reliable source of power.

Furthermore, loads like electromagnetic valve train (EVT or EVA), whose power requirements are a function of engine speed would put high stress on the power network in a different way. Loads such as windshield defrosters, on the other hand, demand power at higher voltage for efficient operation. Electrical loads on the automobile are modeled using power law relations for which nominal power coefficient and exponent are laboratory characterized. For example, an EVA system for an internal combustion engine has approximately a constant torque characteristic, similar to a conventional valve train where the nominal power at idle is P_0 = 275 for V8-4V, 125 for V6-4V and perhaps 80 for I4-2V and xV represents the number of valves/cylinder:

$$P_{EVA} = P_0 \frac{n}{500} \quad (W) \tag{4.6}$$

$$P_{load} = P_0 (\frac{V}{V_0})^\alpha \quad (W) \tag{4.7}$$

In equation (4.6), the EVA actuator power is normalized to idle speed. In all other electrical loads, we model the power demand as a function of the distribution system voltage as a power coefficient and exponent. This power law characteristic captures the load variability as system voltage changes due to temperature stabilization of the regulation characteristic. The exponent further characterizes the load in terms of its inherent character. For example, for a resistance load such as windshield or mirror defroster or heated seats, the exponent is 2.0. For electric incandescent lights the exponent becomes 1.6 and for fan motor loads the exponent is 2.3 to 2.7 (theoretically it would be 3.0, but practical limitations, air bypassing the blades, type of fan whether axial or radial [cage fans] dictate its power characteristic).

Thus, such load demands have resulted in the need to scale up the onboard vehicular power level. Increasing the power with the same voltage level would lead to issues such as increased wiring harness and insulation that would lead to higher cost and weight. Also, the system losses would increase. A possible solution for this would be to increase the voltage level.

Considering these aspects, several decades back, the voltage was raised from its earlier 6V level to the present day 12V level and, now with an ever-increasing demand forecasted into the future, there is a need to switch over to the much higher voltage level of 42V. The rationale today is however not at clear cut as it was in 1955. There is no single issue as pressing today as ignition failures were back then. If there were, the case for 42V would not be delayed as it is and we would have far more 42V vehicles on the highways. Notwithstanding these caveats, 42V is appearing in the market place and it will continue to make inroads into the luxury passenger car and sport utility and light truck fleets.

Such a change in the system network voltage and power brings about a major challenge in terms of the alternator, the heart of the network. The conventional claw pole synchronous alternators have reached their maximum limits of generation capacity. An attempt to generate higher power would lead to heavy losses and unacceptable cooling requirements, resulting in low efficiency. Thus, there is a need for a new alternator system for such high power generation [14].

The new alternator that would be introduced requires higher torque demand than the conventional production alternator in order to generate higher power and closely converges with the torque requirements of the present starter motor used to crank up the engine [15]. This opens the possibility to integrate the starter and the alternator as a single machine called the integrated starter/alternator (ISA), which would perform the combined operation of both. This integrated system that will be designed and developed will be able to supply high power of 4-6 kW at 42V in the near future. The high power capacity of the system would also enable it to lend itself to other multiple applications that would result in higher performance, better fuel economy, and reduced emissions.

4.5.2 Starter

The starter motor is used in automobiles for cranking the engine. These starter motors were first developed around 1910 to replace the hand crank. Since then, the starter motor has gone through multiple upgrades in terms of engagement methods, gear reduction, changing over from 6V to 12V, size and weight reduction, and improving its reliability and performance [14], [16].

Series starter motors used field windings initially. These motors have high starting torque and high no-load speed characteristics, which are ideal for engine starting applications [17].

The permanent magnet (PM) starter motor replaces the field winding with the PM, which improves performance of the machine. The absence of the rotor winding lowers the losses and the high flux density of the permanent magnet provides higher output and a reduction in size of the machine for the same rating when compared to the series starter motor. However, compared to the series starter, this PM starter has lower starting torque and no-load speed for the same performance at the rated condition, which would be harmful to the engine. The cost of these machines is also high compared to the series motors.

The permanent magnet starter motor with auxiliary poles gives the combined advantage of higher performance of the PM starter motor and the high starting torque and no-speed of the series motor [17]. Here, two parts contribute to the magnetic flux: the permanent magnet and the iron. Thus, the amount of permanent magnet required is reduced compared to the PM starter motor.

The starter motor is similar to other machines except that it has a magnetic switch that controls the engagement of the pinion on the rotor to the ring gear on the crankshaft. It is usually a field winding excited machine. When the field is excited and the motor rotates, the magnetic switch is activated and, with the help of a rotary lever mechanism or Bendix assembly, the pinion is coupled to the ring gear. At the end of the starting mode, the magnetic switch is deactivated, which de-energizes the motor and the engine now overspeeds the pinion, causing it to return to its retracted position. A spring at the end of the Bendix assembly facilitates the return of the pinion. During engagement, inertial forces on the pinion due to starter armature acceleration actually throws the pinion into engagement while simultaneously compressing the spring.

Starter motors of small size and high speed versions with planetary gear arrangements are also used for start-up. Here, the torque and the speed are transformed using the gear arrangement.

Since the starter is usually placed next to the transmission, it is always preferred that the starter be small in diameter and axial length. Thus, some starters have configurations in which both the pinion and the ring gear are placed within the hollow rotor.

Conventionally, the ICE is started using a single starter motor, which are gear coupled to the crankshaft. The starter motor brings the speed of the engine up to the 150-200 rpm at which the fuel is injected and ignited. The support of the starter motor is withdrawn and the engine climbs up to the idle speed on its own. This starting method results in the use of a large amount of fuel, most of which is wasted due to incomplete combustion, and also results in a high emission, particularly hydrocarbons, during starts. The lead-time to start the engine is high and the accompanying acoustic noise is high, mainly due to the coupling gears. However, these starter motors are able to provide the sufficient torque of up to 150 N-m for starting.

A second method of starting an ICE involves using the starter motor to crank the engine up to its idle speed of about 600-800 rpm and then injecting the fuel for combustion. This does have the advantage of fast and smooth engine start-up but loses the combustion stabilization benefits at such high speeds. In this case, the power required by the starter would be very high and such high power drawn over a very short interval of 1-1.5s would reduce the battery life considerably.

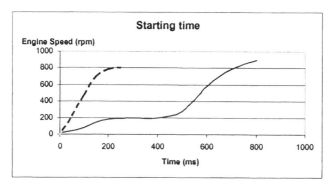

Figure 4.10 Engine cranking by conventional starter and crankshaft mounted ISA.

A third technique of starting an engine involves the use of two starter motors, starter 1 and starter 2. Starter 1 is used to start the engine from 0 rpm and crank it up to 400 rpm. At 400 rpm, starter 1 is disengaged and starter 2 is engaged to raise the crankshaft speed further to 600-800 rpm. At this point, the fuel is injected for combustion and starter 2 is simultaneously disengaged. Starter 1 is essentially of high power and is used to overcome the high inertia and hydrodynamic friction while starter 2 is used only to raise the speed of the crankshaft. Making use of two starters ensures a reliable and efficient cold starting. However, during warm starting, the use of starter 1 may not even be necessary. Starter 2 alone would be able to start the engine successfully. The drawback is that two electrical machines are used instead of one, which brings about two major issues – their packaging and space requirement from the design point, and the additional cost for the second starter.

Another method that is similar to the above makes use of one starter and an alternator during starting. The alternator is operated as a motor and performs the function of starter 2 during starting. The starting torque available from the alternator would be less as it is coupled to the crankshaft, usually through a belt. Here, the alternator primarily adds momentum to the crankshaft and transfer torque. A chain or gear coupling enables higher torque transfer to the crankshaft and the loss due to the belt slippage can be avoided. Thus, the cost of the second starter motor is eliminated and also the noise levels are less if the alternator coupling via belt is used.

4.5.3 Alternator

The electrical power demand in automobiles has been rising considerably over the past 25-35 years. This is primarily due to the introduction of various safety and infotainment loads. In addition, the increasing need to change over the mechanically driven loads to be electrically driven for improved performance has resulted in higher electrical power demands.

Among these loads, the in-car entertainment loads draw a very low power. However, loads such as lighting and heated back lights form a major power consumer of about 57% of the total power generated, while safety controls demand only around 10-11%. There are also loads whose power requirements change with speed, such as the electromagnetic valve train, where the power required at idle is only about 275 W. At cruising speed, it is 2 kW, and it increases to 3.3 kW at wide open throttle (WOT).

The electrical power requirements also vary considerably depending on the season. In winter, more power is consumed, or even on a dry day and a wet night, where the variation is nearly 75%. Even under maximum load periods such as in winter, the power required is 65% less than the maximum alternator capacity.

The alternators in automobiles are designed to match the power demand and the alternator output at idle speed. This typically has resulted in over-specifying the conventional alternator. Besides, the need to attain a lower idling speed in vehicles for higher fuel economy has led to over-sizing of the machine.

The losses in the alternator have to be taken into account as they directly reflect on efficiency figures and, consequently, impact the fuel economy. The losses in the alternator may again depend on the speed of the machine, such as iron losses and friction and windage losses, which increase with speed, while the copper losses are more constant. The nature in which these losses increase might be important too. The friction and windage loss for instance increase gradually to a certain speed, beyond which the losses increase abruptly. On the other hand, copper losses, even though constant with speed, start to increase with temperature during the course of operation.

The rotational loss, which is lower than the copper loss at low speed, becomes higher than copper at high speeds. Thus, more fuel is expended to generate the same required power at high speeds.

The design of the automotive alternator should thus take into account these factors and should involve a design concept that highlights developing an efficient machine and improved design methodologies for various electrical components and identifies operating regions and techniques that lower the losses.

The iron loss in the alternator depends also on the magnetic characteristics of the laminations, electrical frequency, and magnitude of flux density. In order to reduce the iron loss, if the flux density is reduced, then the size of the machine would increase. The core loss may be reduced further by using thin laminations to minimize eddy currents; however, the fragileness of the laminations becomes high, leading to manufacturing problems. Another option would be to use low loss material such as Transil 315/335 at the expense of high cost.

The conventional Lundell alternator is a claw pole synchronous machine used to generate 14V output. It consists of a rotor mounted on the two end bearings of the machine frame, a stator, six diodes, and a regulator. The generated AC output from the Lundell is rectified to a single DC voltage and then regulated to 12V [13], [14]. The rotor is made of iron pole pieces and many turns of fine wire (~300t), which are mounted over the shaft of the machine. The rotor coil leads are traced out through slip rings and brushes to the external circuit. When the rotor coils are energized, an electromagnetic field is produced which magnetizes one set of six teeth claw poles set at magnetic South and the other set at magnetic North.

The stator of the alternator consists of a three-phase winding, which may be connected either in star or delta. The output of the stator is fed to the three-phase rectifier made of six semiconductor diodes to give a DC voltage (Figure 4.11). The rectifier diodes, generally press pack types, are mounted into a collar for heat sinking.

The regulator is used in the system to regulate the output voltage after the alternator reaches cut-in speed, which is approximately 1000 rpm at its shaft. Similarly to the DC generator the output voltage of the alternator increases linearly with speed. A Lundell alternator outputs nominal voltage at idle speed; at high speeds, if the voltage is not regulated, it would result in a very high over-voltages of n_{alt}/n_{ci} *14V, where n_{ci} is the alternator cut-in speed of 1000 rpm. Thus, the regulator functions to limit the output voltage to a pre-defined value. This regulation is achieved here by controlling the field voltage by turning it on and off alternatively [14].

These alternators have been developed and re-engineered over the decades for improving the performance and are used in vehicles for charging the battery and to supply the auxiliary loads. Thus, for the alternator being the main element in the vehicular power systems, it is required that it should have high performance and reliability.

The claw pole structure used in the Lundell alternator produces high windage losses, and the use of high field ampere-turn in the rotor leads to high rotor losses, altogether lowering its efficiency. The magnetic circuit efficiency is also low due to the use of the large number of stator windings and the losses associated with it. Due to the high permeability of the claw poles, the power density output of the alternator is low and the pole face losses are high. Also, the claw pole structure makes it very sensitive to airgap flux variation and is the cause for the high noise level [13], [18].

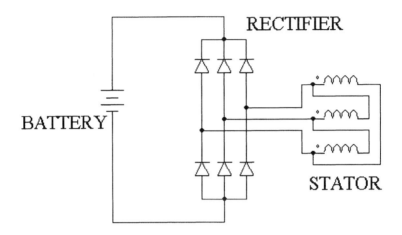

Figure 4.11 Electrical circuit layout of Lundell alternator.

The alternator also suffers from high rotor inertia, leading to high belt slippage. The Lundell alternator, although being a simple machine, has design constraints in terms of length-to-diameter ratio. Many of the avoidable losses in the alternators can be reduced by speed boundary operation of the alternator. Optimizing the duty cycle and using a step or continuously variable reduction system can do this. The speed of the alternator is raised at low engine speed and is lowered at high speeds, thereby operating the alternator within a speed boundary. Avoidable losses up to 150W are eliminated at almost each operating point of the speed boundary.

The automotive industry assigns a high level of importance to every component size and weight that goes into the vehicle. In the alternator, a given power requirement can be attained by varying the length/diameter ratio in accordance to whether there is room for length or diameter. This is because power is proportional to stator length and diameter squared.

In general, the length of the alternator is usually constrained. In a conventional alternator, the end windings extend beyond the stator length and are responsible for a further increase in length of the machine even though they do not make any contribution to the power generated. This is a major concern when alternators of very low depth are required, for example when they have to be placed into the clutch belly. In such cases, triangular winding alternators can be considered, which reduce the end windings. This, in turn, reduces the copper winding and resistive voltage drop by about 25% compared to the conventional alternator. Among triangular shapes, the isosceles triangular winding configuration provides the option for the most compact system.

However, as a consequence of having lower depth of stator slot, designing a poly-phase configuration where the crossing of the windings over the length is essential rather than at end winding and the magnetic field produced that cannot be confined to the plane, have to be considered during the design. It can make use of triangular or rectangular magnets, but with the former, the magnetic flux would have the same sign. Moreover, with triangular winding the magnetic flux is half that of the conventional machine irrespective of the magnet shape. However, the magnitude of the EMF is the same due to the presence of a compensation factor. Besides, a machine with triangular winding and rectangular magnets tends to have a higher rms value of EMF in the windings.

Automotive alternators may be of electrically excited salient pole type. Here the magnetic field is provided by the field windings and can be used up to a speed of 5000 rpm. Beyond this, the stress levels under the pole teeth tips increases considerably and would result in mechanical instability. It has an efficiency of 17% higher than the conventional Lundell alternator.

Hybrid excited salient pole alternators can also be used. The excitation in this machine is provided by both the field winding and the permanent magnet. In this design, the rotor weight must be properly distributed by arranging the magnetic and electrically excited poles appropriately. Here too the mechanical speed is constrained to a maximum of 5000 rpm due to the stress developed below the poles surfaces and at the edges of the magnet slot. The efficiency of this machine is 33% higher than the conventional Lundell alternator; but, the cost is very high due to the use of the magnets.

4.5.4 Integrated Starter/Alternator

4.5.4.1 Integrated Starter/Alternator Requirements

The integrated starter/alternator (ISA) to be developed should not require maintenance. It should have an integrated semi-intelligent system to meet the prime objectives of the future. The ISA would essentially be a four-quadrant drive and an active element having multiple functional modes. The engine operating conditions would dictate its operation as a generator or motor. The ISA should also have wide constant power speed characteristics to be able to deliver at least 4 kW over the entire speed range of 1:10 during generation (Figure 4.12). Developing the controls for such operation is a major task. The ISA is also able to provide the necessary torque and speed for engine start-up. Even thought there is no specific minimum torque required for the engine start-up, it is usually preferred to be greater than 150 N-m [15].

The speed of rotation of the present starter motor used for cranking the engine is about 100-150 rpm. On the other hand, the ISA is able to crank the engine at a speed of 400 to 700 rpm for fast engine start-up. This enables the

ISA to reap the benefits of reduced fuel consumption and lower emissions. However, there are a few factors that delay the cranking operation, such as lags in microprocessors used in the control circuit, large electrical time constant of the machine, and power electronic components. In machines such as the induction type, the time required for building up of the magnetic field contributes to the delay. However, if the cranking speeds are very high, above 500 rpm, it would lead to prolonged time for start-up as the combustion stabilization benefits are reduced and the hydrodynamic friction components increases [16]. Furthermore, the DC output from the machine has violent oscillations due to non-optimal machine parameters such as slip.

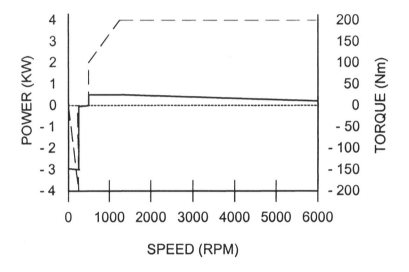

Figure 4.12 Power and torque vs. speed characteristics for ISA.

The ISA also needs to address the issue of cold starting of the engine. For instance, at very low temperatures, the engine oil viscosity is very high, leading to high hydrodynamic friction. The initial crankshaft position also affects cranking due to the non-uniform distribution of hydrodynamic friction. Another issue is that, at low temperatures, the battery voltage falls and thus the maximum cranking speed attainable is also reduced. However, this factor depends on the machine that is used as ISA [16].

As previously mentioned, the increasing safety loads in vehicles driven by electric power necessitate a reliable power source. Thus, the ISA to be developed should have high fault tolerance and, in an extreme case resulting in a fault, it should still be able to deliver power under reduced performance.

Automotive Power Systems

When ISAs are mounted to the crankshaft, they are exposed to extremes of harshness in terms of high temperatures and vibrations. Therefore, the system needs to be as rugged as possible, and an effective cooling, either air or liquid cooling, should be provided. The ISAs should also have low noise levels, particularly at speeds below 3000 rpm, because above this speed the engine noise level dominates.

The ISA, when operating as a generator, should be able to generate high power at high efficiency. Also, in the generation mode, there are the issues of load dump situations that have to be accounted for. For the Lundell alternator, the load dump is the worst-case scenario; however, it may not be the case for other machines. The present Lundell, which operates at 14V, faces 40V on load dump, limited primarily by semiconductor clamping action and specified as the upper limit by SAE, a near 285% higher voltage [16]. On the other hand, an ISA operating at 42V would face 60V on load dump, which is just a 50% higher voltage (Figure 4.13) [11], provided its design adheres to the ISO draft specification for 42V PowerNet.

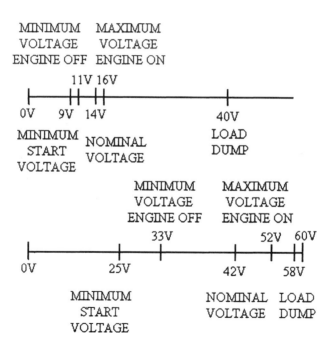

Figure 4.13 14V and 42V system voltage profile.

Since the ISA operates under different modes, but primarily either as a motor or generator, there is a need for a smooth transition from one mode to the other. This may be done by disconnecting the ISA from the power system for a momentary period. It may also be required to be done over a certain speed ranges, for example, after the engine starts up at about 400 rpm. The transition could be made at any speed between idling speed and 500 rpm.

As for the mechanical structure of ISA, it should be simple, lightweight, compact, have low rotor inertia, and low cost. Low rotor inertia requirement is of particular interest for offset coupled ISAs to reduce belt slippage. It would also be desired that the system should be able to deliver high torque at starting and have high power density.

4.5.4.2 Integrated Starter/Alternator Applications

The ISA is primarily used for starting and generating power, thereby performing the function of starter and generator, respectively. However, the ISA to be implemented as a new technology would require that it address a major issue that would enable it to be used in the automotive industry. This would be the start-stop and regenerative braking application, which would bring reduced fuel consumption and lower emissions, thereby providing OEMs with an option of achieving and compiling with the increasingly stringent CAFÉ and emission standards.

The start-stop application primarily requires the engine to shut down a few seconds after the vehicle comes to a halt and the engine is idling. Once the gas pedal is depressed, the engine starts up seamlessly due to the high speed cranking capabilities of the ISA. This reduces the fuel being consumed during idling condition. The regenerative breaking makes use of the kinetic energy of the vehicle during the braking to generate power, thereby eliminating the fuel consumption to generate that power.

ISA also provides the advantage of acceleration support. The present starter motor cranks up the engine and the ignition takes place before the idling speed. This results in wasted fuel and high emission due to incomplete combustion of fuel (A/F is over-enriched). However, in the case of ISA, it cranks the engine, the ignition takes place, and the ISA continues to crank the engine up to the idle speed. This brings reduced fuel consumption and lower emissions due to the complete combustion of fuel. Also, since with ISA, it is possible to carry out the ignition at higher speed levels, the hydrodynamic frictions will be less and, at the same time, can benefit from better combustion stabilization [16].

The ISA can also provide active engine vibration damping, thus offering a smoother vehicle drive. This can be achieved by providing torque pulsations that cancel out the pulsations of the engine [19]. Similarly, the ISA can also be extended to provide torque pulses during coasting with engine shut-down. When the engine is shut down, during the coasting period, jerks are produced in the vehicle due to slow compression and expansion and also due to a vacuum being

Automotive Power Systems

created during each expansion. Thus, by providing torque pulses at particular crankshaft or piston positions, the jerks can be eliminated, thereby providing a smooth ride.

ISA is a high power machine that is connected to the crankshaft, so its application may be further extended to provide a low speed, short distance electric drive, similar to an electric vehicle. However, this would not be possible with an offset coupled configuration.

4.5.5 Machine Types

For starter/alternator applications, the machines identified as strong candidates for performing efficiently are permanent magnet (PM) and switched reluctance machines (SRM), whose design characteristics and control features are discussed below. However, the induction machine (IM) is also discussed for the characteristics that it offers for this application.

4.5.5.1 Induction Machines

Induction machine (IM) technology is quite mature and is very well known for its high torque and good starting torque characteristics. It is a low maintenance, highly reliable, and inexpensive machine. Due to its limited flux weakening, it is difficult to attain constant power speed performance. Alternate methods, such as using switches to switch winding or relays to change over converters can be used to improve the constant power speed range, but this would lead to reliability issues due to the use of switches and mechanical contactors.

IM also has a low lagging power factor when lightly loaded, and the starting current is very high [20] in line start applications. The high starting would lead to over-rating the power semiconductor switches, leading to increased converter cost. In IM, there are also problems of controlling the speed of the machine. Thus, for a vehicular drive profile, IM would have many challenging issues to face.

The efficiency of the IM is very high, about 91% at low and high speeds. The output power of the IM is limited by the frequency, especially above the constant power region due to the inductive reactance of the machine. The inductive reactance is also responsible for high electrical time constant, thereby increasing the response time and increasing the starting time in the engine.

The noise levels of IM are very low compared to the other machines and so are the torque ripples, as sinusoidal waveform is used for operation.

4.5.5.2 Permanent Magnet Machines

Permanent magnet (PM) machines make use of a permanent magnet on the rotor to produce the necessary magnetic field. The PM machines that are potential candidates for ISA applications are the surface permanent magnet

(SPM) machine and the interior permanent magnet (IPM) machine, among other types.

The SPM has magnet blocks glued on to the surface of the rotor and wrapped over a fiber sheet so as to keep the magnets intact at high speeds. Since the magnets are used to produce the magnetic field, the field coils, slip rings, and brushes are all eliminated (Figure 4.14). In addition, the absence of field windings reduces the copper losses in the machine.

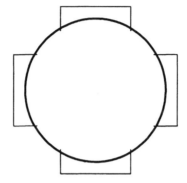

Figure 4.14 Four pole SPM rotor.

These SPM machines have high power density due to the lower reactance as a result of low magnetic permeability. Also, radial magnetic field tends to enhance the power density. Overall, the magnet circuit of SPM is highly efficient, which raises the possibility of lowering the stator windings and stator copper losses with powerful magnets. The SPM can produce very high torque levels and, in particular, has high starting torque compared to the Lundell alternator. This feature helps in reducing the size of the machine.

The SPM offers high design flexibility, in the sense that the length and diameter parameters are independent of each other and thus enable the increase of either of them (independently) for increased efficiency. This also brings flexibility in the manufacturing of the SPM, where the diameter can be kept the same and the length can be varied for various ratings, permitting the use of the same fixture parts such as stator laminations, magnets, bearings, brackets, and bolts, which altogether provide an option to reduce the machine cost. The low sensitivity to airgap variation due to the low magnetic permeability also increases the flexibility in manufacturing. All of these flexibilities arise due to the two-dimensional nature of the magnetic circuit of SPM.

The rotor, which carries the magnet, can be designed to be hollow, lowering the rotor inertia and thereby reducing the belt slippage in offset coupled operation [16]. The durability of the SPM rotor under rotor speed tests have shown that they can operate without failure up to a speed of 24,580 rpm, although the alternator usually operates under 22,000 rpm [18].

Automotive Power Systems

The major drawback of the SPM is the difficulty of achieving constant power over a wide speed range. This is because the mmf of the magnet is constant, offering a constant flux causing back-emf produced to increase linearly with speed. Also, the flux weakening is low due to the very low inductance. Therefore, to maintain a constant output, special converter designs have to be developed, which would turn out to be very expensive.

The presence of magnets in the SPM brings thermal issues, because, at high temperatures, there is lower remanence and coercivity and demagnetization occurs. In addition, at high temperatures, deterioration of the stator windings leads to the failure of the machine [22]. Therefore, air or liquid cooling would be essential for the machine.

IPM machines are differentiated from their close SPM counterparts by the presence of the magnet inside the rotor [15]. These magnets are placed into the slots within the rotor. Here, the torque levels are high, similar to the SPM, but the torque is generated not just by the magnet, but also by the iron saliency. Thus, there are two components that contribute to the torque generated. This feature provides the designer with ample flexibility to design the system. Also, the high torque levels in IPM enable the size of the machine to be scaled down by nearly 15% when compared to IM [14].

The presence of the magnet inside the rotor permits the rotor to be used for much higher speeds of about 24,500 rpm [18]. In IPM, if the rotor saliency is increased, it directly reflects on the stator inductance ratio (L_q/L_d), thereby giving a means of reducing the magnets that are used in the machine [23]. This also reflects upon reduced magnet induced back-emf, cost, and higher speeds. This high rotor saliency can be achieved by using axially laminated rotors, but such designs would lead to manufacturing problems altogether, resulting in lower cost advantage. The IPM offers a wide constant power speed range, which gives the advantage of less complicated converter designs [15].

The presence of a smooth rotor lowers the noise to exceptionally lower levels at low speeds. If hollow rotors are used, at high speeds, the vibration of the air passing through the holes in the rotor tends to produce very high noise; yet, this can be reduced by changing the dimension and size of the holes. However, for the ISA applications, this is no major issue because the engine noise dominates at high speeds.

Thermal compatibility in embedded magnets is also a concern. IPM involves complicated manufacturing with installation and magnetization processes. Another drawback is that the high magnetic saturation in the rotor leads to degrading of the stator winding and starting torque [15]. Other challenges for the IPM would be its electromagnetic design optimization and its performance under fault conditions.

The IPM machine can make use of convention block magnet, sheet magnet, or injection moulded magnets.

As another option, powdered metal magnetic machines essentially have rotors made from pressed power magnets. The use of such a rotor results in nearly 10-20% higher current carrying capacity at high temperatures compared to laminated core machines. Furthermore, the core is found to be about 10-50% cooler [14]. The use of these rotor types would largely depend on their structural capability for ISA applications.

Axial flux permanent magnet (AFPM) machines essentially consist of slotless stators placed between two PM rotor discs [24]. They thus have two independent electromagnetic structures. The stator is torroidal and being slotless is made up of low cost iron strips with thin laminations that make it very light in weight and compact without loss in efficiency. The elimination of the stator slot reduces the stator teeth losses as well as rotor losses at high frequencies. There is also reduced torque pulsation and acoustic noise.

The windings wound on the stator are similar to those of a toroid and are connected in a way to obtain a three-phase winding. The stator is encapsulated within a fiber reinforced epoxy resin, which helps in mechanical strength and heat dissipation. The rotor carries the PM, as in an SPM machine, with the difference being that the magnets in an AFPM machine are placed in a way to provide an axial flux. The rotor may be provided with pits for the magnets to increase the mechanical integrity of the rotor. When placing the magnets on the rotors, magnets of the same polarity are made to face each other.

The type of magnetic material and number of poles mainly characterize the designing of the machine. The electrical parameters and other design characteristics would be based on these two factors.

The AFPM, which would be operated in the speed range from idle to 6000 rpm, produces a high frequency output that leads to increased stator losses. To reduce these losses a good choice of stator material is vital and the choice would be based on the hysteresis and eddy current losses, magnet permeability, and saturation flux density. Based on these characteristics, the best candidates would be iron, sintered, and amorphous materials [25]. Depending on the required output frequency range and characteristics, the optimal one may be chosen.

The stator winding may be Litz wire or multiplier construction. The use of Litz wire involves increasing the number of wire strands, which leads to a decrease in copper and an increase in insulation for a given fill factor. This reduces the eddy currents, but the resistance increases. This is an ideal solution to reduce the skin effect and eddy current losses during flux leakage, as these losses increase considerably due to irregular current distribution in the wire cross-section in a normal winding arrangement. Also, at high frequency, these losses are significantly high, since there are functions of frequency [25].

The axial length of the stator coil has low resistance and reduced power losses. During the operation of the machine, there are two sets of flux linkages, one on either side of the stator surface. The flux takes the path from the PM on the rotor, airgap, circumference of the stator core, airgap, and to the backside of the magnet on the rotor. When the current is passed through the stator coil, a torque is developed tangentially which rotates the rotor. The diameter of the stator and rotor determine the power level of the machine. Therefore, the AFPM machine is ideal for direct drive ISA automotive applications. It is lightweight and compact with short axial length and has high efficiency.

Other types of machines used would be those which have rotors made of only magnet. For these kinds, there are major issues of mechanical stability and temperature. In addition, the power levels of the machine may be small.

4.5.5.3 Switched Reluctance Machines

SRM has a single rotor construction, essentially made up of stacks of iron, and does not carry any coils or magnets [26]. This feature of the SRM enables it to achieve a rugged structure and provides the machine with the advantage to be used at high speeds and better withstand high temperatures. SRM has salient rotor poles and the stator windings can be wounded externally and then slid on to the stator poles. This gives the SRM a very simple and easy manufacturing process and the cost of the machine will be low. SRM achieves high torque levels at low peak currents by using a small airgap. However, for ISA applications, which expose the machine to a high vibration environment, it is required that the airgap length be increased, which is a drawback for the SRM.

The choice of the number of poles to be used is important in the SRM due to the vibration that is produced. A structure such as 12/8 provides lower mechanical vibration compared to the 6/4 structure.

SRM machines are well suited for high-speed operations and tend to have higher efficiency at high speeds [27]. SRMs are also capable of wide constant power speed output, which is an essential requirement for starter/alternator applications where the speed ratio varies up to 10:1.

SRM technology has been well established and used in the aerospace industry for quite some time at high power and speed (1500 hp, 48000 rpm) [28]. Its extension for the automotive application as a starter/alternator would be challenging, however, as its application would be at lower voltage and power levels and the torque requirements would be different. Also, the ISA in the automotive industry would have applications beyond generating and starting.

The SRM is a highly reliable machine as it can function even under fault condition with reduced performance [28]. One of the reasons for this is that the rotor does not have any excitation source and, thus, does not generate power into the faulted phase. Therefore, no drag torque would be produced under motoring mode and there are no sparking/fire hazards due to excessive fault currents. Also, the machine windings are both physically and electromagnetically isolated from one another, reducing the possibility of phase-to-phase faults. The SRM as a system with converter involves two switches and a winding in series per phase. Thus, even in a case of both switches being turned on at the same time, no shoot-through fault would occur, unlike in the case of AC drives, which lead to shorting of the DC bus.

The SRM is a machine that would be run by a non-sinusoidal voltage waveform, thus, resulting in a high torque ripple. This also leads to high noise levels: in keeping with the standards, to have torque ripples of less than 15% in non-sinusoidal excited machines, the number of phases in the SRM has to be increased, which would reflect on the system cost due to the increased part number, in this case the switches [14]. However, this torque ripple in SRM can be exploited to provide active engine damping.

SRM converters have very high efficiency of almost 46% at low speed and 89% at idling at the time of starting the engine. During generation, the efficiency values are remarkably high, around 93% over the entire alternator speed range [28].

The relative merits of each of these four major electric machine technologies for application to automotive systems can be visualized by considering three attributes: (1) how the machine is excited, whether magnetization is fully internal as with SPM, fully external as with IM or SRM, or somewhere in between as with IPM; (2) a description of constant power speed ratio (CPSR), whether it is limited to angle only control as SPM, or amenable to field weakening as IM, IPM and SRM are; and (3) what the control implications on the power electronics are in terms of real and reactive power flows. The IM has low kW/kVA performance during start-up. The SPM has excellent kW/kVA performance and the IPM is somewhat in between. The following figures illustrate these attributes.

Each of the machine types in Figure 4.15, when compared using the attributes of excitation source, inverter burden, and CPSR, show various mixes. Each attribute reflects some specific system performance capability or lack thereof. For example, high kVA/kW reflects more inverter burden hence higher inverter cost. A fully internally excited source, for example, may appear to be a notable benefit but from a manufacturer's perspective it reflects higher capital cost that fully externally excited because the machine excitation must be paid for by the manufacturer whereas for fully externally excited the excitation costs are paid for by the operator. CPSR is a highly sought attribute in automotive systems because it means that transmissions or other gearing mechanisms are not needed or are minimized in complexity.

Automotive Power Systems

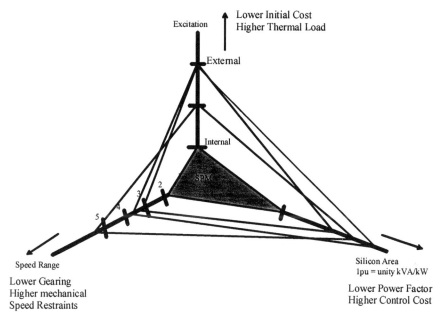

Figure 4.15 Attributes of automotive electric machines for starter-generator, surface permanent magnet machine, SRM shown.

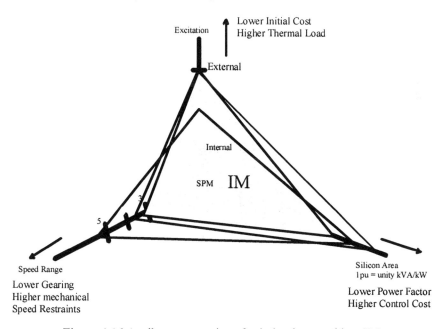

Figure 4.16 Attribute comparison for induction machine, IM.

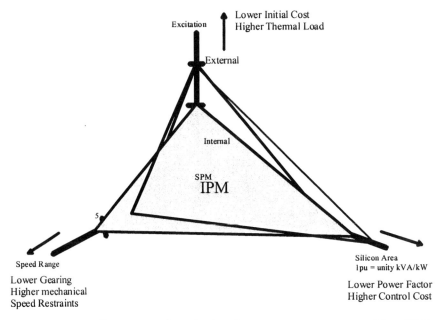

Figure 4.17 Attribute comparison for interior permanent magnet machine, IPM.

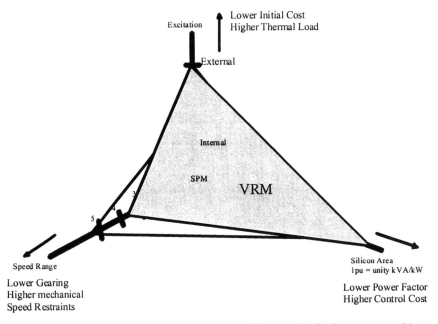

Figure 4.18 Attribute comparison for variable/switched reluctance machine, VRM or SRM.

4.5.6 Control

4.5.6.1 Induction Machines

IM, when used as an ISA, can be controlled by making use of a control strategy that employs a voltage input from the battery and current input from the inverter. Here the objective is to hold the current as constant as possible during the starting mode by appropriately selecting the frequency and voltage. During starting, as the speed increases, there is a fall in the current level due to the torque delivered, which is compensated for by increasing the frequency of the inverter and voltage proportionally. This control is executed continuously until the end of the starting mode. IM has high starting current and this can be avoided by controlling both frequency and voltage.

During system transition, when the system operation is changed over from starting to generation mode, the controls have to be changed as well. This may be done by selecting a reference speed at which the change takes place and, as a result, no speed sensing would be required. During the transition, the frequency of the inverter would be low for smooth transition.

The generation of the power in IM occurs when the speed of the rotation of the rotor is higher than the stator frequency and the output power increases with slip. In the generation mode, the output voltage is kept constant by continuously varying the inverter frequency. Thus, the voltage is constant irrespective of the speed and load conditions.

During generation, the output is controlled by modifying or manipulating the load connected to the machine. This could be done by providing capacitive load. This can be done at the time of transition too. The drawback here is that the speed bandwidth would be severely affected. Another option for control could be by controlling the slip between the stator and rotor. In doing so, controlling the electrical frequency of the stator would be a better option. The output power of the machine is, however, limited by the frequency beyond the constant power region due to the high inductive reactance of IM.

Power may be generated from IM by using fixed capacitor self-excited IM configuration. However, in places of variable speed over a wide range and a varying load condition, it cannot be used. An IM generation system involving capacitors on the load side to provide reactive power to the machine and a discharge resistor and switch on the battery side to dissipate excess power in order to maintain constant DC link voltage for stable operation may not be applicable in the area of vehicles due to the low efficiency and high cost.

The induction motor has a complex dynamic model that can be simplified if the rotor flux vector is aligned along the d-axis, resulting in a zero flux along the q-axis. This leads to decoupling of the flux and the torque components. Thus, the IM can be easily controlled by rotor flux orientation. Since the torque and speed are vehicle parameters, the flux component becomes the flexible parameter for obtaining the desired control. By controlling the d-axis flux or current, the system efficiency may also be maximized while meeting the load demands.

However, rotor parameters such as rotor resistance vary with speed and temperature and have to be overcome by gain schedulers that fine-tune the machine parameters during its operation. The rotor parameter variations are coupled with slip gains and this slip gain is tuned to achieve a linear relationship between the output torque and q-axis current. At the same time, a scaling factor between q-axis and output torque commands has to be established. The dependence of these gain schedulers on factors such as speed, d-axis current, and mode of operation have to be also considered in the rotor flux orientation control.

Power generation by IM, where the IM is coupled with a PWM inverter and a battery as a source, is possible. However, in such cases, the battery life is compromised due to the high ripple currents and also the kVA rating of the inverter would need to be high. Power may be generated using an IM by having a high voltage stator split winding that is supplied with excitation from the inverter such as a PWM inverter. The inherent advantage is lower inverter rating due to the lower current levels and, hence, lower inverter cost.

Alternatively, IM may be directly connected with a rectifier to charge a battery and, at the same time, have a PWM inverter with inductive filters and a capacitor voltage source for excitation. The cost of this system will be higher than split winding type, but it is quite lower than an IM directly connected with an inverter and battery.

The filter inductor is used in order to reduce the high frequency harmonics being injected into the machine while the capacitors act as voltage sources. The indirect field oriented control involved here compares the reference and actual DC bus capacitor voltage, and the error voltage signal is used to generate q-axis control command to the PWM controller. The voltage difference between the rectifier output and its reference is used to generate a d-axis control command for PWM control. The slip of the IM is determined from the generated phase currents and the calculated rotor angles are used for obtaining d and q-axis components of current to generate a PWM duty cycle in the inverter for voltage regulation. The d-axis current is used for regulating output voltage while q-axis current is used to regulate the voltage source of the inverter.

The system efficiency is higher than the Lundell alternator about 20% over a wider speed range. The output power and the power factor, however, drop at high speeds. The output voltage of the IM can be well regulated within a bandwidth of 0.2V from the already established control and converter techniques.

4.5.6.2 Permanent Magnet Machines

In the PM machines, the presence of the PM gives a constant mmf, resulting in a low output voltage at low speeds and high voltage at high speeds.

Thus, the output has to be regulated. Regulation is achieved by incorporating PM hybrid excitation. This involves some changes in the machine construction, where the rotor would have both PM and excitation windings.

Here the excitation by the magnet would form the main part while that from the field windings would form the assistant or auxiliary part [29]. This hybrid excited starter generator (HESG) can be made brushless to avoid losses. The main part would function exactly as a PM machine while the objective of the auxiliary part is to provide the necessary field excitation in order to achieve a constant output voltage. This control strategy is feasible due to the fact that there is no phase difference between the two emf voltages developed by the main and the auxiliary part and they are additive quantities.

If U_n and I_o are the rated output voltage and the rated load current with no excitation from the auxiliary, then the curve to be followed would be E_0. If the load current increases to a higher level I_1, there is a fall in the output voltage. Thus, to increase the voltage to the rated value, current is supplied to the windings of the auxiliary part to provide additional field excitation. This raises the output voltage, and the characteristics shift to resemble those of curve E_1. Similarly, for a fall in the load current I_2, the magnitude of the voltage tends to increase, which is then regulated by reducing the excitation current, and the curve shifts to E_2. This is shown in Figure 4.19.

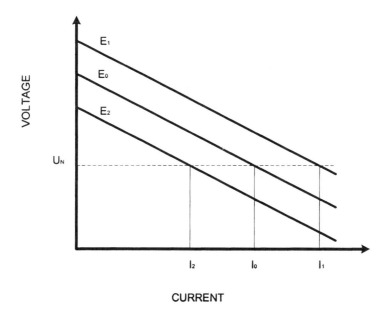

Figure 4.19 Voltage regulation principle.

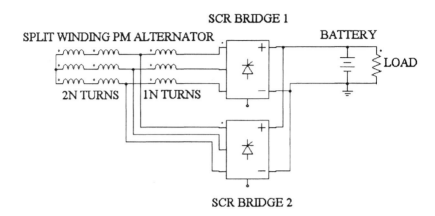

Figure 4.20 Power generation system.

The major design issue related to this control technique involves determining the rated load current, which would be the average of the sum of the minimum and maximum load currents. Designing the assistant part is essential to determine the maximum voltage value that needs to be regulated, which can be determined by knowing the extent of the HESG voltage regulation and the range over which the load current varies. For linear voltage regulation, the magnetic circuit would need to be linear.

Another technique to regulate the output voltage would be to use a split winding dual SCR bridge converter (Figure 4.20) [18]. Here the machine windings can be split into a ratio of 1:3, with one converter connected to 1/3 of the winding and the other converter connected to 2/3 of the winding. Therefore, at low speed operation, the output is taken from the converter that is connected to the larger number of windings; at higher speeds the output is tapped from the converter that is connected to fewer windings. Thus, the output profile is improved, as can be seen in Figure 4.21.

The major design issue here would be the proper control of the phase angle and delay of the SCRs. It would involve the use of a PLL based delay angle generator, which uses the alternator AC signal for synchronism with the alternator frequency; the delay angle would be achieved from the DC voltage signal and the temperature dependent limiter (Figure 4.22).

The other control method involves increasing the number of poles to increase the output voltage at low speeds [30]. This would be the only option at low speeds, because the magnetic flux cannot be increased as mmf of the magnet is constant. Any attempt to increase the number of windings would result in increased machine size, which is not permissible.

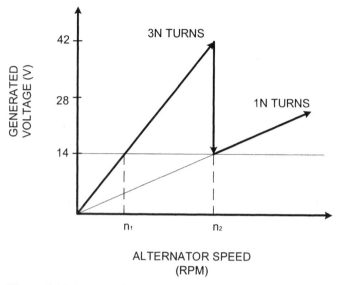

Figure 4.21 Generated voltage (v) vs. alternator speed (rpm).

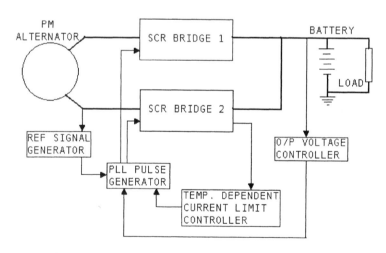

Figure 4.22 Block diagram of dual SCR bridge control.

At high speed, the output voltage of the machine is high and can be reduced by using a claw pole structure, which tends to increase the demagnetizing effect at high speed [17], [30]. This also ensures that the voltage rises gently and that the leakage reactance is high. In traditional machines, d-axis armature reaction has less influence on the airgap flux, mainly because the magneto-conductivity of the magnet and the air is the same. However, by using the claw poles whose magneto-conductivity is high, there will be a tendency to create a higher impact of the d-axis armature reaction on the airgap flux. Thus, as the load current increases at high speeds, the d-axis armature reaction also increases, thereby, reducing the output voltage.

A magnetic notch is another option that could be used to regulate the output voltage [30]. A magnetic notch introduces increased flux leakage, which reduces the output voltage. Here the fall in the voltage is a function of angular velocity. Thus, at high speeds, the output voltage is reduced to a large extent, thereby regulating it, while, at low speeds, there is a faint fall in the voltage magnitude. As a result, the output, at low speed, is not affected much. Therefore, an improved performance is obtained (Figure 4.23) [30].

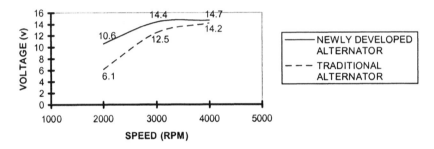

Figure 4.23 Traditional and newly developed alternator.

The output voltage can also be regulated by using flux weakening algorithms in which the instantaneous current in a synchronous time frame is rotated towards the negative d-axis, which would enable operation under the constraints of available voltage, current, and rotor speed [23]. Here the flux weakening can be made to automatically adapt itself to the bus voltage.

In the case of AFPM machines, the low phase reactance makes it difficult to attain wide constant power output over the entire speed range that is required for ISA applications. The AFPM has high efficiency as shown in Figure 4.24, but, at low and high speeds, for constant power output, its efficiency reduces [25].

A constant power output from AFPM machine can be obtained by flux weakening. This can be done by injecting current into the stator and partially offsetting the magnetically driven flux. It can also be done by using a slotted stator with slots covered with soft material bars, which increases the inductance and, thus, reduces the net magnetic flux in the machine. The flux weakening by stator reaction may result in a very high hazardous voltage at the terminal of the machine if the control circuit fails. Therefore, one alternative to the above two options might be a mechanical control.

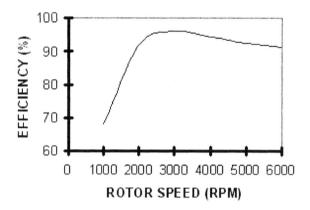

Figure 4.24 AFPM efficiency vs. speed.

A mechanical control can be used to reduce the peak airgap flux density by increasing the airgap length, which would result in reduced flux linkage, eventually leading to flux weakening [24]. This is possible in AFPM but would involve energy to overcome the attractive forces between stator and rotor.

An optional method is to provide angular displacement between the two rotor discs, as shown in Figure 4.25. In this case, there would be only half of the magnetic flux that would link and, thus, result in flux weakening. The advantage here would be that no torque or energy is required to achieve this, and that the airgap flux density remains the same.

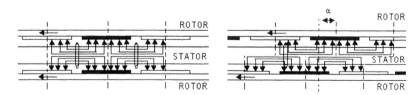

Figure 4.25 Aligned and angularly displaced rotor discs.

When there is no displacement, the flux linkage is maximum and, on displacement, the flux linkage reduces. The emf induced in the coils during the operation adds up to give the total emf, which is represented as E_t in the vector diagram of Figure 4.26. In the unaligned position each rotor disc is displaced by an angle α in the opposite direction; the resultant total emf is a cosine function of the electrical displacement angle (electrical displacement $\alpha_e = P\alpha$, P is number of poles). Since the magnitude E_t is a function of both speed and cosine of electrical displacement, the magnitude can be kept constant as the speed changes (Figure 4.27), i.e., increasing α_e when speed increases and vice versa. This would be done under the generation mode, above the idling speed.

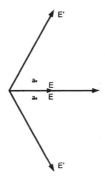

Figure 4.26 Induced coil emfs due to rotor angular displacement.

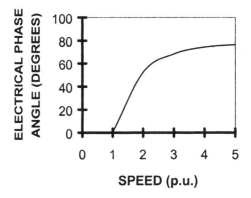

Figure 4.27 Electrical phase angle vs. speed.

This is achieved by using a rotor hub that comprises the rotor disc, synchronizer, and cam-spring governor. The cam-spring governor operates in accordance with the speed of the machine and thereby controls the position of the treadle that is rigidly connected to the one of the rotor discs. The operating mechanism of the cam-spring is based on the restrain force of the spring and the centrifugal force of the cam. The synchronizer ensures that the rotor discs are displaced by α degree in the opposite direction (Figure 4.28).

Automotive Power Systems

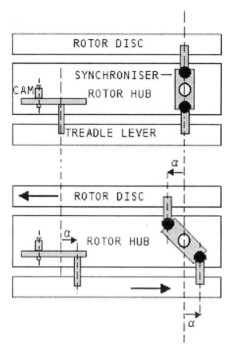

Figure 4.28 Initial and displaced position for synchronized motion of rotors in the opposite direction.

This control method provides constant output power and even performs well under transient speed changes. However, the output does have some harmonics. This method lowers the strain on the power converter systems.

AFPM machines have better performance in terms of output power and switch utilization when they are fed with square wave than when they are fed with sinusoidal wave for a given converter current limit. In addition, low cost hall effect sensors can be used instead of high resolution encoders for synchronizing phase emf and current at starting and also for disenabling inverter switches at ICE ignition speed.

4.5.6.3 Switched Reluctance Machines

SRM provides constant power over a wide speed range and is highly dynamic with speed. General SRM control layout is shown in Figure 4.29. Electromagnetic torque in the SRM is produced by the tendency of the salient rotor poles to align with the excited stator poles and attain the lowest reluctance position. The torque developed depends on the relative position of the phase current with respect to the inductance profile (Figure 4.30). If the current falls on the negative slope of the inductance profile, then the machine is in the generating mode. The torque does not depend on the direction of the current, because torque is proportional to the square of the current. The back-emf developed depends on the magnetic parameters of the machine, rotor position, and the geometry of the SRM.

The current waveforms for the motoring mode and the generating modes are mirror images of each other [27]. During generation, initial excitation has to be provided from an external source since it is a single excited structure. The stator coils are turned on when they are in aligned position and then turned off before un-alignment position. The turn-on and turn-off periods determine the peak value of the current. During motoring mode, the peak value of the current depends on the turn-on time. These timings greatly help in designing optimal and protective control.

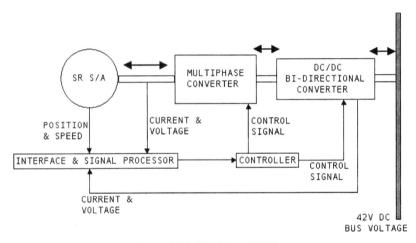

Figure 4.29 SRM based ISA.

To increase the robustness of the SRM, the sensorless operation approach can be used [26]. The magnetic parameters of the machine carry the position information that can be retrieved. But, since the dynamics of SRM change with speed, the algorithm have to be modified or adjusted for each operating mode – standstill, generating, motoring, reverse motoring, and dynamic braking.

At standstill, the prime objective is to determine the correct phase for initial excitation, for which fixed, short voltage pulses are applied to all the phases, and then the magnitude of the current is compared to select the most appropriate phase [27].

During low speed operation, the back-emf is neglected since the errors are small; back-emf is insignificant, since the SRM is operated mostly at high speeds. At low speed, the unaligned position can be detected. The rotor positions are obtained from electrical speed, and the speed can be obtained from the frequency of the previous phase excitation. The detection is done by applying short, fixed frequency pulses to an idle phase, and the magnitude of the current from the rotor movement is determined, with the maximum current corresponding to the unaligned position, which is the position of least inductance. However, frequency and duty cycle need to be properly selected to avoid negative torque effects and sluggish behavior and to attain a good level of resolution.

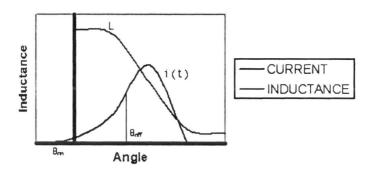

Figure 4.30 Current pulse and inductance profile at generation.

At high speeds, the back-emf levels are high and need to be taken into account. There are significant currents at unaligned and aligned positions during motoring and generating, respectively.

In SRM, closed loop control is stable and reliable, as open loop control leads to instability. For torque control, the electromagnetic torque is estimated and commutation angles as functions of current and conduction band length at significant inductance slope are determined. Here the PI controller would be used to compensate for mismatches and the feed-forward controller would be used to accelerate convergence.

In the case of negative incremental loads, there is destabilization, which can be reduced by using capacitors. Also, in SRM, by changing the turn-on and turn-off angles, the system may be opted to operate under maximum efficiency, minimum torque ripple, or minimum rms DC link current.

During the motoring mode, the SRM can provide high torque of up to 160 N-m and the absence of dead zone in the torque characteristics enables it to provide good starting torque. The SRM torque varies over speed (130 to 20 N-m between 100 to 1000 rpm), which highlights its ability to provide acceleration support. Also, the extended constant power speed ratio enables lower power requirement during the motoring mode.

SRM can be current controlled for both motoring and generating modes of operation [31]. During motoring, the current is controlled by adjusting the firing angle Ψ and applying the current during the magnetization period Φ_p. If, during the current control, there is overlapping of the phase currents, it leads to an increase in the maximum torque level.

During the generation mode of operation, the torque must be fed to the machine when the inductance level is decreasing, i.e., when the rotor is moving from the aligned (L_a) to the unaligned position (L_o) (Figure 4.31). The generating operating domain can be plotted on a plane of Ψ and Φ_p, representing average output current. Cutting the current with zero battery current would give a generating zone, which can be represented by the difference of Ψ and Φ_p, e.g., $\Psi - \Phi_p = 70$, for maximum output current. However, this condition changes with the inductance profile (Figure 4.32). In Figure 4.32, profile no. 1 is $\Psi - \Phi_p = 80$, and profile no. 2 is $\Psi - \Phi_p = 60$. To attain a constant current, the PI corrector can also be incorporated, which requires Φ_p as a function of speed and Ψ as a function of speed and desirable output current.

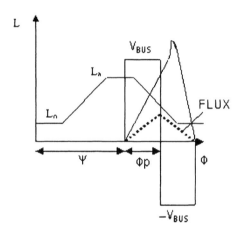

Figure 4.31 Supply voltage and control parameters.

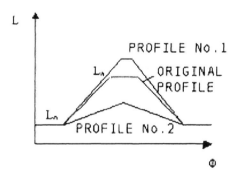

Figure 4.32 Inductance profile of 6/4 SRM.

Another phase current control scheme involves prediction of the phase current within an interrupt routine. Here the back-emf for the reference current is determined, provided the phase is turned off at the very next execution of the routine. If the phase current decreases before it reaches the reference value, then the turn-off is delayed. If the back-emf increases beyond the maximum applied voltage, the turn-off is done immediately.

Electromagnetic torque is developed in the SRM as a result of the tendency of the polarized rotor pole to align itself at the lowest reluctance position, as mentioned above. Thus, the torque characteristics would very much depend on the rotor position. Torque pulsations will be produced, which can be reduced by bell shaped torque characteristics and by using appropriate commutation techniques. The major source of pulsation is improper commutation, mutual coupling between phases, and hysteresis control of current.

The torque pulsations of the SRM can be used to effectively cancel the torque pulsation produced by the ICE. This would be based on matching the frequency of the oscillation of the engine and the SRM. In the control, the engine torque is introduced as a disturbance and the T_{error} is determined (Figure 4.33) [27]. By matching the timings of the maximum and minimum torque developed by the SRM and the engine, the T_{error} can be reduced. For example, at starting, the maximum torque of SRM and engine are displaced by 180 degrees.

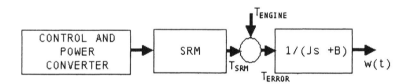

Figure 4.33 Block diagram for engine torque cancellation.

Online monitoring and control of SRM may also employ torque pulse cancellation, which involve unsymmetrical control of the stator phase, resulting in reduced efficiency. The total torque developed by SRM is made to follow the engine torque closely to achieve cancellation. In such tracking controls, SRM with high poles/phase is preferred.

Switched reluctance generator (SRG) has low radial vibrations, since the stator phases are excited at the unaligned position, and also there is absence of significance phase current. The presence of large machine time constant also brings small radial forces. The use of an anti-vibration configuration helps to further reduce the vibration. The radial forces being position dependent, the magnitude of attractive forces is lower and the absence of sudden change in the rate of change of radial force in SRG reduces the vibration to a large extent.

Current profiling greatly reduces the noise level in SRG, but it leads to lower performance levels.

4.5.7 System Cost

The cost of the ISA along with the drive that performs all the required functions is shown in Figure 4.34 [14]. The converter-to-machine cost is about 10:1 for each machine type. That is 95% of the system cost is from the converter and only 5% is from the machine. The cost of the SPM and SRM are higher than the other two, mainly because of the presence of magnets and a larger number of converter components. It is also seen from the graph that the cost of the IPM is higher than that of the SPM, even though it uses less magnet. This is mainly due to the complicated manufacturing of the IPM. However, the converter cost of SPM is higher.

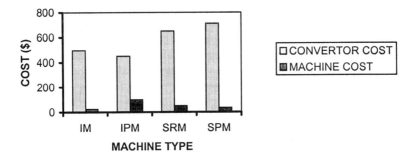

Figure 4.34 6kW ISA cost.

In general, the cost of the ISA should be lower than $500 for its commercialization. For this figure, the closest candidates are IM and IPM. One of the ways of reducing the system cost is lowering the cost of the converter, which should be possible with the further development of components in the coming decade.

The system cost can be estimated in various ways for the ISA. One method would focus on the cost per watt of the converter (an estimation similar to that being done in the UPS industry), in which case it is estimated that the cost should be lower than 0.10 $/watt. System cost evaluation may also be based on cost per component or favorable performance/price ratio. Irrespective of the cost model used, the converter forms the major portion of system cost.

Numerous studies have been performed that break down the costs associated with ISG systems for automotive applications. Consider the power electronics or inverter unit consisting of power stage, control card, power supply, communications, gate driver, heat sink, and bus bars and package. The

Automotive Power Systems

relative costs associated with these subsystems in just the inverter can be broken down as shown in Figure 4.35 and 4.36.

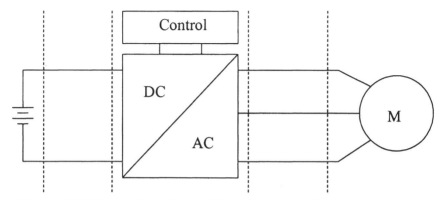

Figure 4.35 ISA installation in a vehicle and corresponding costing partitions (battery, cable/connection, inverter, cable/connection, and electric machine).

PNGV's Automotive Integrated Power Electronic Module (AIPM) Program

Component	% Total $
Control/Communications/Power Supply	13%
Gate Drive/Modules/Cable	30%
Die Cast/Heat Sink/Connection	33%
Bus Burs/Current Sensors/Capacitors	24%

Figure 4.36 Inverter cost breakdown by major group and comparison to PNGV targets for each group.

The procedure we follow to determine the cost breakdown for the power electronics unit can be summarized as follows:

- Constrain the power silicon junction temperature, $\Delta T_{jc} < 40°C$. This will effectively establish a durability criterion for the inverter and result in durability specific power silicon size.
- To the simulation, input the desired power throughput and system voltage level.
- Calculate the device currents for the machine technology in use.

- From the calculated device currents, determine the device losses that match the thermal constraint and from this calculate the resultant device area (iteration). Then calculate the device current density and costs.
- Compute the costs of all the remaining components in the four groups for each power level and voltage selected.
- Tabulate resultant losses, efficiency, and costs
- Repeat the process for the next semiconductor technology, i.e., MOSFET, IGBT, etc.

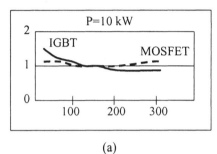
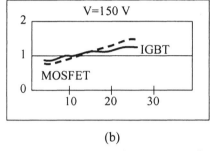

(a) (b)

Figure 4.37 Per unit cost of inverter ($pu) versus (a) system voltage (V) and (b) power level (kW).

We notice from Figure 4.37 that below 150V the MOSFET is clearly more cost effective than an IGBT, whereas above approximately 10kW of power the IGBT would be preferred. The incremental costs are $18/kW for the IGBT and $14/kW for the MOSFET ISG inverter. Note that these are the incremental costs, not absolute specific costs. If the throughput power were zero the cost associated with defining an inverter, package, control, power supply, communications etc. would establish a baseline cost. This can easily be gathered by extrapolating the cost versus power curves in Figure 4.37 to zero throughput power. In that case the cost of all the housekeeping logic, communication, chassis, and heatsinking would define entry level fixed costs. Table 4.1 tabulates the costs associated with the remaining components in the partition of Figure 4.35 when the complete inverter-controller is taken as a single module. Notice that the inverter and power electronics clearly dominate the cost picture at 10 kW regardless of voltage level. The electric machine, an integrated starter-alternator (ISA) in this case, represents just a fraction of the cost picture. The battery, Pb-Acid in the case described, is of valve regulated technology and relatively low cost. If an advanced battery such as nickel-metal-hydride (NiMH) were used, the cost breakdown would shift so that battery, power electronics, and everything else would be an approximate equal split to the total cost.

Table 4.1 ISA system cost breakdown by voltage when P=10kW

Component	42V %	150V %	300V %
Battery	21	25	30
Inverter	49	42	37
ISA/Electric Machine	25	25	23
Harness/Connectors	5	8	10
Total	100	100	100

In an ISA system having an advanced battery such as NiMH the cost of the battery would be on par with the cost of the complete power inverter system. Therein lies the dilemma with ISA applicability to automotive idle stop and mild hybridization systems. The cost of entry at low power levels precludes affordability of ISA. If the power levels were greater than 20 kW the situation would change and overall benefits such as fuel economy of ~10% or higher improvement versus ~5% for 10 kW ISA would yield a much more favorable value equation and business case.

4.5.8 ISA Coupling Configurations

The ISA can be coupled to the ICE in two ways, either by direct coupling, which involves placing the ISA on the crankshaft, or by offset coupling, which is very similar to the way the present alternator and starter are connected to the crankshaft. Each coupling method highlights its own individuality in terms of design constraints, performance, and cost.

4.5.8.1 Offset Coupling

Offset coupling is very similar to present day coupling techniques and, thus, it can be easily incorporated without much change near the engine. Offset coupling may be done either by way of the chain, gear, or belt. Figure 4.38 shows an offset coupled ISA using a belt drive [16].

Chain coupling involves the use of the chain to run over the pulleys. This coupling requires the ISA to be placed near to the transmission. The chain drive gives the advantage of using chains that have smaller width than when a belt is used, due to the high durability, strength, and longer life [16]. The durability and chain life can be enhanced by pressured oil lubrication. One of the drawbacks of the chain drive system is the high noise level, which could be reduced by using silent chains, but this would result in increased cost.

The gear drive coupling also involves coupling at the transmission side. A major issue to be considered here is that the high speed, not high torque levels, affects the gear drives performance. At high speed operation, scoring of the gear teeth occurs, resulting in damage of the gears. This can be reduced by using materials of high grade, but there are issues of high material and manufacturing cost due to the need for grounded teeth of high accuracy and surface finish. Thus, the gear drive is preferred to be driven at low speeds with multiple stage arrangement. This would require that the transmission case be modified.

In the case of the ISA with a belt drive option, there is a need to increase the width of the belt compared to the width that is currently used in order to accommodate increased load capacity. The belt drive offers silent operation and a low cost solution. It also offers more packaging freedom and involves no lubrication, unlike other offset coupling methods.

The belt drive gives the option of using a single or two separate belts for the ISA and accessories. The former provides the advantage of using a single belt, but the higher torque and power transfer to the ISA would demand a wider belt. This would involve the need to replace all pulleys with a new type. The latter allows the use of the same pulleys for the accessories but would require an extension of the crankshaft length external to the engine to accommodate an additional pulley for the ISA. This would result in a reduced cost but would put increased stress on the bearings of the shaft due to overhanging.

When using the belt drive, other accessories such as a bi-directional torque tensioner, keyed and locking nuts to fasten pulleys to the ISA, and high durability connecting belt(s) have to be properly selected, and very importantly there may be a need for dampening materials in the hub of the ISA and/or engine crankshaft pulley due to higher inertia and torque ripple in the ISA. The choice of the belt usually depends on the behavior of the belt at high speeds, weight, width, wear, centrifugal force, and noise. Chain drives have also been attempted in offset coupled ISA but here again the problems of high reflected inertia, ISA and engine torque pulsations, and rapid engine rpm slew rates during downshifting have been known to snap even steel link chain belts.

Automotive Power Systems 117

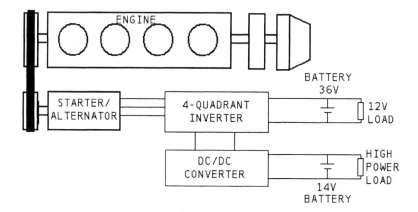

Figure 4.38 Belt driven offset coupled ISA.

The offset coupling offers the advantages of being easily incorporated as well as of being easy service to the system. With offset coupling, power generation of up to 5 kW is possible, beyond which there would be heavy stress on the belt and mechanical parts and the system would be unable to deliver high power [14]. Thus, the drawback with offset coupling would be the mechanical reliability issues of the chains, gears, and belts, not just at high power but even at low power levels.

4.5.8.2 Direct Coupling

Under direct coupling, the ISA is directly incorporated into the drivetrain of the vehicle and, therefore, it is referred to as an integrated starter/alternator (Figure 4.39). The ISA can be placed either between the transmission and the engine or on the accessory side. However, the latter would require the extension of the shaft external to the engine, thereby, posing a problem leading to the wear of the bearings. Furthermore, the size of the machine/engine has to be re-configured, as the space is available at a premium, especially if the engine is transverse mounted. The former configuration involves packaging issues, as the drivetrain has to be modified to accommodate the ISA.

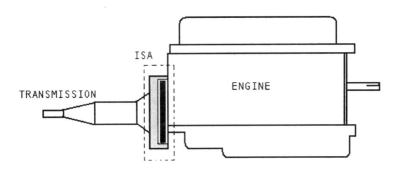

Figure 4.39 Direct coupled ISA.

This coupling offers silent start, high power generation of greater than 5 kW (and up to 12-15 kW), and the lower speed operation of the alternator [32]. This enables the removal of the pulley, tensioners, and belt. Depending on the design, the flywheel, whose function would be served by the rotor of the machine, may also be removed. Being directly coupled, the ISA can lend itself to provide vibration damping of the engine; such ISAs are referred to as integrated starter/alternator dampers (ISADs).

Under the direct drive configuration, the ISA is placed in a 1:1 speed ratio for generation and starting. At present, a coupling speed step-up ratio of typically 2.5 is being used. This results in a need for a larger and heavier machine that would be able to provide all the necessary starting torque. The increased size would be compensated for by the fact that a single machine is being used instead of two. Direct coupling also places the ISA under severe environmental conditions of high temperature and vibration. Thus, due to these issues, more effective cooling would be required, and a machine with a larger airgap would need to be specially designed and developed.

We compare the fundamental physics of direct and offset coupled ISAs in Figure 4.40 by noting that at the rotor surface of either machine the tractive force per unit area, also referred to as the surface traction or pressure, would be identical. There may be some offsetting factors such as higher electric loading possible in the direct coupled ISA due to its larger diameter, but in general we can assume that electric and magnetic loading are approximately the same. The difference is that direct coupled ISAs have rather large rotor diameters typically 2.5 times that of an offset coupled ISA, or a conventional alternator for that matter. Now, the offset coupled ISA will have a pulley ratio (assuming no internal gearing or transmission that shifts depending on cranking or alternator mode) of 2.0 to <3.0 for the same reasons given for conventional alternators. Primarily ISAs are used as gearing for starting, yet sufficiently low such that rotor burst limits are not exceeded, even in the case of engine overspeed such as

Automotive Power Systems 119

occurs during fast downshifting or in faults such as loss of the engine controller rev-limiter logic. The pulley ratio of virtually all belt connected, or B-ISA's, on the market is on the order of 2.5:1. We will refer to direct coupled ISA as C-ISA for crankshaft mounted.

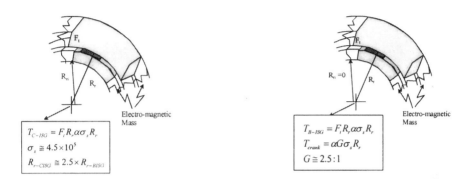

Figure 4.40 Fundamentals of B and C-ISA.

What is interesting to note from a first principles comparison of direct coupled C-ISA or ISG and offset coupled B-ISA or ISG is that the lever arm ratio of C:B-ISA is about the same as the pulley ratio of B-ISA:C-ISA or 2.5:1 since C-ISA by definition is 1:1 on the crankshaft. This leads to some rather interesting results when they are compared, as is done in Table 4.2.

Table 4.2 Attribute comparison of B and C-ISA.

Physical Limits	B	C	Typical
Mechanical Surface speed	=	=	200 m/s
Magnetic Iron sat.	=	=	1.7 T
Electric Slot currents	Higher	Nominal	$<2*10^7$
Thermal Temp. rise	Must liquid cool	Air/liquid cool	$<60\ °C$
Gearing Gear ratio	2.6:1	Direct drive	$<3:1$
Machine size Rotor radius	1	2.5x	50 mm
Cost of mat'l E-M volume	1	2.5x	

In this table, attention should be focused on the last two rows, which compare material content and relative cost of B-ISA versus C-ISA. For the same torque delivered to the engine crankshaft the C-ISA will have 2.5 times the electromagnetic volume (mass) as the B-ISA and its cost will be approximately the same ratio. We continue this rather approximate comparison of B-ISA with C-ISA by noting the impact each has on reflected inertia presented to the crankshaft, which is a truly performance limiting attribute when it is excessive. Start by assuming, as noted in the table above, that the average electromagnetic density is the same for each machine. This is not unreasonable since each rotor/stator contains the same materials in about the same relative proportions, including slot fill and other volumes that are air only. We also note that active rotor length, L, will be approximately the same in both cases (about 50 mm). Start with the definition of polar inertia for this composite rotor.

$$J_r = \frac{\pi}{4} <\rho> LR_r^4 G_r^2 \quad (kg-m^2/rad) \tag{4.8}$$

Define the ratio of C-ISA to B-ISA reflected inertia as follows, noting that R_{rx} is the rotor radius and G_{rx} is the gearing ratio between ISA and crankshaft for each case where x=B or C as appropriate.

$$\frac{J_{rC}}{J_{rB}} = \frac{<\rho_C> R_{rC}^4 G_{rC}^2}{<\rho_B> R_{rB}^4 G_{rB}^2} \tag{4.9}$$

Using the values noted above for B-ISA and C-ISA in equation (4.9), we find that the inertia ratio equals 6.25. We can validate this using some known rotor inertia values for C-ISA and B-ISA. A C-ISA having an approximate 120mm rotor radius and 50mm stack will have a polar inertia, J_{0C}=0.07 kg-m^2/rad, from which we would predict using equation (4.9) that a corresponding B-ISA would have J_{0B}=0.011 kg-m^2/rad, which is entirely reasonable. A production 14V, 120A Lundell alternator for comparison has a polar inertia $J_{0\text{-Lundell}}$=0.0046 kg-m^2/rad or roughly half the value of the more powerful B-ISA described here.

In conclusion, with the increasing power demand in automobiles, the ISA stands as a strong prospective solution to the increasing power demand in vehicles. By careful consideration of the characteristics of the PM machines, IM, and SRM, as well as their various alternative drive configurations, an ISA can be developed whose application goes much beyond the primary starting and high power generation, to lower emissions and to enhance performance as well as fuel economy.

4.6 Automobile Steering Systems

Due to the growing demand of cars, trucks, and sport utility vehicles (SUVs) in today's society, there becomes a need to create automobiles that are optimized in different aspects. In current cars, many mechanical systems are

Automotive Power Systems

robbing the conventional internal combustion engine of valuable horsepower. For instance, the steering systems of most later model vehicles could be observed to be one of these resource-robbing systems because of their inefficient use of either the rack-and-pinion system or the re-circulating ball system. In either case, they are both mechanical systems that incorporate the use of hydraulics (power steering), which prove to be very inefficient systems. The implementation of the electrically assisted or steer-by-wire systems brings hope for creating a steering system for most vehicles, which will still be equipped with power steering. This is a much safer and more efficient choice for the future production of cars.

The two most common steering systems used today are rack-and-pinion and the re-circulating ball system, which have been in use for several decades. Both of these manual steering systems incorporate similar parts, which include a steering wheel and column, a manual gearbox and pitman arm or a rack-and-pinion assembly, linkages, steering knuckles and ball joints, and the wheel spindle assemblies. Although these are equipped in most modern-day automobiles, it should also be noted that over 90% of all domestic vehicles are also equipped with power steering, or to better describe it, assisted manual steering due to its ability to allow for manual steering in the case of a power steering failure. In short, power steering has almost become a standard equipment item on many automobiles. Although this conventional system of steering has been considered to be dependable, times have changed and modern technology has allowed for the creation of better steering systems. Systems which are much more efficient, reliable, safer, and cheaper to produce and maintain than what is in use today.

4.6.1 Conventional Steering Systems

The rack and pinion steering systems could be observed to be the most common type of steering found on most cars, small trucks, and SUVs. This system in actuality is a very simple system that could be broken down into its individual parts, as shown in Figure 4.41. To begin with, the rack and pinion gearset is enclosed in a metal tube with both of the ends sticking out. A tie rod is then attached to these overhanging ends on the rack. The tie rod is then connected to the steering arm, which is directly connected to the automobiles wheel, which allows it to turn in the desired direction. The pinions gear is directly attached to the steering shaft, which is directly connected to the steering wheel. As a result, when the steering wheel is turned, the pinion is turned, which cause the rack to move either left or right and, as a desired outcome, it causes the wheels to turn in the specified direction.

The rack-and-pinion gearing system actually only accomplishes two goals. It converts the rotational motion that the driver applies to the steering wheel into a linear motion, which is needed to turn the wheels. In addition, due to the large size of the steering wheel in comparison to the small pinion in the rack and pinion system, a gear size reduction is created, which makes it much easier to turn the wheels. Although rack and pinion steering system could be considered to be the most commonly found steering system, it is not the only system used.

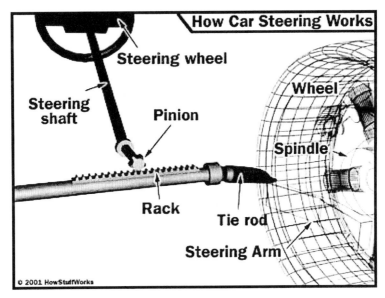

Figure 4.41 Rack-and-pinion steering system (Courtesy of HowStuffWorks, Inc., http://www.howstuffworks.com/).

The re-circulating ball system, as shown in Figure 4.42, is also a very popular system, which could be found on many trucks and SUVs. It is significantly different from the rack-and-pinion system; however, it could still be broken down into its base components, which are steering wheel, re-circulating ball gearbox, and pitman arm. The steering wheel, directly connected to the re-circulating ball gearbox, turns the pitman arm. The pitman arm is connected to the track rod, which connects to the tie rods directly controlling the wheels they are connected to. They are located at both of the ends on the track. The box could be observed as two parts. The first piece is a block of metal with a threaded hole drilled inside of it and gear teeth cut on the outside of it. The teeth located on the outside of the block are what engage the gear that moves the pitman arm. The steering wheel connects to a threaded rod that sticks inside the hole of the block of metal. When the steering wheel is turned, it turns the bolt, which is held in a fixed position, therefore, causing the block to move up and down the rod, which results in the gear turning the wheels to move.

The re-circulating ball steering gearbox contains a worm gear. Instead of having the shaft make direct contact with the threads in the block, all the threads are filled with ball bearings. These ball bearings circulate through the gear as it

Automotive Power Systems

turns in the box. The result of using these ball bearings is that they reduce friction and wear in the gear system. They also help to reduce slop, which is a free movement feeling that is experienced when changing the direction of the steering wheel.

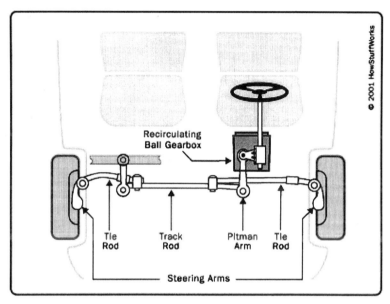

Figure 4.42 Re-circulating ball steering system (Courtesy of HowStuffWorks, Inc., http://www.howstuffworks.com/).

4.6.2 Conventional Power Steering Systems

Power steering is very similar for both the rack-and-pinion and re-circulating ball steering systems. The power steering system, as shown in Figures 4.43 and 4.44, primarily consists of a rotary valve, a steering gear, and a pump that is powered by a belt driven pulley. In order to obtain the hydraulic power that is for the steering, a rotary-valve pump is used. This pump is driven by a belt and pulley system driven by the car's engine. Inside the pump is a set of retractable vanes similar to propeller blades on a fan that spin inside an oval casing. As these vanes spin, they pull low-pressure hydraulic fluid from the return line and force it out at an outlet as a much higher pressure. This amount of flow that is created by the vanes is determined by the car's engine speed. As a safety precaution, a relief-valve is part of the pump system to make sure that the pressure does not get too high, specifically when the engine is revving hard causing a large amount of fluid to be pumped out.

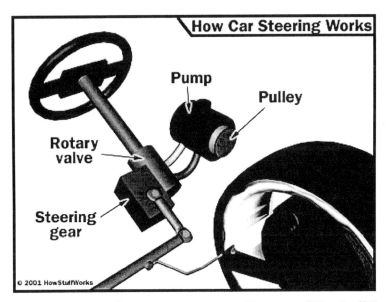

Figure 4.43 Conventional power steering system (Courtesy of HowStuffWorks, Inc., http://www.howstuffworks.com/).

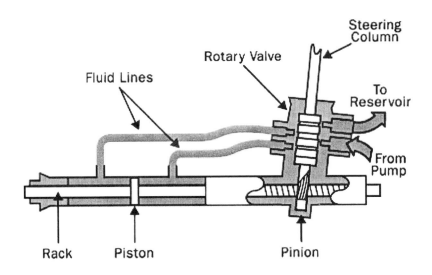

Figure 4.44 Typical hydraulic assisted steering layout (Courtesy of HowStuffWorks, Inc., http://www.howstuffworks.com/).

Automotive Power Systems

In these power steering systems, a cylinder with a piston in the middle is attached to the rack (rack and pinion) or track rod (re-circulating ball). On the piston are two fluid ports, located on the opposite sides of the piston. These ports are connected to the lines that provide the flow of hydraulic fluid needed to the specified side of the piston forcing the piston to move, as a result, creating assistance to the steering. As a safety precaution, it should be noted that in the case of a power steering malfunction or failure, manual steering would always be available.

4.6.3 Advancements in Power Steering Systems

With the increase in size and weight of automobiles, power steering has been developed to assist motorists with slow speed maneuvering such as parking. A power steering system provides most of the torque necessary to overcome the friction between the wheels and the road. As was explained, the most common power steering system in use today is hydraulic power steering. A simplified diagram for such a system is illustrated in Figure 4.45.

The major components in Figure 4.45 include the hydraulic pump and reservoir, rotary valve, steering wheel, torsion bar, rack and pinion, hydraulic piston, and the hydraulic lines that connect the system together. The hydraulic pump and reservoir, powered by the engines drive-belt, provide pressurized hydraulic fluid to the system through Line 1. The rotary valve directs fluid flow from Line 1 to Line 2, Line 3, or Line 4 depending on the desired steering operation. Steering commands are inputted through the steering wheel to the torsion bar. The torsion bar has two functions: one is to convert torque to lateral force via the rack and pinion, the other is to signal the rotary valve to assist the steering maneuver.

For a right-hand turn, the driver turns the steering wheel clockwise (directions are referenced to the driver in the driver seat). That spins the torsion bar in the same direction, causing the rack to move laterally towards the driver side. At the same time, the torsion bar signals the rotary valve to allow fluid flow from Line 1 to Line 3 and from Line 4 to Line 2. That forces the hydraulic piston towards the driver side, providing most of the force needed to make a right-hand turn. Similarly, to make a left-hand turn, the rotary valve allows fluid flow from Line 1 to Line 4 and from Line 3 to Line 2, causing the hydraulic piston to move towards the passenger side.

Whenever the wheels are in the centered position, no power steering assistance is needed. The hydraulic fluid simply circulates through Line 1 and Line 2, resulting in a major inefficiency for the hydraulic power steering system. In fact, the hydraulic pump is always working, even when it is not needed for steering assistance.

In an aim to improve the efficiency of a hydraulic power steering system, the engine drive-belt can be replaced with an electric motor. The electric motor provides power to the hydraulic pump only when turning of the wheels is desired. This greatly improves the efficiency of the power steering system, but it adds cost and time to the manufacturing process along with weight to the system. Therefore, a further step has been taken to develop a new type of power steering system, namely electric power steering.

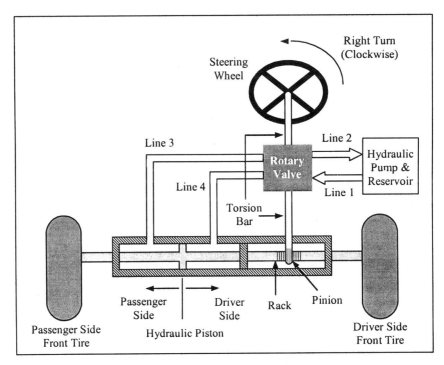

Figure 4.45 Hydraulic power steering system.

4.6.4 Electric Power Steering Systems

It has been shown that, in all the current conventional steering systems, the steering wheel and the column count as a major source for injury to the driver in automobile front-end collisions. As a result, a numerous amount of energy-absorbing and non-intrusion designs have been developed. Energy-absorbing columns were developed in order to serve two purposes. To begin with, they must stop the wheel and column from being pushed to the rear of the car in the event that the car is crushed from a frontal impact. Secondly, the energy-absorbing column must be able to provide the driver with a tolerable amount of resistance as he/she thrusts forward and strikes the wheel with his/her chest. Although the idea does not seem to pose any alarm or concern, one major problem that has risen is that the collapse of the column due to a frontal impact should not obstruct its ability to provide a proper "ride down" for the driver's chest.

Automotive Power Systems

The problem with the power steering system is that the rotary-valve pump must provide a sufficient flow of hydraulic fluid when the engine is at an idling state. As a direct result of this, the pump moves a much larger amount of fluid than is needed when the engine runs at faster speeds. It is noted that, when a vehicle is being driven and the steering wheel is not being turned, it is necessary for both the hydraulic lines to provide the same amount of pressure to the steering gear. When the steering wheel is turned, a spool valve is turned in one direction or the other and ports open to allow the flow of the high-pressure fluid needed in the appropriate line.

Electric type of power steering system completely does away with all hydraulic components. It provides steering assistance with an electric motor that directly assists steering maneuvers only when turning is desired. The electric motor may be mounted to assist lateral motion, as shown in Figure 4.46 (a), or to assist circular motion, as shown in Figure 4.46 (b). The control system for the electric motor consists of the typical components for an electric motor drive. The controller uses torque commands and current/voltage feedback to control the power electronic converter. The converter then outputs the voltage needed to carry out the desired steering operation.

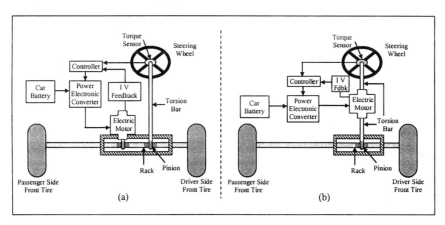

Figure 4.46 Electrically assisted power steering system.

4.6.5 Steer-By-Wire

Steering systems in today's modern vehicles have adapted many new ideas and technologies, which has ultimately directed them to experiment with the idea of steering by wire. The following is the chronological order of the essential steps and breakthroughs that were taken that created the path for the discovery of the steer-by-wire steering system.

Due to the constant pumping of fluid in the power-steering pump on most cars today, much valuable horsepower goes wasted. This wasted horsepower translates into wasted fuel that everyone could do without. The variable-displacement power-steering pump is a device that is already utilized in a few of the cars in production today. This pump reduces the amount of fluid that is being pumped at higher speeds, when it is not necessary to have the assistance of power steering, thereby causing a reduction in the amount of power that is consumed from the engine.

Taking the next step forward came the development of electro-hydraulic systems. In this system, an electric motor with variable speeds would be used to power the steering pump. This would allow the motor to turn off when the engine is running hard, thus causing a reduction in the amount of power that is being consumed.

Once again, if we were to take the technology of steering one step further, the development of electric power steering would be explored. In this system, all the hydraulic equipment would be eliminated completely. Instead of having hydraulics assist in power steering, an electric motor would be directly mounted on the rack to assist in steering. Electronic sensors would be strategically places to the steering wheel, sending signals down to a control system that controls the electric motor on the rack. This would allow the electric motor to provide the proper amount of assistance.

The gradual evolution and breakthrough discoveries in automotive steering finally lead to the discussion of the steer-by-wire system. In this system, the mechanical connection that is placed between the steering wheel and the steering would be eliminated completely. As a replacement, there would be a purely electronic control system. This form of steering would contain sensors that would send signals to tell the car what the driver wants the wheels to do. It is even possible for motors in the steering system to provide the driver feedback on what the car is experiencing.

4.6.5.1 Comparison of Conventional Steering Versus Steer-By-Wire

The major benefit with the current manual steering systems (rack and pinion, re-circulating ball) is that we are using technology that has been well developed and known; however, the major disadvantage to this is the high steering effort that is due to the increasing from-axle loads and tire widths. To resolve this problem, hydraulic power steering was developed. Although hydraulic power steering reduces steering effort and allows for manual steering in case of a hydraulic pressure decrease, there is the issue that it requires the use of hydraulics and the large amount of energy that is consumed because of the directly driven boost pump. Generally, traditional steering systems are no longer acceptable due to numerous reasons. To begin with, the steering column is a major inconvenience when pertaining to crash structure development. The steering assembly also heavily influences the engine compartment package. It is important to note that the disposal of hydraulics becomes an important issue. The driver support system also only allows for longitudinal interventions

Automotive Power Systems

(brakes-ABS, engine moment-ASC, brakes and engine movement-DSC). Lateral interventions through steering would greatly improve active safety.

With the development of electric power steering, there becomes the ability for variable steering assistance, which results in less energy being consumed. As a major added bonus, there is no longer the need for hydraulics. The problem with this system, however, is that there is no driver independent steering assistance and there is still the presence of the inconvenient steering column. This system would not be considered as a by-wire system due to the mechanical connection.

A steer-by-wire system with mechanical or hydraulic backup is also a steering system that could be observed. The benefit to this system is that there is driver independent steering assistance; however, there is the disadvantage of needing a safety case for switchover and there is no guarantee in the function of the backup. When having a mechanical backup system, the hydraulics are eliminated; however, the steering column as well as the clutch assembly will still be present; this would then result in package difficulties. If we were to choose to have a hydraulic backup, this would allow for the elimination of the steering column, but we would now require hydraulics and a boost pump. This would ultimately lead to the electric redundant steer-by-wire system. In this system, there would be no steering column and hydraulics, which therefore would eliminate the need for a hydraulic or mechanical switchover mechanism. This system would take advantage of all the by-wire technology. As a result, there would only be the need for one system technology instead of three (standard, backup, and switchover mechanism). The only problem with this option would be the requirement of a safety case for the redundant electric system.

4.6.5.2 Components for the Steer-By-Wire System

Steering systems have always had physical mechanical connections, which will in the future be replaced by those in which an electronic signal communicates the driver's intention to turn also known as the steer-by-wire system. The most obvious benefit to switching to the steer-by-wire system is the reduction of parts and the elimination of a hydraulic system. In the steer-by-wire system, as shown in Figure 4.47, electric actuators are placed at each wheel which do the work of the current hydraulically assisted steering, thereby eliminating the need for the vacuum booster, master power, hoses, clamps, and hydraulic fluid.

In the steer-by-wire system, there are a few primary components. Directly connected to the steering wheel is a steering sensor. The sensor is very similar to the type that would be found on a driving simulation game. The steering sensor is designed to interpret what the driver is choosing to do to the vehicle and pass the message down to the controller. Similarly, the steering sensor also reads feedback information that is sent to it from the controller. For instance, the controller will determine how much resistance is felt on the steering wheel or what kind of responses should be experienced to the driver. In short, the controller will sort out the information that is not desired from the information that is desired for the driver to experience. For example, road vibrations and engine vibrations are undesired road and manners which will be eliminated to create a much more effective drive. From there, a signal is transferred down a wire to a controller box. The purpose of this controller box is to send the message to the electrical components on what they should do. The controller box acts as the brain for the systems, telling each specific component of the steer-by-wire system to perform their specific duty at a specified time. From there, a steering gear, which is placed at each of the steering wheels then receives the message from the controller to determine whether the wheels should be turned left or right. The steering gear system contains a motor, which performs the actual work necessary in turning the vehicle. It is an independent motor that can be almost compared to the conventional hydraulic power steering. The motors are there to take what the driver has inputted and amplify the effort a numerous amount in order to turn the wheels in the correct direction.

In the current mechanical system, it is believed that the driver either experiences positive feedback in which the driver feels everything or negative feedback. The truth to this matter, however, is that mechanical steering lies somewhere in the middle. For an example, when a tire nears its handling limit, the torque felt through the steering wheel decreases by an amount so miniscule that most drivers will not even be able to detect this point. It is only when the tires begin to squeal and the amount of steering lock increases that drivers notice that the tires have reached their handling limit. In the steer-by-wire system, if a driver were to add or subtract too much steering lock for the situation, the steering system would independently of the driver's input, change the angle of the front wheels based on the information from the vehicles stability system. This would result in a steering system that is much more precise and eliminate any unnecessary wear on the vehicle.

There are many ideas that could be experimented and/or tested in the steer-by-wire system. An example would be a speed-independent variable steering ratio. Using this idea, when a vehicle is traveling at low speeds the steering wheel would be turned with minimal effort; small turning efforts inputted to the wheel sensors from the driver could create large changes in the angle of the front wheels. As a direct result of this, parallel parking will become a much simpler task due to the fact that the wheel may only require a 20-30 degree turn in either direction. Conversely, as the velocity of the vehicle increases on the road, the relationship between the input and response would be modified toward the more conventional values. As a result, at high speeds, all vehicles will have handling and steering capabilities similar to those of a high performance sports car.

The primary goal of the steer-by-wire system is to enhance safety by separating drivers from routine tasks and assisting the driver when responding to a critical situation. This technology also creates a great potential to make cars

Automotive Power Systems

much more environmentally friendly and cheaper to produce. By removing many of the mechanical parts, the direct result would be that there is a better use of the materials and that there becomes more freedom in choosing how the interior of the vehicle should be laid out. Since intelligent electronic systems are used, there is a guarantee that the systems will become much more precise and accurate, causing less wear on the engine, better fuel economy, and much less required maintenance.

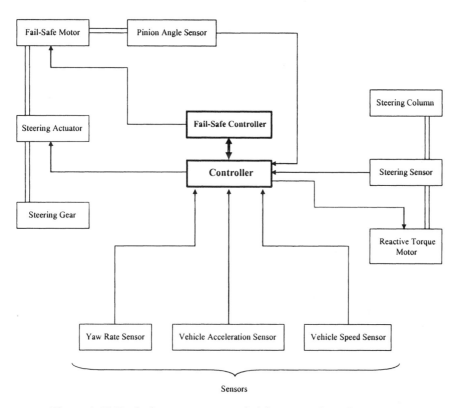

Figure 4.47 Typical components needed for a steer-by-wire system.

4.6.5.3 Benefits of Steer-By-Wire

When a steer-by-wire system is used, there is no longer a need for the steering column. This means that there is an increase in passive safety/crash performance (no intrusion, airbag position). It creates production benefits due to the ability to easily convert left and right hand vehicle steering. In addition, the system would only use "dry" actuators and only local hydraulics (less oil, environmental benefits). As a direct result of this, there will be a drastic reduction in the amount of energy that is consumed. Finally, by having this mechanically decoupled steering wheel, there is no longer any unintended feedback/road disturbance to the driver.

Overall, the steer-by-wire system not only creates free space in the engine compartment by eliminating the steering shaft, but it would also reduce the amount of vibration felt in the car and create a much safer environment for the driver. The technology also possesses a potential to make the production of cars much more environmentally friendly with a significant cut in the vehicle's price. Due to the removal of many mechanical parts in the steer-by-wire system, results include better use of materials, more freedom in the interior of the vehicle while, at the same time, there becomes greater precision and accuracy from the intelligent systems, which will result in less engine wear, better fuel economy, and less maintenance. Despite all these listed benefits, there is still a potential for other advantages in the future.

4.7 Semiconductors for Automotive Applications

The application of semiconductors to automotive systems has enabled dramatic improvement in function and durability of historically mechanical systems. The first application of semiconductors to an automotive application was the use of the bipolar junction diode in the alternator rectifier, first and for only a brief period as a replacement for the generator commutator, but most importantly as the rectifier in the Lundell alternator. With this innovation the earlier dc generator was replaced within a matter of just a few years by the more efficient and higher power density alternator. Following rapidly on the heels of this introduction was the replacement of the generator regulator with a bipolar transistor regulator, transistorized ignition, and microprocessor controlled engine functions of spark and EGR. During the past two decades the use of electronic controls, principally engine and power train controls, have increased rapidly. Attention in recent years has been on electrification of the vehicle power train and the move to a higher voltage EDS, known in the industry as 42V PowerNet. But what have been the semiconductors of choice for automotive application? The following sections explore the selection and rationale.

4.7.1 The Role of Power Electronics

It has been shown that the choice of system voltage has been one of the leading concerns facing electrical system designers for the past two decades[2].

---

[2] SAE Dual and Higher Voltage Committee, a grass roots effort from 1988-1995 led by engineers in the big three auto companies, made recommendations for a 48V battery, 56V nominal system J2232. The MIT-Industry Consortium (1996-present), having global representation of automotive manufacturers and suppliers, reached consensus on a 36V battery, 42V nominal system that was

Automotive Power Systems 133

Early efforts with semiconductors were concerned with their survival in the harsh automotive environment, particularly understood. As system complexity increased, and as designers better understood the electrical distribution system voltage variations and transients and methods of limiting, major semiconductor manufacturers established reliability levels adequate for automotive use [41], [42]. This motivated the EDS system designers to increase the applications for electronic controls for switching high currents in applications such as ignition coils (8A) and alternator regulators (3.6A) and at reliability levels so high that the days of frequent distributor points replacement were all but forgotten except by the most nostalgic car buffs. Power electronics and smart-power devices continue to replace previously mechanical systems in all aspects of the vehicle so that today the value of semiconductors in the average automobile is several hundred dollars ($200 in 1998) and the electronic content is approaching 30% of the cost of producing the car.

The principle hurdle to power electronic introduction in automotive systems has been the cost challenge for those systems that interface directly with the vehicle's EDS and power supply (without over voltage protection or a secondary circuit). Power semiconductors must have guard bands in voltage so that transient over voltages such as alternator load dump, which is a destructive transient, are survivable. Present 14V nominal EDS voltage is now clamped to 40V maximum transient from its earlier 80V limits. The proposed 42V PowerNet is even more restrictive of transients (58V maximum) so that its guard band is less than 1.5 times the working voltage of 42V. For semiconductor device reliability the stress level imposed by a more restricted guard band means higher durability. Guard band, or the ratio V_{cc}/V_{max}, is a critical determinant in semiconductor cost to meet reliability levels. Guard banding has significantly improved from 0.175 when 80V transients were present, to 0.35 today and 0.72 in 42V PowerNet standards. With these levels of guard banding, 40V rated devices for today's 14V system and 60V devices for PowerNet, there are no reliability issues. In higher power applications such as 1 kW electric assist power steering or 10 kW ISA, the semiconductor devices should be rated 75V to allow sufficient guard band for not only system voltage transients but module and package unclamped inductance induced voltage steps on the device metalization. When temperature variability of device avalanche and tolerance are included the guard band may reach 80V or even 100V. A very important point to be observed for 42V PowerNet is that its specification is consistent with popular 75V and 80V process technologies. Figure 4.48 repeats the specification for 42V PowerNet and the consequent requirements for semiconductors [42].

Nominal operation of the 42V PowerNet is from 30V to 48V. Allowance for generator ripple extends the high end of the working voltage to 50V. The

---

compatible with all international standards for touch safety and is now a draft standard within ISO.

maximum over-voltage of 58V is permitted for 400 ms or less. Voltage dips to 18V for 15 ms and to 21V for 20 s max are also permitted to accommodate vehicle starting. Central reverse polarity protection (alternator rectifier diodes) limit the reverse polarity to –2V for 100 ms or less. As already noted, the semiconductor devices should have voltage ratings of 75V or higher. If we assume a nominal guard band of 30V then the required silicon area for automotive applications is as shown in Figure 4.49 (diamonds or red trace). The squares or blue trace shows the dramatic reduction in chip conductance required (hence chip area reduction). As a result, semiconductors for 42V systems are on the order of 20% the size of their counterpart 14V device.

Definition of 42V PowerNet:

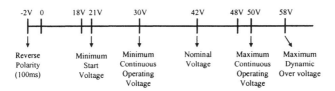

Definition of Semiconductor Requirements at 42V:

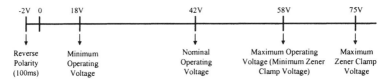

Figure 4.48 Specification for 42V PowerNet (DIN, ISO) and semiconductor rating specifications.

4.7.2 Semiconductor Device Technology

It was noted in section 4.5.7 regarding system cost that the choice of semiconductor device technology was highly dependent on system voltage and power level. Introduction of higher powered loads such as electromechanical engine valve actuation, electric driven fans and pumps, electric assist power steering, and braking, as well as electrified chassis functions such as active suspension and active roll represent kilowatt level loads [43], [44]. For power levels of 3kW to 10kW as representative power levels of B-ISA and C-ISA, it is not readily evident whether or not the power MOSFET is the best choice or if an IGBT might not be a more appropriate selection. Detailed investigations on a 3kW, 100V and 50V power electronic inverter for ac motor control aimed at a determination of minimum system cost have been performed [45]. The comparison was made for the same power throughput and losses. In this study off the shelf 100V and 200V rated MOSFETs were used for the 50V and 100V dc bus cases respectively while 250V IGBTs, the lowest available voltage, were

Automotive Power Systems 135

used for both 100V and 50V operation. It was found that at 100V dc bus the IGBT inverter cost was approximately half the cost of the MOSFET design. When the bus voltage was dropped to 50V (this was several years before 42V PowerNet proposals) the advantage the IGBT had at 100V was lost, and its cost depended on the particular load conditions and PWM frequency employed.

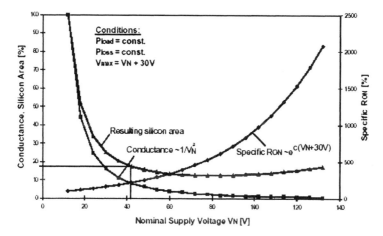

Figure 4.49 Drastic reduction in chip area, hence cost, of semiconductor devices for 42V systems compared to 14V as the baseline case (Courtesy Infineon Technologies AG, Munich).

Several conclusions can be made regarding the choice of MOSFET or IGBT for such low voltage inverter applications. We note that in the case of the IGBT inverter that the same freewheel diode was used. In addition, and for both cases, the inverter current to the three phase ac permanent magnet ac motor was regulated in the stationary frame using a hysteresis band regulator. For the same motor torque at 50V as for 100V, the phase currents were doubled, so hysteresis band window was doubled as well. From this several conclusions can be made:

- For high load currents at 50V bus the dramatic reduction in MOSFET on-resistance resulted in approximately the same conduction losses as for 100V bus. However, MOSFET switching losses were doubled due to slower switching speed, but total inverter losses increased by only 20%.
- In contrast to the MOSFET, the relative ineffectiveness of paralleling IGBT die to reduce conduction loss resulted in the IGBT having nearly double the conduction losses as the MOSFETs even when 8 IGBT die were connected in parallel.
- The benefit of fast switching IGBTs with low switching loss did not benefit the situation because the switching frequency was low (2.4 kHz to 12 kHz) due to the high currents involved and the relatively wide current

regulator band. At switching frequencies above 5 kHz the IGBT would tend to have lower total losses than the MOSFET.
- For low levels of load current the relatively high switching losses of the MOSFET completely overshadow its conduction losses so that only 2 IGBT die in parallel result in lower total inverter losses.
- It was found that 8 of the 250V size 5 IGBT die have approximately the same total silicon area as 6 of the size 6 power MOSFETs (IRFC260), so the IGBTs would have the same delta-T increase (same losses). At high currents and if 8 IGBT die are used, the relative cost would be 1.4 times that of the MOSFETs whereas if 6 IGBT die were used the cost would be the same.

Some of the differences noted can be attributed to the fact that the intrinsic body diode of the MOSFET has a poor reverse recovery characteristic (without electron irradiation or ion implant carrier lifetime reduction). As a consequence the speed at which it could be switched was deliberately slowed down, especially during turn-on when its complementary phase leg inverse diode was undergoing reverse recovery. The use of good quality free wheeling diodes across the IGBT permitted its switching characteristics to remain. Lastly, at low currents the conduction loss in the MOSFET were quite low because there is no threshold voltage to overcome, and importantly for this comparison, the MOSFET's could be operated in a synchronous rectification mode in concert with their body diode conduction.

When the two systems were compared for equal losses, it required 2.5 of the size 5 IGBTs to equal the losses of 6 size 6 MOSFETs. This results in the IGBT inverter costing 50% of its MOSFET equivalent at 100V bus voltage. Switching speeds were such that peak over-voltages, dV/dt's and di/dt's were all comparable.

4.7.3 Integrated Power Modules

Continuing electrification of power train and vehicle systems in general is enabled by innovations in higher levels of power electronics integration. Early designs relied on discrete, standalone modules that were function specific. For example, discrete devices were integrated in the early engine control microprocessor for solenoid, relay, and fuel injector control. With higher integration levels these actuator drivers were integrated first into multichip modules (DIP packages) and later into functionally integrated packages. For AC motor control the trend has been the same, but at a slower pace. For automotive applications that demand quality, durability, and performance all at lowest cost, the demand for higher integration is evident. Higher levels of integration yield higher system reliability and performance is improved through better thermal management and more flexible packaging. It is desirable to integrate not only the power electronics devices and their attendant gate drivers and sensors but

Automotive Power Systems 137

the pre-driver and control stage as well. Trends to smartpower are advancing this trend in automotive applications.

Environmentally the automobile can be split into five zones (three zones at minimum): underhood, chassis, exterior, interior/cabin, and trunk. Each of these zones represents a unique packaging environment for electronics including temperature variation, humidity, and exposure to automotive fluids, salt spray, water immersion and splash, dust, sand and gravel bombardment, altitude, vibration and shock, electrical transients, and EMI and ESD. Electromagnetic interference, EMI, is particularly problematic in the automotive environment because of proximity of sources and sensitive modules, open wire harnesses, and minimal shielding. Electrostatic discharge, ESD, can be a major issue when harness and module pins are exposed to touch potentials of kV such as might occur by an occupant or mechanic sliding from a cloth seat in cold weather and touching metal areas of the vehicle. All automotive electronic modules are ESD protected, and underhood modules are capable of withstanding direct exposure to spark plug wire arcing to their pins.

The worst case packaging location for electronics in the automobile has always been in the engine compartment and on the engine/transmission as depicted in Figure 4.50. In fact, simultaneous temperature and vibration present the most demanding packing application challenges of any industry. The evolution of electronics, including power electronics, integration in the automobile is illustrated in Figure 4.51. Note that the ultimate goal is control electronics integrated with its associated actuator and only termination needs for power and communications. In some cases these goals are being met; integrated electronic throttle control in which the communications and control electronics are integrated with the power drivers and packaged on the engine throttle body have already been put into production.

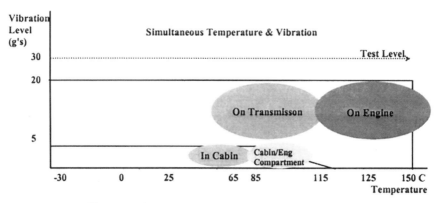

Figure 4.50 Automobile environmental conditions.

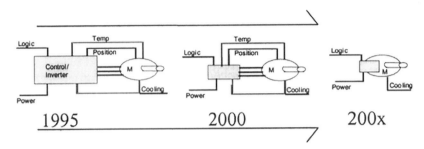

Figure 4.51 Evolution of automotive electronics integration.

A discussion of automotive electronics integration would be incomplete without mention of durability due to thermal shock and thermal cycling, particularly in the presence of vibration. Thermal shock is defined as a temperature cycling test where the rate of change of temperature is equal or greater than 10°C/min. A thermal cycle test is generally performed at a rate of temperature change that is less than 5°C/min. Damaging effects of thermal shock include de-lamination of circuit cards, cracking of ceramic substrates, and interconnect system failures. Thermal shock is most severe when a module is operated intermittently at low temperatures. In automotive systems, power cycling at low temperatures with a relatively low thermal mass heatsink results in large temperature excursions of the electronics chip and consequent mechanical stress. The result is generally material fatigue leading to catastrophic failure. It is for this reason that electronics that must be packaged underhood are attached to the engine intake manifold since it has a very substantial thermal mass and its temperature variation is limited by the presence of engine coolant circulated within. It is temperature cycle range and rate of change that is most damaging to packaged devices and modules.

Thermal cycling is performed as a test, typically over the temperature extremes of -40°C to +140°C, with ramp up and down times of 15 min and hold times at the extremes of 15 min. Thermal cycling provides an assessment of an integrated modules durability. Thermal expansion mismatch of materials used in the module, such as ceramic substrate to chip heat spreader to silicon chip, contributes to deformation and accumulation of strain. Thermal cycling results in an accumulation of deformation, which leads to stress induced fatigue failures. Fatigue failure in solder interconnections and chip attach methods are strongly influenced by the duration of time spent at temperature extremes leading to creep because strain energy increases with time for a given temperature. During the ramp changes in temperature, plasticity is evident because of the resultant deformation that accumulates which is dependant on he ramp rate of change in temperature. Because plasticity and creep resulting from temperature cycling are so cycle dependent it is very important to specify the temperature cycle appropriately to best reflect the intended service location in the automobile where these modules will be used.

4.8 Automotive Communication Networks and Wireless Techniques

4.8.1 Necessity of Networks in Automobiles

The automobile industry is fast becoming a technological field where application of high-end state-of-the-art electronics is becoming a necessity rather than a luxury. This is due to the fact that this industry is driven by consumer demands, which demand more and better convenient features inside an automobile as time goes by. Coupled to this is the fast growth of computing solutions to many modern automotive problems that not only simplifies the solution, but also creates the possibility of integrating a host of functions aimed at satisfying the consumers demand for convenient features and services. The cost of implementing such electronics amounts to almost a quarter of the total cost of manufacturing an automobile [47]. While electronic systems averaged $100 in an automobile way back in 1977, they averaged around $1800 up to until recently [47]. The following figure represents graphically the growth in wiring in the recent years [48].

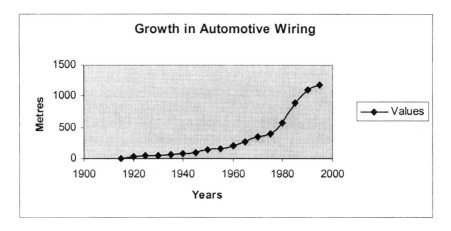

Figure 4.52 Growth in automobile wiring.

With such electronic systems comes the necessity for wiring up all these systems. This would increase the functionality of the systems by integrating services provided by individual systems. Older systems just consisted of a point-to-point wiring between two components. This caused a lot of problems as far as the wiring was concerned because point-to-point wiring resulted in complexity of connection and addition of weight as well as reduced reliability and functionality. However, with the modern systems, all the wires are connected to a central control unit. This results in more functionality and reliability. Also, with the development of various protocols and bus architecture for communication purposes, the task of wiring became simpler. Some of the protocols that have been developed, for example, x-by-wire, employ the use of components such as actuators and sensors to provide the feedback required for making adjustments based on the information obtained. An example of this system is the steer-by-wire column. The details of the x-by-wire system would be provided later on in this section.

The implementation of these networks not only solves the problems of traditional wiring systems, they also provide scalability; they leave open opportunities for adding new components in the future without any major changes to the network. They also make it possible to integrate different networks together. This would increase the functionality of systems already existent. The high data rates that are available today vary from a few kilobits/sec to a few megabits/sec depending on the network used, making it possible to implement such seemingly simple networks for real time control systems. To implement these networks in safety critical systems of a car would take some time because a failure of these networks would result in a loss of human lives, which is a matter of very grave concern. However, such networks are already in use in many places of a car such as multimedia, doors, seat adjustments, and trunk release.

We shall now proceed to study such automotive networks that are used. Various automobile companies who developed those networks have already implemented a few of the networks that are going to be mentioned. Possible examples are given of the practical implementations of these networks.

4.8.2 Controller Area Network (CAN)

It was in 1983 that Robert Bosch GmbH in Germany started an internal project to develop an in-vehicle network that would serve to connect all the electronic components in a car. In 1986, they introduced the concept of the serial bus based Controller Area Network (CAN) for the first time at the Society of Automotive Engineers (SAE) Congress.

The CAN protocol is based on the principle of broadcast transmission technique. It is basically a communication technique in which a data to be transmitted on a network is sent to all the nodes including the one to which the data is intended. The nodes look inside the data packet to see if the message was meant for them. If not they simply discard the packet. The node to which the data was intended then downloads the data and processes it.

Some of the features that have made CAN so popular are speed, data length, and the event triggered mechanism. The speed of data transmission in a CAN network can go up to 1Mbps. This is very helpful in real time control systems that can afford very low latency. Thus, due to the high speed offered by CAN, low latency and, hence, time efficient control can be achieved. CAN frames are also short in length due to which there is minimal delay in the reception of messages.

CAN is basically an event triggered mechanism. This means that the transmission of data is prompted only when a specific event happens. For example, data transmission may take place when a button is pressed or a lever is pulled. Due to this property, the bandwidth available is made maximum use of, due to the minimum load put on the bus system.

Therefore, the combination of high speed and short message length results in a low delay, so CAN can be implemented in control systems that have a very low tolerance for delays.

It must be noted that the addressing system is not based on the physical addresses of the nodes in the network; rather, the transmission on a CAN network is message oriented. Every message that is to be transmitted is assigned a particular message identifier. This message identifier is unique within the network. If it is not unique, then the network would be swarming with messages having the same identifier and the nodes would not be able to resolve which one is old and which one is new. This message identifier actually defines both the content of the message as well as its priority. This priority is resolved at the bit level. Every message identifier is made up of a binary value. The message identifier with the lowest binary value gets the highest priority and vice versa. The message identifier gives the definition for the priority as well as for the content of the particular message. For transmission purposes, a node forwards the message to its CAN controller along with the message identifier to be used for that particular data. The CAN controller then formats the data into a format that can be used over the CAN network and transmits it on the bus. Once this data has been transmitted, all the other nodes then function as receivers and receive the message that is transmitted. If it is not intended for them, they discard that particular message.

4.8.2.1 CAN Message Format

As shown in Figure 4.53, every CAN message [49] starts with a start-of-frame (SOF) field to indicate the beginning of a new message. It is followed by an arbitration field. It contains both the message identifier and remote transmission request (RTR) bit used to differentiate the data request frame and the data frame. The control field follows the arbitration field. It contains the information necessary to differentiate between a CAN frame and the extended-CAN frame. The data field that follows the control field can hold up to 8 bytes of data. The CRC field is used to detect any errors in the messages. CRC stands for cyclic redundancy code. It is an error detection scheme used in most networks. The ACK (acknowledgement) field is used by the receiver to acknowledge that the receiver had received the data correctly. In this way, the transmitter can know that a receiver received the data that it transmitted correctly. Finally, the end-of-frame (EOF) acts as a frame de-lineater. It signals to the entity involved that the current data frame has ended.

| SOF | ARB. FIELD | CTRL. FIELD | DATA FIELD | CRC FIELD | ACK FIELD | EOF |

Figure 4.53 CAN message format.

4.8.2.2 CAN Message Arbitration

The procedure invoked for allocation of the bus to various nodes is the CSMA/CD, which stands for Carrier Sense Multiple Access/Collision Detection. It is a kind of arbitration method [50] in which the nodes wishing to transmit first "sense" the bus to see if it is idle or busy. If the bus is busy, they defer transmission of the data by a certain amount of time. If not, they go ahead and transmit the data. This is done so that a collision between packets can be avoided on the bus. This would lead to waste of the bus available bandwidth.

The resolution of the priority between nodes for transmission is based upon the identifiers that they transmit. They not only transmit the identifiers, but they also continuously keep listening to the status of the bus. The method of bus arbitration can be well understood by considering the following example. Let us say that a node transmits with identifier of logical value "0" and another node a logical "1". It has been mentioned above that the node with the lowest binary value would be given the higher priority. In this case, it is the node with the identifier "0". Therefore, the node with the identifier "1" does not transmit. Also, the bus state is now "0". It is for this reason that bit "0" is known as the dominant bit and bit "1" is known as the recessive bit. After the node has finished transmitting, the node with the recessive bit as identifier tries to obtain control of the bus by the arbitration process described above.

4.8.2.3 CAN Error Detection

Error detection and correction is an important part in any network. Faulty nodes, faulty transmission channel, or simply a noisy environment could be one of the several reasons that could be attributed towards errors in networks. The error tolerance level of networks differs from network to network. Some networks are more fault tolerant than others. For example, the fault tolerance level of a simple network linking a few computers may be higher than that of a network implemented in an automobile. It depends on the kind of network implemented as well as the application for which the network is implemented. The errors that occur in a network may be at either the bit level or at the frame level. Fortunately, CAN is equipped to handle both kind of errors [49]. The major types of errors that can be handled by the CAN protocol are bit, stuffing, CRC, form, and acknowledgement errors.

When a node starts transmitting, it simultaneously keeps track of the status on the bus. If for example, a node transmits a "0" but on the bus it is read as "1", a decision can be reached that an error has occurred. Such an error is known as a bit error.

Most networks use a method known as bit stuffing when long sequences of the same bit are very likely to occur in the data field. In this method, if a particular bit occurs consecutively for five times in the data field, an extra bit is added at the end of the consecutively occurring similar bits. This extra bit, called the stuffed bit, is the opposite of the 5 similar bits. For example, if a run of 5 consecutive "0's" occur, the bit following the 5 consecutive "0's" would be a "1". This is to indicate to the receiver that there need not be any reason to worry if a run of the nature described above occurs and, for the receivers, to resynchronize themselves to the master clock. However, sometimes the stuffed bit may get corrupted and instead of five consecutive similar bits there would be six consecutive similar bits. This kind of error is known as the bit stuffing error.

As mentioned before, the cyclic redundancy check (CRC) is employed to detect any errors occurring during transmission. Here both the receiver and transmitter would have stored a mutually agreed polynomial P(x). The transmitter then represents the data to be transmitted by a polynomial G(x). G(x) is then divided by P(x). The remainder is then transmitted in the frame to the receiver. The receiver then performs the same division to see if it gets the same remainder. If not, then the receiver can infer that an error has occurred.

Form error occurs when the size indicated by the data length code (DLC) and the actual size of data in the data field do not match.

When a node has received a message correctly, it sets the value in the ACK field of the data to a dominant bit. If the transmitter does not detect a dominant bit in the ACK field, then it concludes either that the receiver did not receive the data correctly or that the ACK field is corrupted.

The CAN specification also details the implementation of "error counters", which are implemented in the hardware. The transmitter and receiver each maintain two error counters. The values in these error counters are incremented by 8 when an error is detected and decremented by 1 when a message has been transmitted without any error. Under normal circumstances with few errors, a CAN node is said to be operating in an "error-active state". If an error is detected at this stage, the node indicates it by the transmission of an active error frame. However, once the counter reaches a value of 128 the node enters the "error-passive" state. There is a limit to the value that the counters can reach up to; it is 256. Once the counters reach this value, the bus goes into the "bus OFF" state, shutting down all transmission. The bus can again start transmission after being reset by the CAN controller.

The CAN controller consists of the following:

- CPU interface logic (CIL) handles the data transfer on the bus.
- Bit stream processor (BSP) handles the streaming of data between buffer and bus line.
- Error management logic (EML) is involved with error management.

- Bit timing logic (BTL) is responsible for the synchronization of bit streams.
- Transceiver control logic (TCL) handles error detection and correction, transmission, and reception of the data and arbitration.
- Message buffer memory is used to store message for future transmission.

Basically there are two kinds of CAN controllers: full CAN and basic CAN. In the full CAN controller, spaces in the memory buffer are reserved for message with certain identifiers. If an incoming message does not have the required identifier number, it is rejected. However, in a basic CAN, all the messages are stored in the memory buffer. Therefore, in the basic CAN, it is the software that decides whether or not to accept the message.

For simplicity and transparency, the CAN protocol is made of three layers where each layer provides a service that is utilized by the layer directly above it. The three layers of the protocol stack are CAN object layer, CAN transport layer, and CAN physical layer, as shown in Figure 4.54.

The application layer is mainly concerned with the implementation of the specific application, which is implemented by the hardware. The object layer is designed for message handling and filtering. It is responsible for the selection of appropriate messages for transmission as well as filtering of unwanted messages. The transfer layer is responsible for functions such as error detection and transmission of an error frame upon detection of an error. It also deals with the issue of bus arbitration as well as framing issues. The physical layer is entrusted with the responsibility of the transmission of data frames. There are standards that have been established by the CAN consortium for the physical and application layers, written down in the CAN 2.0 Specification. For example, the physical medium is a pair of twisted wire 40 m long. For the protocol implementation, the CAN can take up to a maximum of 2032 identifiers. Recently, a variation of the CAN protocol, called the CANopen, has been developed [50]. It is a protocol that has been developed keeping the basic CAN application layer protocol as the foundation. Through this it is possible for various devices of different types to communicate with each other. This is made possible by a procedure known as "profiling". By this method, it is possible to build the specification of the CANopen protocol on the specification of other protocols. Therefore, it is possible for real-time data transfer and control as well as synchronization between the devices. This CANopen protocol incorporates the seven-layered structure of the ISO/OSI model into it. Thus, using the CANopen protocol provides us with a network solution where different networks running different protocols can be integrated to form a single network.

Some of the advantages offered by the CAN network are high speed, low latency, flexibility, and security. Ironically, there exists a particular problem due to the CAN specification. The CAN rule specifies that the last 7 bits of a frame should be recessive bits. As far as the receiver is concerned, a message that is received is said to be correct if there is no error present till the penultimate bit.

Automotive Power Systems

The last bit is ignored completely. Therefore, even if the last bit is a dominant bit, the message is resolved to be error free.

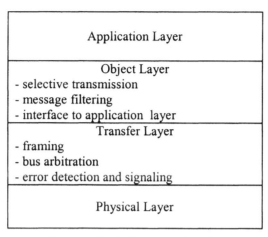

Figure 4.54 CAN protocol stack.

Another problem that exists in a CAN network is that sometimes duplicate messages are created when one node receives a message correctly, but another node does not receive the same message properly. When this happens, the node, which did not receive the message correctly, sends a re-transmission request to the transmitter that is promptly obeyed. Therefore, this leads to spurious duplicate messages.

There are many ways in which CAN can be implemented in automobiles. For our case, we shall consider the various ways in which the Jaguar car company has implemented CAN in its XK8 sports car [51]. The applications in this car are the engine control module (ECM), antilock braking system (ABS), transmission control module (TCM), instruments and driver information module (INST), J-gate illumination module (JGM), and inter-suspension data transfer system.

4.8.3 Time Triggered Controller Area Network (TTCAN)

Communication protocols can be divided into two classes: event triggered and time triggered. Both these methods vary much as far as their operating principles are concerned [52]. The event triggered model is known as an asynchronous model whereas the time triggered model is known as a synchronous model. In the even triggered model, data transmission takes place when a certain event takes place, for example, when a button is pressed or a lever is pulled. Here data transmission takes place at random on the time line. However, in the case of a time triggered model, data transmission takes place at specific intervals on the time line. In this mode, all the nodes are synchronized to a master clock so that all of them have the same sense of time. Each node is allotted a slot time. It is only during its allotted slot time that the node can transmit. This is something akin to the time division multiple access scheme.

As mentioned before, transmission of data takes place due to the progression of time. But, how do the nodes know when to start transmitting their data? In the static scheme, nodes are allotted slots in time. But, in the time triggered CAN, this problem is solved by the transmission of a special frame known as the reference frame. This is distinguished from other frames by virtue of its identifier. This reference message has a bit called the start-of-frame (SOF) bit. This reception of this bit denotes the instance of time at which the data transmission can take place. The TTCAN has two levels of operation defined in the specification [54]. Level 1 is by virtue of its property of the reference message. This reference message ensures the time triggered operation of the CAN. Level 2 ensures the global time synchronization of the nodes in the network. The difference is that for Level 1 implementation, the reference message carries one byte of information necessary for control purposes whereas, for Level 2 implementation, the reference message contains about four bytes of control information, with the other 4 bytes being usable for data transfer.

The advantage that it enjoys over CAN is that during its operation cycle, TTCAN permits the transmission of the regular time – triggered messages as well as event triggered messages. Certain time slots are reserved for the event triggered messages. For example, x-by-wire systems, which integrate sensors and actuators for performing certain critical real time control functions, need the TTCAN. This leads to a predictable behavior of the network, which is very important due to the distributed architecture of many networks. A distributed network consists of many subsystems organized into sub networks. Each sub network may be implementing a different protocol. Therefore, a protocol, which clearly defines the transmission procedure, to be followed at well-defined time slots would clearly reduce the complexity of the network as well as utilize the available bandwidth fully. In addition, it would reduce the time required, however small, for the arbitration procedure in the CAN that grants the nodes access to the bus for transmission of messages. It would also reduce the probability of a message with low priority from being refrained for a long time from transmitting that message.

As mentioned before, the fundamental transmission principle of TTCAN lies in the reference message. The time difference between two consecutive reference messages is known as the "cycle time" [53]. The period in between is known as the "basic cycle". The basic cycle is divided into various time slots called "exclusive windows". Data transmission takes place during the time slot of each exclusive window. Exactly which node is supposed to transmit and receive during which exclusive window is decided by the system design engineer. These exclusive windows support not only time triggered messages but they sometimes also allow the transmission of event triggered messages as well, which is the inherent property of TTCAN. These exclusive windows can be of varying size and duration to facilitate the transmission of variable data. Typically, during a basic cycle, multiple exclusive windows are incorporated so that utilization of bandwidth is kept maximum. The combination of several basic cycles constitutes what is known as a "matrix cycle" or "matrix system".

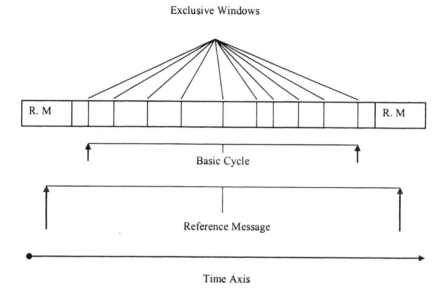

Figure 4.55 TTCAN transmission.

As mentioned before, event triggered messages can be sent too in the TTCAN. Special windows called "arbitration windows" are used for this type of transmission. The bitwise arbitration method decides which node is to be granted access to the bus during the arbitration window. Multiple messages can be sent in the arbitration windows. Therefore, it is up to the application software to decide which messages should be sent and which messages should not be sent. Automatic retransmission of corrupted messages is not possible. Another distinct feature permissible here is that, when the system is being designed, it is possible that a few windows be left empty. This leaves it possible for new nodes, which may be added in the future, to be integrated into the system mechanics without much trouble.

Unlike conventional network topologies that use routing tables, which are used by individual nodes to store the routing information from a sending node to receiving node, the nodes in TTCAN are given the information that is specific to that node only. The information supplied pertains to the time slot during which the node is supposed to transmit or receive. Since a node need not know any other information about the matrix cycle, a lot of memory is saved.

As has been mentioned before, several basic cycles make up a matrix cycle. The matrix cycle essentially describes the transmission schedule of the entire network. It defines which node should be transmitting at any given instant of time. In this way, each node knows when exactly it is supposed to transmit. The matrix cycle is repeated over time. The concept of the matrix cycle has been depicted in Figure 4.56.

Each basic cycle consists of several exclusive and/or arbitrating and free windows. As can be seen from the figure, this structure is highly column oriented. Each column in a matrix cycle is known as a "transmission column". All the windows belonging to a single transmission column are equal in length. A counter Cycle_Count is used to indicate the number of the current basic cycle. It is incremented by 1 every time one basic cycle is over.

Apart from the exclusive and arbitrating window, another window known as Tx_Enable is used. This is contained within the exclusive and arbitrating windows. This is used to indicate to the node the time at which it is supposed to start transmitting. This is done so that the message following the current one may not be delayed due to the delayed transmission of the current message. The Tx_Enable window concept is shown in Figure 4.57.

Another feature of the TTCAN system is that two arbitration windows can be merged to form a single arbitration window. However, there is a constraint that is applied to this situation. This is done only after it has been verified that the message for which the merging is being done fits exactly within the two arbitration windows.

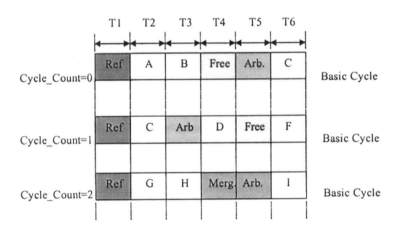

(T1, T2, T3, T4, T5, T6 – Transmission Cycles)

Figure 4.56 Matrix cycle in a TTCAN system.

Automotive Power Systems

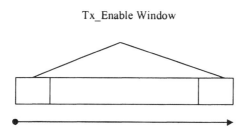

Figure 4.57 Exclusive/arbitration window.

The start of every exclusive and arbitration window is denoted by something called a "time mark". A time mark that precedes an exclusive window which is to be used for transmission purposes is known as the Tx_Trigger. It consists of the following:

- A pointer that points to a specific message that is to be transmitted.
- The basic cycle during which the message is to be transmitted.
- The transmission column during which the transmission is to take place.
- The position in a transmission column at which a re-transmission has to take place. If not required this is de-activated.

Similarly, a time mark that precedes a window that is to be used for reception of a message is known as an Rx_Trigger. Both the Tx_Trigger and Rx_Trigger are actually registers in which the required information needed for the transmission and reception of messages are stored. Also, a time mark is made up of a base mark and a repeat count information. The base mark contains the number of the basic cycle that follows the basic cycle during which the node is supposed to transmit/receive the message. The repeat count information contains the number of basic cycles between two consecutive transmissions or receptions.

The time unit that is used in any network transaction in the TTCAN network is known as the network time unit (NTU). It can either be a CAN bit duration in the Level 1 TTCAN or a fraction of a second when implemented in the Level 2 TTCAN.

A message state counter (MSC) is attached to every message that is transmitted or received. An error to this message is responded by an increment in the MSC. A node then flags an error if either the MSC reaches a value of 7 or the difference between 2 MSCs is greater than 2. Due to certain offset effects, the nodes loose their sense of "global time". Therefore, some mechanism should be designed so that all the nodes are synchronized in unison to the master clock.

To compensate for the drift in time, the node measures the time that it sees locally. When a master node sends a frame synchronization pulse, it also sends its global time value. The node, upon receiving this frame synchronization pulse, captures the global time contained in it. It then calculates the local offset as the difference between measured global value and the measured local time value. During the next basic cycle, the global time is then corrected by a factor equivalent to the local offset, i.e., global value = local time + local offset. However, each node may have a drift that is different from other nodes. To compensate for this effect, a unit known as the time unit ratio (TUR) is used. The difference between two consecutive frame synchronization pulses is measured, both globally and locally. The quotient of these two gives the correct TUR.

Since the entire TTCAN architecture depends on the operation of the master node, fail-safe operation of the master node is essential. To ensure this, a couple of nodes are identified to be potential master nodes with corresponding identifiers. If a potential master node sees no traffic on the bus or sees no reference messages, it transmits its identifier and assumes the role of a master node. It then starts transmitting the reference messages. If it sees another potential master node arbitrating for the role of a master node, but with an identifier of higher priority, it synchronizes itself to that particular reference message. Higher utilization of bandwidth, lower jitter and latency, higher security for messages, and greater flexibility and deterministic behavior to automotive networks are the main advantages.

In summary, with the addition of a session layer to the protocol stack of a CAN network, it is possible to have a time triggered model of the CAN which lends flexibility as well as a deterministic behavior to the network. This fact can be exploited by systems such as x-by-wire systems in the future. Slowly, such networks would be added to the existing mechanic and/or hydraulic back-ups to ensure complete safety of the operating control system.

4.8.4 Society of Automotive Engineers (SAE) J1850

The J1850 protocol defined by the Society of Automotive Engineers (SAE) is a standard protocol defined for automobiles. The main advantages are that it is economical in implementation and that it has an open architecture. This ensures that it can be implemented on a wide scale during the manufacturing process of automobiles.

The SAE defines three specifications for the network protocols that are generally used in automobiles: class A, class B, and class C. A Class A network has a data transfer which has a maximum of less than 10kbps. It is basically used for actuators and smart sensors. According to the SAE, a Class B system is defined [55] as "a system whereby data (parametric values) is transferred between nodes to eliminate redundant sensors and other system elements. The nodes in this form of a multiplex system already existed as stand-alone modules in a conventionally wired vehicle. A Class B network shall also be capable of performing Class A functions". It has a data transfer rate of around 100kbps. It is basically used for applications that do not need any real time control. The J1850 protocol belongs to this group of data communication. Class C is the last class of data transfer defined by the SAE. It has a data transfer rate of around

1Mbps. Due to this high data rate it is basically used for real time control systems. The controller area network (CAN) belongs to this class of network.

Another distinct advantage that it enjoys over its counterparts is that it has no master-slave architecture. It is completely masterless. The bus access scheme is actually a Carrier Sense Multiple Access with Collision Resolution (CSMA/CR) [56]. This means that any node can transmit when it senses that the bus is free. If there were a collision, an arbitration mechanism would decide which node could transmit. It is basically based on the same identifier scheme that is used in CAN. The contention algorithm allots control of the bus to the nodes based on the priority of the identifiers. If there is more than one message to be transmitted at the same time, the algorithm keeps sifting through all the identifiers until only one transmitter remains while all the others have been eliminated due to lower priority. This transmitter then continues to transmit. Here the transmitter receives the message that it transmitted. This is to check whether the message that appears on the bus is the same as that the one it had transmitted. Once it has finished transmitting it relinquishes control of the bus. The other transmitters then try again to gain control of the bus. Therefore, an open architecture is created which enables the addition of new nodes or the removal of old nodes without many changes to the network. Also, due to the CSMA/CR scheme that is used for bus access, messages are not lost due to the collision. This makes it possible to save a lot of time, which could be lost due to re-transmission of messages. This leads to higher bandwidth utilization.

The SAE J1850 protocol has two different approaches that are commonly used in the automotive environment: a high speed, 41.6 kbps, two-wire approach that uses pulse width modulation (PWM) and an average speed, 10.4 kbps, single-wire approach that used variable pulse width (VPW) modulation.

There are a lot of differences between these two approaches. To put it in a nutshell, representation of bits in terms of voltage transitions per bit differs in the two schemes. The PWM approach uses two voltage transitions to represent one bit whereas the VPW approach uses just a single voltage transition. Incidentally, General Motors developed the VPW approach. It uniquely defines a transition according to four main characteristics: active or passive state and long or short pulse. The VPW modulation takes a unique approach towards representation of bits. In this approach, a "1" bit does not mean a transition from a low state to a high state. Instead, a "1" bit or a "0" is determined by the amount of time that the bus stays at a particular voltage level. According to VPW J1850 protocol definition, a "1" bit is a bus driven at a high potential (active) state for 64μs or low potential (passive) state for 128μs. Alternatively, a "0" bit is defined as a bus in high potential (active) state for 128μs or low potential (passive) state for 64μs. A high potential is usually any voltage from 4.25V-20V. A low potential is any voltage from ground voltage to 3.5V.

When a node wants to transmit, it first senses the bus. When a bus is idle, it is usually at the ground potential or low voltage. The node wishing to transmit then can either leave the potential on the bus as it is or pull it up to a high voltage using pull-up transistors. Therefore, even if there was a case of two nodes trying to simultaneously transmit data with one node allowing the bus to remain at low potential and another trying to pull up the potential, the node trying to pull up the potential would win and the other node would have to defer its transmission and try to gain control of the bus at a later time.

The signals can be transmitted using the technique of multiplexing. There are basically two types of multiplexing: frequency division multiplexing (FDM) and time division multiplexing (TDM). In FDM, each transmitter is allotted a particular frequency band from the entire bandwidth for transmission, whereas in TDM each transmitter is permitted to transmit at different instances of time on the same channel. The 10.4 kbps model uses the variable time TDM.

Since the bus access scheme used in the SAE J1850 protocol is CSMA/CR, there are no well-defined instants of time when any node can start transmission. The nodes transmit as and when they can sense an ideal bus. It is for this reason that data transmission in the SAE J1850 is asynchronous. As mentioned before, this architecture does not follow the master-slave architecture used in many protocols. Furthermore, the messages are broadcasted in this protocol. This means that the message to be transmitted is not only received by all the other nodes in the network, but it is also received by the transmitter like an echo. This is done to ensure that the message present on the bus is the same as the message transmitted by the transmitter.

Here, the node that wishes to transmit first senses the bus for a fixed amount of time. If it senses that the bus is idle, it starts transmitting. If the bus is busy, it waits for some amount of time before trying to start transmission once again. Let us assume that a node wishing to transmit transmits a passive symbol. If there is no challenge, it goes ahead and transmits the message. It has to be noted here that even during transmission each bit is checked to make sure that there is no other node with higher priority waiting to transmit. This is called bit-by-bit arbitration. This process is carried out until only one bit is left. However, before transmission, if there is a node transmitting an active symbol, the node with the passive symbol looses control of the bus and starts functioning as a receiver. The node with the passive symbol then is said to have "lost arbitration". Thus, we see why the arbitration scheme used is called Carrier Sense Multiple Access/Collision Resolution (CSMA/CR). It is called multiple access because it is possible for multiple nodes to gain equal access to the bus.

Once the transmitting node has finished transmission, the nodes that had lost arbitration again try to gain access to the bus and the whole arbitration procedure is repeated again. Invariably, it is seen that nodes that have more active symbols in the starting of the message compared to others gain access to the bus earlier than the other messages. However, it must be stressed that, due to the nature of the bus access, it is impossible to specify instants of time at which certain nodes can transmit or receive, very much unlike the TTCAN.

The diagram in Figure 4.58 shows the format of the message frame that is used in the SAE J1850 protocol. SOF stands for Start Of Frame. It appears at

Automotive Power Systems

the beginning of every frame. It is used to indicate the beginning of a frame. SOF is usually a "high potential" on the bus that lasts for around 200µs. A header field follows this. The header field contains information regarding the message that follows the current one, for example, how many bytes are contained in the header of the next message or how many bytes are contained in the entire message frame. Usually, the size of the header field varies from a single byte to three bytes.

The date field contains a sequence of 0's and 1's. This is the part where the data to be transmitted is placed. Its length varies from a single byte to eleven bytes. However, it must be remembered that both the data field and header bytes put together must fit inside the standard J1850 protocol message frame.

As mentioned in CAN and TTCAN, CRC stands for cyclic redundancy check. Here, the entire message is divided by a predetermined polynomial. The 1's complement of the result is then attached to the message to be transmitted. A similar calculation is performed at the receiver's end. If there is no error, the result is usually a constant (C4 hex) [57].

EOF stands for End of File. This is to indicate to the receiver that the message ends at that point. The receiver then can choose its response, whether it wants to transmit a message of its own or transmit the in-frame-response (IFR). The IFR is a way adopted by the receiver to acknowledge to the transmitter that it received the message correctly. The EOF is indicated by a low potential on the J1850 bus for 200µs. Figure 4.59 shows SAE J1850 message with IFR.

SOF	HEADER FIELD	DATA FIELD	CRC	EOF

⬅——————————— Message Frame ———————————➡

Figure 4.58 SAE J1850 message format.

Figure 4.59 SAE J1850 message with IFR.

4.8.4.1 SAE J1850 Specifications

The total length of the network wiring is restricted to 40m. This included 35m of on-vehicle networking and 5m of off-vehicle networking. The wiring itself is a single wire in the 10.4 kbps case and double wires in the case of the 41.6 kbps model. The maximum number of modules or nodes, including off-vehicle equipment, is 32. This makes the assumption that a single node is seen as a unit load, which corresponds to 10.6KΩ and 470pF. Off-vehicle minimum resistance should be 10.6KΩ and the maximum capacitance should not exceed 500pF. Another major specification is that a "0" bit symbol is the dominant symbol and the "1" bit symbol is the recessive symbol. The pulses on the 10.4 kbps model are all wave shaped. The advantage of using this shape is that it rounds of the sharp edges of the bus transition, thereby, removing unwanted high frequency components. Table 4.3 shows VPW DC parameters.

4.8.5 IEEE 1394 Protocol

The IEEE 1394 is a standard data communication protocol that has been developed for multimedia applications. The main feature of the IEEE 1394 protocol is its ability to be able to scale with increasing number of nodes. Also, it has an ability to scale with increasing number of services for which the bus serves as the communication protocol. The physical layer is based on optical fibers. Since this can allow implementation of services, which would require a high bandwidth rate, it could be used to offer an entire gamut of embedded solutions in an automotive environment. The bandwidth offered by an IEEE 1394 serial bus is usually from 2^i – 98.304 Mbps, with i taking on the values of $\{i = 0,1,2,3,4,5\}$ [58].

Table 4.3 VPW DC parameters.

Parameters	Symbol	Min	Type	Max	Unit
Input High Voltage	V_{ih}	4.25	-	20.00	Volts
Input Low Voltage	V_{il}	-	-	3.5	Volts
Output High Voltage	V_{oh}	6.25	-	8.0	Volts
Output Low Voltage	V_{ol}	0.00	-	1.5	Volts
Absolute Ground Offset Voltage	V_{go}	0.00	-	2.0	Volts
Network Resistance	R_{load}	315	-	1575	Ohms
Network Capacitance	C_{load}	2470	-	16544	pF
Network Time Constant	T_{load}	-	-	5.2	µs
Signal Transition Time	T_t	-	-	18.0	µs
Node Resistance (unit load)	R_{ul}	-	10,600	-	Ohms
Node Capacitance (unit load)	C_{ul}	-	470	-	pF
Node Leakage Current	I_{leak}	-	-	10	µA

The IEEE 1394, like the SAE J1850 protocol discussed earlier, is masterless. However, the similarity ends here. A feature of importance in the IEEE

Automotive Power Systems

1394 is that it has a distributed architecture. This means that various functions have been distributed throughout all the nodes. Therefore, in case of some event in which a part of the system or serial bus is cut off from the rest of the network, the sub-system can still function independently of the rest of the network.

As mentioned before, the IEEE 1394 has a distributed architecture. The management functions are distributed throughout the nodes. The protocol stack consists of the physical layer, link layer, transaction layer, and a serial bus manager. Each layer has some services or functions to perform. Figure 4.61 depicts the protocol stack architecture.

IEEE 1394 supports two types of data transfer [59]: asynchronous data transfer and isochronous data transfer. Asynchronous data transfer is mainly used for guaranteed data delivery, which some protocols like the SAE J1850 protocols offer (the in-frame-response). The messages are short and they are mainly used for control and setup purposes.

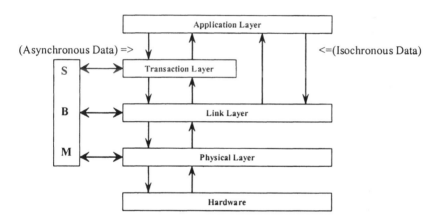

Figure 4.61 IEEE 1394 protocol stack architecture.

Isochronous data transfer is used for mass transfer of data. However, reliability is not a major criterion in this case. Constant speed and bandwidth are the major factors taken into consideration for this kind of data transfer. Messages are sent in the form of packets which are sent out every 125µs. A typical application is video streaming.

Both these methods utilize a common physical as well a data link layer. The link layer performs functions such as addressing, error checking, and message fragmentation and re-assembly. The physical layer is mainly concerned with the conversion of binary data into electrical symbols that can be transmitted over the communication channel. Sometimes it can also be used for bus arbitration in case multiple nodes want to transmit at the same time. Data is transferred directly to the link layer in the case of isochronous data transmission, whereas for asynchronous data transmission an additional transfer layer is required. The data, which are in the form of packets, have a header containing information regarding the address of the destination as well as the type of data. Error checking is implemented in the form of the cyclic redundancy check (CRC). The physical layer, link layer, and the transfer layer exchange messages with the serial bus manager (SBM), which is used for the maintenance of the serial bus. Applications that need vast coding could be difficult for the user to code. Therefore, to make the job of coding an application easier, the IEEE 1394 has an additional layer that is called the application programming interface (API) layer. Another layer known as the hardware abstraction layer (HAL) is also implemented. This is done to make the API, transaction layer, and the serial bus manager independent of the link layer controller so that they can be used on various embedded systems.

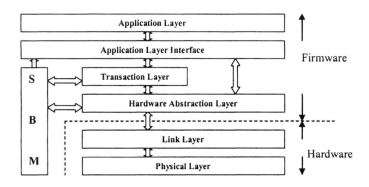

Figure 4.62 IEEE 1394 protocol stack for embedded systems.

The addressing is implemented by the way of a 64 bit addressing system [60]. Out of these 64 bits, the higher order 16 bits represent the node ID. 10 bits of these 16 bits are reserved for the bus ID. Therefore, we have a total of 1023 buses possible ($2^{10} - 1 = 1023$). The other 6 bits represent the actual physical address of the bus, which makes it possible for accommodating 63 nodes ($2^6 - 1 = 63$). The rest of the 48 bits are utilized for publishing addresses for memory space.

The configuration process consists of 3 steps [61]: bus initialization, tree identification, and self-identification.

A feature that is supported in the IEEE 1394 is the plug-and-play mode of operation. When a node is added, a reset is generated in the network. When this happens, all the nodes start to re-identify themselves. Therefore, branch nodes, root nodes, and leaf nodes are identified. Once this process is over, the root nodes start issuing IDs to every other node. Slots of 125μs are reserved for data transmission. The root node, which issues a cycle start packet, indicates the start of every cycle.

Main applications include the avionic industry to handle CNI waveforms, industrial cameras that are for industrial inspection systems for quality control, and multimedia applications in automobiles.

4.8.6 Media Oriented Systems Transport (MOST)

The development of the MOST network [62] stemmed for the need for a peer-peer network that had high data speeds but was cost effective. The MOST network was essentially to be used for multimedia applications in the automotive environment. But, due to its efficiency, it is increasingly being used for home networks too. Also, the implementation of optical fibers at its physical layer ensures high reliability under adverse circumstances. Some of the key features of the MOST networks are ease of use, cost effectiveness, availability of synchronous and asynchronous data transfer mode, flexibility, and the wide range of applications.

As pointed out before, the MOST network is a peer-peer network. The point-to-point network connection can be established using either of the ring, star, or daisy chain topologies. Two approaches can be used for administrative tasks: centralized and decentralized approaches. In the centralized approach, a single node handles all the tasks. In the decentralized approach, all the administrative tasks are distributed throughout all the nodes.

A MOST network is essentially made up of the MOST interconnect, MOST system services, and MOST devices. The MOST interconnect is essentially concerned with the establishment of connection between the devices at the start-up phase. It is also during this phase that device addresses are distributed throughout all the devices. Synchronization between all the nodes by means of a bit pulse is achieved during this phase.

The MOST system services are composed of low level system services. Functions such as data routing, channel allocation, fault detection, or delay detection are performed by the low level system services.

The application socket and the basic layer system services make up the MOST netservices, which is basically concerned with the transmission and reception of different kinds of data as well as providing standard interfaces to access various network management functions.

MOST Devices can vary from simple displays to complex applications. They have the capacity of bandwidth allocation as well as packet and control data capacity handling capability. Figure 4.63 illustrates the implementation of a MOST device in a typical configuration.

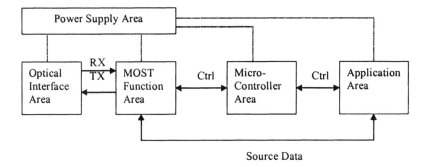

Figure 4.63 MOST device in typical configuration.

The data types supported by MOST are control data and bursty data. All the transactions taking place on the MOST network involve the generation of a synchronous frame by a frame generator or timing master. Each node has an internal timing device that locks onto this signal by means of a phase locked loop (PLL). By locking onto this synchronization frame, all the nodes are synchronized to the timing master.

Sixteen frames make up a block. Each frame consists of 5 sections. There are sections reserved for administrative data. Sixty bytes are allocated for data transfer and 2 bytes for control messages. The structure of the MOST frame is shown in Figure 4.64.

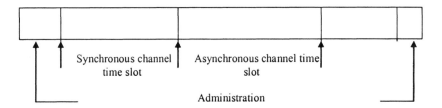

Figure 4.64 MOST block and frame structure

MOST networks are designed to use fiber optical cables as the medium of transmission. Therefore, the network is designed to accommodate all the possible complex topologies that can implement applications based on the fiber optical cables. This provides enhanced capacity as well as security features. This also enables the MOST network to support the plug-and-play mode of operation. The signal lines that compose the physical structure are RX-receive data and TX-transmit data. Other optional lines are power, ground, and wake-up. As the name suggests, the wake-up line is used for waking up the target device.

Automotive Power Systems

In this network, each node has a set of functions that are to be carried out. The most important ones are:

- Synchronization
- Managing data flow
- Decoding addresses
- Detection of system start-up
- Power management.

These are the functions that are implemented in each node. However, the software too has many functions that are important. Some of them are:

- Physical addressing
- Channel allocation
- Control data transfer
- Packet data transfer
- System monitoring

Since this network can support real time data transmission, it can be used for networking CD audio drivers, set-top boxes, and TV sets. Current MOST technology can support over 12 uncompressed stereo CD quality audio channels at the same time or 12 MPEG1+ audio channels or several MPEG2+ audio channels. It can also be used for networking multimedia at homes, for example, multi-room audio and video devices.

4.8.7 X-By-Wire

The automobile industry is always concerned with safety related issues. It is always trying to develop methods that keep increasing the safety standards, for example, intelligent driver assistance. However, such systems need to be computer controlled to deliver maximum efficiency. With this comes the need to replace all the mechanic or hydraulic backup with electric/electronic components. This can be done only when it has been ascertained that the systems that are replacing the mechanical or hydraulic backups are very safe. Such systems are known as "x-by-wire" [63] systems. A consortium comprising of Daimler-Benz, Centro Ricerche Fiat, Ford Europe, Volvo, Robert Bosch, Mecel, Magneti Marelli, University of Chalmers, and Vienna Institute of Technology has carried on work on this field. Examples of this kind of system are steer-by-wire and brake-by-wire systems. Figure 4.65 illustrates a typical application of an x-by-wire system.

The system comprises of actuators and sensors connected to electronic control units (ECUs). The sensors and actuators are used for taking measurements, which are in turn fed to the ECUs for driver feedback. Based on the measurements taken, the driver can make suitable modifications which are then relayed to the actuators for implementation. It must always be ensured that the measurements taken are always accurate. If the readings taken by the sensors are not accurate or the response of the driver given to the actuator is not accurately interpreted, it could lead to very serious losses. Another problem with implementing this technology for mass production is its economic feasibility. In addition, associated with this technology are problems such as reliability and maintainability.

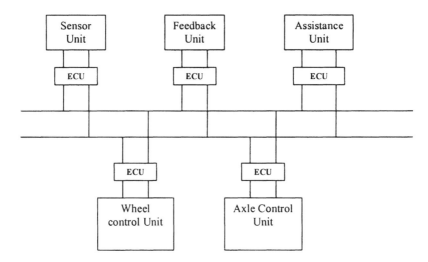

Figure 4.65 An implementation of x-by-wire system for steering and braking systems.

The entire architecture of the x-by-wire system is based on a time triggered approach. Here, all the activities are carried out at specific instants of time. Time is divided into slots of equal duration. For each time slot, there is a specific task assigned. All the nodes are synchronized to one global sense of time. This is achieved by the transmission of a synchronization pulse by a master node. This pulse is periodic in nature.

At specific instants of time, all the sensors capture their measurements which are in turn transmitted on the bus. These measurements are used to update certain variables. The new values over-write the old ones. These new values are then used for control applications.

A main feature of the x-by-wire system is what is known as composability. It means that whatever properties are exhibited by sub-systems, the same properties are exhibited by when the various sub-systems are integrated to offer some function or application.

Another important feature is the ability of the system to handle errors which are generated when nodes "talk" during a slot which is not allotted to them. This node is known as a "Babbling Idiot". This error can be avoided because, in a time triggered system, information is known about which node is supposed to transmit at which time interval.

The bus access is controlled by a TDMA scheme. Broadcast topology is used. As mentioned before, information regarding which node can transmit at which interval is stored in the memory. This information is stored in every node locally. Every node is supposed to follow this timetable. The protocol used is the Time Triggered Protocol/Class C (TTP/C).

The node architecture is based on what is known as a "fail-silent" mode of operation. This means that if a node is detected as having an error, it is asked to stop transmission. This is done so that the erroneous node does not interfere with the normal performance of the network. Sometimes, two nodes are used as a pair for fault tolerance. If one node has gone silent, then the other node can continue to perform the task in an error-free way.

This technique has many distinct advantages. First of all, the need for complex circuitry to implement other fault tolerant mechanisms is avoided. Second, combined with software fault detection, the fail-silent architecture would be able to cover the entire range of possible errors comprehensively. Thirdly, this mechanism can be tested to test its full functionality.

Each ECU, as mentioned before, is made up of some sensors and actuators. It must be ascertained that, in a node comprising two ECUs, there should not be any disparity in the decision taken finally. This may mean that each ECU exchanges sensor values with each other on the common bus so that a common decision is reached. An alternative to using the common system bus is using a sensor bus. Also, if digital sensors are used, each ECU will have the same values. If the two ECUs are located far way from each other, then the sensors are connected to the closest ECU. If, on the other hand, they are situated close to each other, then the sensors may be connected to the ECUs by means of a sensor bus as mentioned above.

The software model consists of two layers: system software layer and application software layer. The system software layer is used to provide services to the application software layer. Services such as fault tolerance and detection are implemented in the system software level. The language used to write codes is the ANSI C subset with certain modifications like exclusion of certain error prone constructors and inclusion of exception handling. Efficient design, higher fault tolerance, ability to be tested, and ease of synchronizing are the main advantages.

With this kind of system highly efficient and robust driving aids can be developed for the driver. This could eliminate the use of the current mechanical/hydraulic back-up systems. The combination of sensors and actuators minimize the deviation from the normal values. Thus, intelligent driving aids can be developed.

4.8.8 Local Interconnect Network

Local Interconnect Network (LIN) [64] is a communication protocol that has been established for automobiles. This protocol is based on the SCI (UART) data format with a single master/multiple slave architecture. A consortium formed in 1998, comprising Audi, BMW, Daimler Chrysler, Volvo, Volkswagen, VCT, and Motorola, worked on establishing the specification for this protocol.

The development of LIN was based on the necessity for a communication protocol that was a very cost effective protocol addressing not only the issue of the specification for communication, but also other issues like signal transmission, programming, and interconnection of nodes. It basically takes an all-around approach for the development and consolidation of an automotive protocol. The basic advantage that LIN enjoys is that it is very economical compared to protocols such as CAN. However, this advantage is negated by its inherent limitations such as low bandwidth and performance and the single master topology of the network.

There are quite a few criteria that have to be taken into consideration when a network is designed. Factors such as bandwidth, security, latency, electromagnetic interference, fault tolerance, and cost should be balanced in order to design a network to suit the requirements. There are basically two different approaches that could be taken to design a network.

One approach is to divide the sensors and actuators in a zonal manner, connected to a central ECU. Various main ECUs are connected to each other by CAN links. This extensive use of CAN is to enable the usage of high bandwidth for signal exchanges.

The other approach is to totally abolish the zonal concept. Here, all the actuators and sensors are connected to the central ECU by means of LIN links. This has the advantage of being scalable. No major changes have to be done to the network in order to accommodate additional nodes.

As mentioned before, LIN is based on the SCI (UART) byte word interface. The network has a single master/multiple slave topology [65]. All the slaves have the job of transmitting and receiving. The master node, apart from the task of transmission and reception, has the additional task of maintaining the synchronization in the network. This is done by means of a message header which consists of the synchronization break, synchronization byte, and a message identifier. The message identifier is used by the nodes to identify the messages meant for them. Each identifier is unique to the node. This way the node knows the messages meant for it. It has to be remembered that the message identifier indicates to the content of the message and not to the destination. Once the node has received the message meant for it, it then sends back a response that contains data with the size ranging from 2, 4, or 8 bytes of data with 1 checksum byte. One message frame consists of the header and the response parts.

Each message frame is made up of a byte field, which has 10 bits. These 10 bits include a dominant "start" bit, 8 bits of data, and a recessive "stop bit". The message frame consists of the header sent by the master node and the response sent by the slave node. The header sent by the master node consists of synchronization break, synchronization field, and the identifier field. Each message frame is built upon the 8N1-coding scheme. The synchronization break must be minimum 13 bits in length to ensure proper synchronization. The synchronization field is a string of bits with an equivalent hexadecimal value of 0x55. With this type of synchronization pattern, it is possible for nodes not equipped with quartz stabilizers to re-synchronize themselves. There are basically two levels of synchronization which are defined for the LIN network:

- Unsynchronized: The slave clock time differs from the master clock time by less then +/- 15%.
- Synchronized: The slave clock time differs from master clock time by less than +/- 2%.

There are 4 identifiers reserved for special purposes. Out of the 4 identifiers, 2 are used for uploading and downloading purposes. The only way they differ from the normal data frame is that in these frames, instead of data there are user-defined command messages.

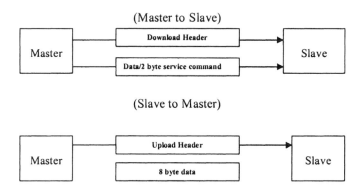

Figure 4.66 Use of LIN identifiers for uploads and downloads.

The other 2 special identifiers are used for ensuring upward compatibility with the future versions of the LIN protocol. These identifiers are called extended identifiers. There are three types of communication modes that are supported:

- Master to slave or multiple slaves
- Slave to master

- Slave to slave – the slaves can talk to each other without routing the transmission through the master

This is illustrated in Figure 4.67. LIN specifications are:

- Medium access control: single master
- Typical bus speed: 2.4-19.6kbps
- Size of network: 2-10 nodes
- Coding scheme: NRZ 8N1(UART)
- Data byte/frame: 2, 4, 8 bytes
- Error detection: 8-bit checksum
- Multicast message routing: 6-bit identifier
- Communication speed: 2.4, 9.6, 19.2 kbps
- Voltage level: 13.5V
- Signal slew rate: 2v/μs
- Termination resistor master/slave: 1kΩ/30kΩ
- Termination capacitance master/slave: 220 pF/2.2nF
- Line capacitance: 100-150 pF

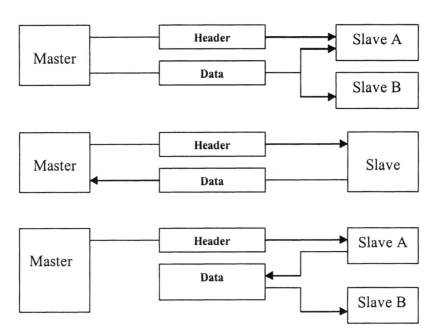

Figure 4.67 Data communication in LIN.

Single wire-transmission, low cost of implementation, and no resonators are the advantages. Low bandwidth and single bus access scheme are the main disadvantages. Implementations include:

- Door control
- Roof control
- Steering wheel and steering column
- Seat control and heating
- Switch panels
- Motors and sensors
- Smart viper motor
- RF receiver for remote control

4.8.9 Bluetooth

Bluetooth [66] is a wireless communication standard that has been developed for networking small devices such as printers, palm pilots, and mobile phones. It was initially conceived by Ericsson in 1994. Since then companies such as Nokia, IBM, Intel, and Toshiba joined it in its research efforts in 1998 to form what is known as the Bluetooth Special Interest Group (SIG). Currently, about 2000 companies are part of the SIG, and they are developing Bluetooth related products.

Bluetooth, which could become one of the defining technologies of the 21st century, has its roots in a town that dates back to the 10th century. Lund, Sweden, a vibrant college town near the southern tip of the country, is where the Swedish cell phone giant Ericsson started the Bluetooth movement in 1994. Ericsson's primary research facility is closely tied to the town's university.

Bluetooth was discovered quite by accident. Researchers wanted to integrate something like an earphone, cordless headset, and a mobile phone with a wireless channel. They eventually settled on small chips that make wireless communication possible between devices that contain them. Thus was born the technology of Bluetooth. What began as simple experiment is now a leading technology that could perhaps change the way transfer of information takes place.

Bluetooth is implemented in devices by way of small chips. It uses a short-wave radio signal that is always on for signal transmission purposes. It creates what is known in the technical terms as an "ad-hoc" network [67]. The participants in this ad-hoc network are the slaves and the masters present in the entire network. Each master and slaves form what is known as a "piconet". It basically represents an area of active participation by one master and one or more slaves. A slave can participate in any number of piconets. However, a master can belong to only one piconet. The topology is also called the scatternet topology. Each piconet has its boundary. We have at least one slave acting as a bridge between two piconets. Each piconet has its own characteristic transmission frequency. Also, since each piconet uses it own transmission frequency, each piconet is uniquely identified by the hopping pattern that is used for transmission. The master of that particular piconet determines the hopping pattern. We shall look into the details of the Bluetooth radio system.

4.8.9.1 Bluetooth Radio System

This is the most important part of Bluetooth. The specifications listed for the radio system are used all over the world for devices that use Bluetooth technology. This is a standard that has been established by the Bluetooth consortium.

a) Radio Spectrum

Since Bluetooth uses a wireless communication channel, the need was felt for a spectrum that could be used all over the world and have public access. Ultimately, it was decided to use the unlicensed Industrial, Scientific, Medical (ISM) band. The spectrum of this band varies from 2.4-2.483 GHz in the United States and from 2.4-2.5 GHz in Japan. The spectrum used in the U.S.A is centered around 2.45 GHz. There are different regulations for this unlicensed band in different parts of the world, but all of them ensure easy access to the public for their usage.

b) Multiple Access Scheme

There are a number of multiple access schemes that are used widely. There is the frequency division multiple access (FDMA), the time division multiple access (TDMA), and the code division multiple access (TDMA). The disadvantage of using FDMA is that its operation is limited by interference which could disrupt data transfer. The interference in the FDMA system is attributed to its inability to perform frequency spreading. If we consider TDMA systems, it would require all the devices to have the same global sense of time. Envisioning a scenario in which there are many devices in any given area, it is not possible for all the devices to be referenced to the same time. Therefore, in Bluetooth systems, FH-CDMA is used, where FH stands for Fast Hopping. This means that, for each transmission, a different frequency is used out of a well-defined set of frequencies. This has been likened to "hopping" frequencies. The advantage of using FH-CDMA is that, at any point of time, only a small part of the bandwidth is utilized, which results in efficient bandwidth utilization and negligible interference.

According to specifications, a total of 79 hop carriers have been defined 1MHz apart from each other in the ISM band. The only exception here is that France and Spain have defined 23 hop carriers here instead of 79. The channel has a dwell time of 625μs. Pseudo-random codes are used for frequency spreading. The "masters" of a piconet determine the pseudo-random sequences

Automotive Power Systems

to be used in that particular piconet. The slaves in that particular piconet use the identity of the master to determine the pseudo-random sequence to be used. Added to these features is the usage of time division duplexing. This means that a node, whether a master or a slave, can alternately transmit and receive. This effectively cancels the interference due to simultaneous transmission and reception by a single node. This concept is illustrated in Figure 4.68. The figure depicts the TDD concept that has been explained.

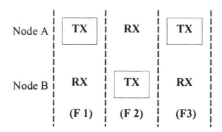

Figure 4.68 FH/TDD channel illustration.

c) Medium Access Scheme

Bluetooth has been developed to handle uncoordinated communication taking place in the same area. With the kind of modulation scheme that is used for the communication channel, each channel had a gross data rate of approximately 1 Mbps. Since 79 carriers are used, we have a gross data rate of 79 Mbps for the entire system. However, due to non-orthogonality of the hop sequences used, the data rate actually implemented is less than the theoretical rate of 79 Mbps.

As mentioned before, each piconet has one master and one or more slaves. The piconet channel is identified by the identity of the master and the system clock. Any unit that wants to participate in the piconet must have its clock synchronized to the system clock. A slave can belong to one or more piconets. However, to maintain efficient utilization of the communication channel the number of slaves that can participate in a single piconet has been restricted to 8. The hopping sequences are determined every time a fresh piconet has been established. Any node can act as a master. This is the essence of peer-peer communication. By default, any node that sets up a piconet is the master.

The master is also used to regulate traffic on the communication channels. The dwell time for each channel is limited to 625µs. This is just enough to transmit one packet. When a master wants to transmit to a slave, it includes the address of the slave. Once the slave has received the message, the master, in turn, polls it to see if it has any data to send. If the slave has some data to send back to the master, it sends the data in the slot following the slot in which the master sends the slave some data. If the master has no data to send to any slave, it polls all the slaves to see if any of the slaves have any data to send. All the decisions are made on a per-slot basis. Collision is avoided by using the ALOHA protocol without carrier sensing, i.e., the nodes start transmitting without seeing if the packet would collide with another packet or not.

d) Bluetooth Physical Layer

The Bluetooth physical layer supports both kinds of links: synchronous connection oriented (SCO) and asynchronous connection-less (ACL). The SCO link is used as a point-to-point link between the master and the slave, and the ACL link is used a point-to-multipoint between master and multiple slaves.

e) Data Communication

The communication concept used in Bluetooth is based on packets. Packets are small blocks of data that have been formed by fragmenting the original data. Due to the standardized dwell time of 625μs for each slot, only a single packet can be sent in a single slot. The format of a standard Bluetooth packet is shown in Figure 4.69.

The access code contains the identity of the master of the particular piconet. When a slave receives a packet, it checks up the access code with the expected code. If it matches, it means that the packet is from the master that is in the same piconet as the slave is. If it does not match, then the packet is not accepted by the recipient. In this way, it is possible to prevent slaves from receiving packets that are not meant for them.

Access Code	Header	Payload
72 Bits	54 Bits	0-2745 Bits

Figure 4.69 Bluetooth packet format.

The packet header contains 3-bit slave address, 1 bit ACK/NACK for the automatic repeat request (ARQ) scheme, 4 bit packet definitions that define around 16 types of payload, and an 8-bit header error code (HEC), which uses the CRC scheme. Four types of control packets [68] are defined:

- ID Packet – consists only of the access code.
- NULL Packet – contains the access code and the payload, used for transportation of link control information.
- POLL Packet – used by the master to poll the slaves.
- FHS Packet – used for synchronization.

The asynchronous link supports packets that may or may not implement the 2/3 FEC scheme. This link supports a data rate of 723.2 kbps. Synchronous link

Automotive Power Systems

supports a data rate of 64 kbps in full duplex mode. As mentioned before, the nodes alternatively switch from transmission to reception mode. The time difference is estimated to be 200μs. This interval defines the variable length of the payload. It also allows the processing of the payload.

f) Modulation Scheme

Bluetooth uses a Gaussian-shaped frequency shift keying (FSK) modulation, with the nominal modulation index "k" being equal to 0.3. Binary "1's" are represented by positive deviations in frequency and binary "0's" as negative deviations in frequency. The advantage of using this kind of modulation is that it allows the easy implementation of low cost-radio units.

g) Connection Establishment

One of the most important things in an ad-hoc network is connection establishment. How do units connect to each other? How do units find each other in the network? The issue of connection establishment deals with these questions. Three services are defined in the Bluetooth protocol: scan, page, and inquiry.

When a Bluetooth module is idle for a long time, it goes into a sleep mode. Even in the sleep mode, the module keeps waking up periodically to check whether or not another module is trying to communicate with it. It does this by implementing a sliding correlator to check the access code which is based on its identity. The window here is limited to around 10 ms. The module scans different frequencies to check whether its access code is being transmitted on any of the frequencies or not. It has a predetermined sequence of 32 hops. All these 32 frequencies are unique and distinct and occupy a spectrum of 60 MHz of the available 80 MHz. The paging unit then transmits the same access code on 16 of those 32 frequencies at a spacing of 1.25 ms. In the 10 ms period, the access code is sent out to 16 different frequencies. If it does not receive any response from the targeted device, it sends out the access code on the next 16 different set of frequencies. The assumption made here is that the paging device knows the hop sequence of the targeted device as well as its access code. Once the targeted device wakes up to the page, it replies back to the paging device by sending back its access code. The paging device then sends back the FHS packet, as described earlier, for synchronizing the device's clock to itself. The inquiry service is used when the paging device does not know the access code of the device it wants to wake up. It broadcasts the inquiry message and all the nodes send back their addresses and their clock information. With this information, the paging device can decide which units are in range and what their characteristics are. Once this has been decided, the paging device then sends an FHS packet to the targeted device.

h) Error Codes

Bluetooth implements forward error correction (FEC) as well as re-transmission schemes. The two different FECs implemented in Bluetooth are 1/3-rate FEC code and 2/3-rate FEC code. The 1/3-rate FEC code essentially uses a 3-bit repeat code whereas the 2/3-rate FEC code uses a shortened Hamming code for error correction. For re-transmission purposes an automatic repeat request (ARQ) scheme is used. When a node realizes that the message that it has just received is corrupted, it sends a negative ACK to the sender. However, the difference between other ARQ schemes and the one implemented in Bluetooth is that the NACK is sent in the slot immediately following the slot in which the data was received.

i) Power Management

There are two major issues to be considered when we talk of power management: the power with which a signal is transmitted and the power consumed when a module is in idle mode. Generally, the power with which a signal is transmitted in Bluetooth is 0dBm. However, the power could go up to 20dBm. The advantage of using a power level of 0dBm is that it results in minimum interference. When a power level of 20dBm is used, a technique known as the received signal strength indication (RSSI) based power control mechanism is used. This is implemented based on the power of the received signal compared to the measured interference. This is done to compensate for propagation losses as well as slow fading.

When a module is idle, it has to be ensured that it draws as little current as possible, just sufficient for it to be able to scan and listen for its access code on the channels. Usually, in this case, the duty cycle is below 1%. However, another state, defined as PARK state, has a duty cycle that is well below the duty cycle in the idle state. In this state, the node listens to the access code and payload for around 126 µs every time. However, to be in PARK state, the node must first belong to a piconet. Yet another state, known as the SNIFF state, has been defined. In this state, instead of scanning at every master-slave slot, the slave scans at larger intervals.

j) Security

The transmission of data in Bluetooth is secured by the modulation scheme that is used: the spread spreading technique as well as its transmission range, which is very low, around 10m. However, additional safety features such as authentication and encryption are provided too. Three levels of security have been provided for the Bluetooth network [69]:

- Non-secure: This mode does not include the authentication and encryption functions.

Automotive Power Systems 171

- Service level security: These security services are not provided until the L2CAP channel has been established. This is used to restrict access to units and services.
- Link layer security: This is provided when the LMP sets up the link. This technique uses a key (128 bit number) for a pair of devices. When the devices want to communicate, they use their keys to see if it is really the same device that is communicating.

4.8.9.2 Bluetooth Protocol Stack

The Bluetooth protocol stack contains some protocols that are part of a standard set of protocols used in other stacks and there are some protocols that have been specially developed for Bluetooth [70]. Figure 4.70 depicts the Bluetooth protocol stack. The various layers shown are:

- RF layer: This layer specifies the receiver and transmitter characteristics and other specifics of the radio modem.
- Baseband layer: This handles the fragmentation and reassembly of data into and from packets as well as the encryption scheme. It also determines the hop selection pattern.
- LMP (Link Management Protocol): As the name suggests, it used for link management. Issues like filtering of messages, traffic scheduling on links, and power control are all handled by the LMP.
- L2CAP (Logical Link Control and Adaptation Protocol): This layer is used for segmentation and re-assembly as well as multiplexing higher level protocol layers.
- Service Discovery Protocol: This protocol is used to discover services offered by other Bluetooth devices in the same surrounding and if necessary makes it possible to connect two Bluetooth devices to offer some services.
- Other protocols: Bluetooth hosts a set of other protocols like RFCOMM, TCS, HTTP, and FTP.

Amongst these protocols, there has been a lot of research on the SDP [71], since it is a very essential part of the Bluetooth network. The SDP in Bluetooth uses a Universally Unique Identifier (UUID), which is a 128-bit number assigned to a service. The Bluetooth device just includes the UUID of the service that it wants in its service discovery request. However, in some cases, just the inclusion of the UUID may not be a feasible option. To solve ambiguous service requests, along with the service requests, certain semantics are included so that even if the exact services are not found, some attributes matching closely with the requested attributes are found. The SDP is very important, whether it is used in a wide area like for example in a shopping mall or in a small area such as in an automobile. Along with the issue of developing a robust SDP, there is the need to consider the mobility issue of Bluetooth nodes.

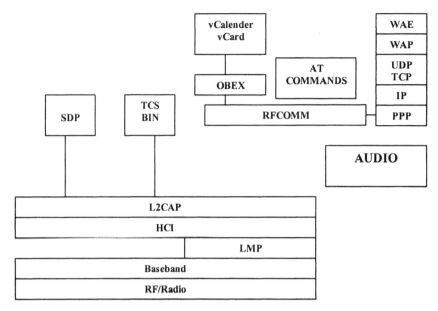

Figure 4.70 Bluetooth protocol stack.

A solution to prevent the loss of data transfer for devices that are highly mobile has been provided by Aman Kansal and Uday B Desai [72]. Let us imagine a scenario where a person with a Bluetooth device, in a shopping mall, wants information about the stores that display a particular product. Also, another assumption made is that access points to the network for Bluetooth devices have been installed at various locations all over the mall. Now, if the person is moving very fast, e.g., in a small golf cart, by the time the access point can transfer data to the Bluetooth device, he/she would already be far away and, therefore, the device has to connect to the next closest access point. This is known in as a "handoff". Traditionally, cellular IP has been used in Bluetooth to support the issues connected to mobility. However, the disadvantage was that the handoff delay was found to be around 5s. The technique developed by the above mentioned authors, which is known as Neighborhood Handoff Protocol, uses "entry points," which continuously scan for devices and, once it is locked on to a device obtains its address and clock information. It then passes this information to the nearest access point, which then, in turn, pages the device. The advantage of using this technique is that it prevents loss of data during handoff as well as detects loss of connection very quickly. The handoff delay has been estimated to be around 1.28s. Thus, we can see that the performance of NHP is at least 4 times better than that of Bluetooth PAC. Furthermore, the time to detect a connection loss has been found to be around 50 ms, in the worst case. Therefore, it has been shown that with the new technique, it is possible for data applications to run on the Bluetooth device without the fear of data loss due to high-speed handoffs.

Let us consider another example. Suppose that a driver is driving by an area relatively unknown to him. He/she wants to obtain information about all the restaurants that are present in that area. If the car is Bluetooth equipped, the device can obtain this information from the nearest Bluetooth access station. It is assumed here that Bluetooth access station is stationary while the car is moving. However, it is also possible for data transfer to take place between two moving vehicles too. One has to consider factors such as fading, signal loss, mulitpath and Doppler effects. Experiments were done to test the maximum time that a Bluetooth device in a car spends in the range of a stationary device [73]. Both Class I and Class II networks were experimented upon due to the outdoor nature of the application. The nominal Bluetooth range of 100m was chosen for the Class I network and a range of 10m was chosen for the Class II network. Table 4.4 gives us the range when the experiment was concluded.

Table 4.4 Bluetooth range test results.

	Maximum Range (m)	Spec Range (m)
Class 1	250	100
Class 2	122	10

Based on these results, the speed for vehicles containing the Bluetooth device has been recalculated. The results are shown in Table 4.5.

Table 4.5 Recalculated speed of vehicles.

Vehicle Speed (km/hr)	Maximum Time in Range (s)
20	90.00
40	45.00
60	30.00
80	13.50
100	18.00

4.8.9.3 Bluetooth Baseband Model

The Bluetooth baseband model is the intelligent core of communication. The functionality of the module depends on the kind of the application. Simple applications can be run on the baseband processor whereas complex applications are run on the host processor. Therefore, the module should very flexible in terms of load sharing; it must be able to accept any load at any point of time. Efficient usage of resources is a prime issue to be considered when designing a module. There are several modules that have been developed, but their usage has been hampered by the fact that they are large due to huge hardware. All the functions are performed by this hardware block instead of the micro-controller. This leads to a waste of the micro-controller. Therefore, the modules that have existed thus far have been inefficient. However, to increase the efficiency of the module, [74] transferred a lot of functionality to the micro-controller with the hardware blocks implementing only essential hardware functions. The basic block diagram of a module is shown Figure 4.71.

The micro-controller is used for controlling other units as well controlling some link control protocols. The baseband unit is entrusted with the task of encoding and decoding bit streams as well as control of the RF module. The USB and UART blocks are used for implementing the Host Control Interface (HCI) and the physical layer. The audio CODEC, as the name suggests, is used for audio coding and decoding as well as supporting both the A-law and the µ-law modulated schemes.

4.8.9.4 Bluetooth in Vehicles

Nothing stands for long in today's world. The automotive industry is an example for this. It is increasingly becoming a breeding ground for high-end electronics where manufacturers of vehicles employ the use of such state-of-the-art electronics to provide sophisticated features to the consumer. Due to this, the manufacturer has to deal with problems arising from the use of wiring of electronic parts: networking, power efficiency, weight factor, shrinking layout space, as well as assembly line difficulties, to name a few. There is a wide range of automotive networks reflecting defined and functional and economic niches. High bandwidth networks are used for vehicle multimedia applications, where cost is not critical. Reliable, responsive networks are critical for real-time control applications such as powertrain control and vehicle dynamics. Not only can Bluetooth be used for control applications, it can also be used to network multimedia devices. Vendors have envisioned a scenario where it is possible to obtain data from a PDA as well as control multimedia operations in a car via a Bluetooth enabled cellular phone [75]. Applications running on voice recognition can also be added on for applications such as banking or shopping online, which can then be implemented using Bluetooth technology. Therefore, using this technology the car can be made into a virtually moving office. As mentioned before, there are many applications that can be supported in a vehicle. Each application demands a different kind of a network for the devices implementing that particular application. The reason is that every application may have varying demands of bandwidth, security, reliability, and so forth. Figure 4.72 shows the kind of networks employed in a car.

Automotive Power Systems

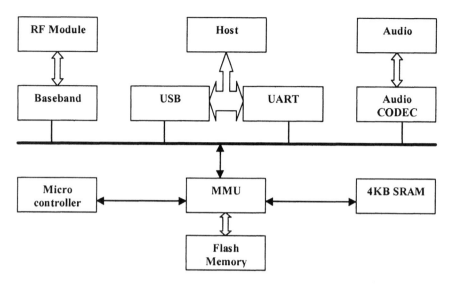

Figure 4.71 Bluetooth baseband module.

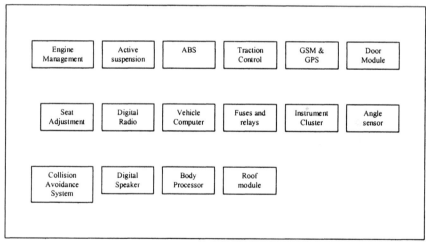

Figure 4.72 Electronic equipment requiring networking.

The definition for the class of networks is shown Table 4.6.

Table 4.6 Different network classes in a car.

Network Classification	Speed	Application
Class A	<10 kbps (low speed)	Convenient features (trunk release, electric mirror adjustment)
Class B	10-125 kbps (medium Speed)	General information transfer (instruments, power windows)
Class C	124kbps-1Mbps (high speed)	Real time control (power train, vehicle dynamics)
Class D	> 1Mbps	Multimedia applications (internet, digital TV, X-by-wire applications)

However, with the advent of wireless technology, cars are no longer isolated systems, since connectivity and connection with the world outside the car enable new applications. Vehicles can make use of various communication networks available outside the car for communication services to applications inside the car. It is at this conjuncture that Bluetooth comes into the picture. The problem with the existing networks for a mobile environment like a car is that there are so many networks involved, where each one has such different data transmission rate and characteristics that it is very difficult to integrate different applications on a common network. With the employment of Bluetooth, it is possible to integrate various networks on the backbone that Bluetooth offers. This can lead to exciting new applications. A mobile environment like a car can then be turned into virtually a mobile office station where office work like e-mails or file transfer can take place, or a mini theatre where a mini screen in the car can play a movie being broadcast with a touch of a button on a keypad that acts like a wireless remote control. Or it could be a simple application like browsing the Internet while driving and booking an airline ticket online. The possibilities are tremendous. What remains to be seen is the efficiency of this technology. With so many devices linked by radio and so many applications, the question of efficient bandwidth utilization and security issues, to name a few, have to be answered. It must be mentioned at this stage that Bluetooth would initially be applied to safety-uncritical applications in and around the car, for example, communication and infotainment applications or applications that involve exchange of pure car-specific information (control and status data of the body electronics). Inside the car, Bluetooth allows the wireless connection of various car embedded electronic devices such as control panels and headsets with other car embedded electronic devices. But, beyond replacing wires between in-car devices, the scope of Bluetooth in the automotive industry is to ensure inter-operability between wireless devices (e.g., mobile phones and PDAs) and car-embedded devices (e.g., car navigation system, car radio, and car embedded phone). The main principle here is to allow devices to co-operate and share resources and in almost the entire time access the car embedded system through mobile devices. This is implemented by the way of a Central Control Unit (CCU) [79]. Apart from controlling various devices in the car, Bluetooth can be used for car-car communication as well as other Bluetooth enabled devices close to the car's environment. The exchange of data between the car and stationary facilities provides a base for various special application scenarios, e.g., access of traffic information or e-commerce applications at designated "access points", or download of software for car electronic systems for service stations. Listed below are some of the applications that are envisaged for Bluetooth in the near future [77]:

a) Anytime Information

With this kind of information service, users receive data via broadcast or multicast. The main advantage of these services is their "always on" nature, such that services can be accessed without any setup effort. Moreover, data carousels allow starting reception at any time. A certain casting service may be offered in large areas or in limited areas only, according to the coverage of the broadcast program. In this way, it is subject to vary along the traveling route of the automobile. The received information can be displayed immediately or can be cached for later retrieval. Broadcast services may include the local weather forecast, updates about the local traffic situation, or advertisements for local events like movie trailers of currently playing cinema programs. Also, entertainment services such as broadcast of regular television programs (for backseat passengers) fall into this service category.

b) Information Push

The main difference between information push and casting is that users must subscribe to push services. Notification of an incoming personal fax, a customized stock market report, or updates of a user's car software are applications for information push. This personalization requires addressing mechanisms for individual users and devices.

c) Data Retrieval

To request information that is not present within the automobile system or not delivered automatically, a bi-directional wireless link has to be set up. Data retrieval can mean broadband applications, e.g., video streaming to view multimedia presentations about a destination. In addition, route information may be requested by the navigation system. An example of a convenient low-bandwidth service is wireless access to music CD recognition systems, which automatically download the track titles of a CD from a database on the Internet. In most application cases, data retrieval has a strongly asymmetric character.

d) Transparent Access to the Internet and Intranets

A transparent IP connection to the Internet is required to make the car part of the Internet. Business people, for example, should be able to access their corporate network for mobile office applications, e.g., e-mail.

e) Telecommunication Services

Common telecommunication services that rely on cellular systems (real-time conversational services) also have to be integrated parts of the services accessible from a vehicle. Telecommunication services will comprehend not only common voice telephony, but also broadband communication services like multimedia video-conferencing. All these service types are well known from non-car-specific communication situations. However, using them in a car environment produces some specific requirements for service execution. Besides the physical availability of the communication infrastructure, services must be aware of some other factors in the vehicular environment. Car applications mostly have varying users, for example, alternating drivers and different passengers using various terminals, especially with corporate car fleets and rental cars. In this case, services should be aware of the user's profile and provide possibilities for its storage and retrieval. Furthermore, the specific driving situation (speed, traffic, time of day/year, and general driving behavior) must be considered when executing a service. Services should thus be aware of the current context. Furthermore, location awareness is an important factor for service provisioning.

f) Car Production

In the car production system, a lot of software is downloaded as a final step in the production line. This is an application where Bluetooth is really well suited [80]. A Bluetooth base station is connected to the product field bus. When the car online gets connected to the Bluetooth base station, it uploads its serial number. The production computer then downloads the software via the field bus to the base station, which is then uploaded to the car. However, this is a dedicated application and no other Bluetooth enabled device should be connected to the cell.

g) Car Diagnostics

Perhaps the first application of Bluetooth technology in the automotive industry would be for car diagnostics [81]. Here, without the need for electrical connection of devices, the factory environment would be able to monitor the car equipment and perform diagnostic experiments. The software for such applications also would be downloadable into the car from any stationary station that is Bluetooth enabled and hosts the software.

h) Toll Payment

Bluetooth has the capability to enhance the existent automatic toll payment methods and to advance the widespread standardized adoption of such systems. When a vehicle already contains a Bluetooth transponder, all that is needed is for the system software to be upgraded, which can be done in the method mentioned above. After the system software has been upgraded, various other applications can be implemented like parking meter systems and parking access systems.

i) Filling Station

An interesting application of Bluetooth can be envisaged at a gas filling station. Here, the make and model of a car as well as its tank capacity and other characteristics can be downloaded to a "robotic" attendant. The filling station then can transmit certain diagnostic parameters such as oil pressure or tire pressure to the vehicle driver along with certain suggestions or, in the simplest case, can transmit certain special offers that the vehicle driver can make use of.

j) Non-Safety Critical Control Systems

Bluetooth can be used for providing control network to various non-safety critical devices such as trunk release, seat movement, door module, and rain sensors to name a few. Additionally, with the help of Bluetooth, cables that traverse moveable interfaces such as those between car hoods, trunks, doors, steering wheels, road wheel, and the vehicle frame may be replaced. Also, complex interfaces like those on the dashboard can be eliminated and customizable user interfaces could be brought into the picture.

k) Intra-Car Communication

Inside the car, there are a host of devices that would be helpful to the driver, e.g., navigation systems. A car driver would like to bring his/her mobile devices and connect them to the car network. For example, the address of a business partner could be stored in a PDA. After the address has been fed into the car navigation system, the system may suggest the route, which may take the shortest time, based on the weather and traffic related data which would already have been downloaded by the navigation system. Also, the sound system and movie screens can be used for various applications.

There are a lot of advantages when Bluetooth technology is used for networking inside an automobile. These advantages are related to cable replacement, seamless connectivity, mobility, and design issues.

With the implementation of Bluetooth technology, there would be replacement of cables, which would contribute towards a major reduction in the overall weight of the car. This, in turn, would lead to improvements in factors such as fuel economy and emissions.

Devices that previously needed wires for inter-connectivity can now communicate with each other via a wireless link, without the need for cables between them. Furthermore, there could be interconnection between devices not only in the same subsystem of the automobile, but also in devices belonging to various other subsystems in the same automobile. This in effect provides a seamless connectivity between devices.

With the ability of the user to plug his/her mobile devices into the network of the automobile, this technology would give a new dimension to the meaning of mobility. While driving the automobile, the Bluetooth enabled device would be able to download the required information, say, for example, the nearest gas station or the nearest restaurant. This would be very helpful to people driving in unfamiliar territory.

Since Bluetooth is a wireless based communication system, it is highly flexible and can be modified too without much difficulty [82]. It is possible to add and remove nodes from the network without any difficulty at all.

Even though Bluetooth is a versatile technology, it has its limitations. There are inherent inadequacies that make practical implementation of this technology seem difficult. In this topic, we shall consider some of the limitations that this technology and the devices that implement this technology possess.

In an environment where data transfer takes place almost all the time, the issue of privacy is foremost in everybody's mind. In the above-mentioned applications, we took the example of a PDA communication and a car navigation system. The possibility of data in the PDA being transferred to an electronic logbook in the car is possible. In such scenarios, almost anyone can access this data. This problem can be overcome in the following way. If any Bluetooth device, in our case the PDA, had an address akin to the IP addresses issued to computers connected to the Internet, the owner then could add that particular address to the piconet environment, so that now the devices know that they are communicating with a legitimate device.

Since this technology involves a lot of electronic devices, the factor of the conditions existing in the operating environment should be taken into consideration. Equipment would be subjected to various mechanical vibrations, temperature swings from -40°C to +80°C, splashes of petrol, and water, and potential mis-wirings can lead to short circuiting of essential systems, load jumps, jump starts, electromagnetic fields, and voltage spikes [78]. These electronic systems must be enabled to tolerate such a hostile environment because failure of these systems could affect the overall performance of the subsystems in the vehicle.

We know that the maximum number of devices that can operate in a piconet is restricted to 8, and around 10 piconets can operate in a given environment. This is good for present day applications; but, it must be increased in the future. In any case, even with an environment where so many devices are communicating with each other, some devices, which are initially switched off, may be triggered by mistake to switch themselves on.

The current data rate is just sufficient for transfer of small data files. However, as traffic increases and the size of the data files becomes larger, provisions have to be made for problems such as congestion control, segmenting, and re-assembly of large data into smaller chunks, to name a few.

Such large scale implementation of this technology requires overhauling of the entire automotive industry because such changes have to be made right from the point where the automobiles are being manufactured. This would involve large scale revamping involving a lot of financial considerations. This would cause skepticism in the automobile manufacturer's mind.

So far Bluetooth has been used for non-critical systems in the car. However, when the situation arises, the seriousness of implementing this technology in critical systems in the car could be tremendous because any failure of this system would lead to a loss of human lives.

The football stadium phenomenon is also a problem that arises when a Bluetooth enabled car is parked for a long time in a parking lot where other Bluetooth enabled cars are present. The Bluetooth device goes into "sleep" mode if it remains inactive for long periods of time. However, due to constant activity in its surrounding environment, it could erroneously be woken up and start transmitting. This could lead to broadcasting secure data, which would be a serious malfunction.

In summary, Bluetooth technology is set to become as commonplace as a household television system. However, it must be remembered that this is still an emerging technology with great possibilities of improvement. Even in its nascent stages, implementation of this technology would require passing stringent and complete tests certified by the automotive industry to demonstrate its safety. However, with the current low cost of a Bluetooth chip, implementation of this system in cars should not be much of a problem. With consumers willing to pay more for invaluable applications, the ability of Bluetooth to provide such applications is exciting. But the problem of lifespan still exists. Whereas cars have an average lifespan of around 10 years, Bluetooth devices have an average lifespan of just a couple of years. This would require constant replacement of devices. How long would it be effective? Only time will tell.

Infotronics will play a large role in the future of automotive electrical and electronic systems. Access to large information networks may be used to enhance vehicle reliability though diagnostics of onboard systems using onboard algorithms or prognostics of full vehicle systems. Prognostics of vehicle systems in the future can be achieved through augmentation of existing onboard algorithms today or by downloading more powerful algorithms in the future. Figure 4.73 illustrates the interconnected nature of automotive systems. It should be noted that all of these modules have excess microprocessor and RAM storage available to various degrees.

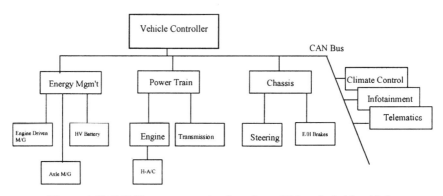

Figure 4.73 CAN bus communications in an ISA or hybrid vehicle.

The large proliferation of electronic controllers on the vehicle, each having accessible RAM and μC processing power, makes it possible to contemplate a distributed diagnostic/prognostic system for assessing the state of health of the vehicle's electrical and electronic systems. There are already movements to migrate high RAM, ROM, and EEPROM/Flash memory from previously standalone and functionally dedicated control modules to the vehicle controller. Although physical partitions (modules) have been presented, it should be understood that in tomorrow's vehicle the definition of functional interfaces will be separate from these physical partitions. For example, the module labeled vehicle systems controller may in fact not be a physical module at all, but the function of vehicle system controller would exist amongst the various modules connected to the communications bus and accessible through it. The presence of this highly interconnected and distributed network means that the distributed computing power can be used to improve reliability.

In summary, the trend to a more connected automobile means that on-board excess processing capability can be used as transient, hence transparent to module applications, residence for powerful diagnostic and prognostic routines that are downloadable from the network. To fully assess such capability it is necessary to understand what types of microprocessors are available in the networked modules, the microprocessor's capability in terms of clock rate and word size, and available memory sizes in each of the distributed modules. In today's environment this is no small task, since automobile manufacturers are unlikely to share such information; even if they were, the modules are designed and the algorithms/code owned by the suppliers, so it is further unlikely that suppliers would share such information. Notwithstanding the issues of understanding the specifics of the code and memory sharing in each module, the hurdles of implementing the downloadable and highly dispersed diagnostic or prognostic algorithms described above would present a major challenge to anyone undertaking such an endeavor. However, the future is unfolding in a manner that suggests that just such an effort should begin soon because the complexity of automotive systems and the need to assess the overall vehicle state of health will soon demand such capability. In the years following the introduction of electronic engine controls, all the manufacturers developed proprietary code and self-defined diagnostic test routines and codes for module archiving. This meant that repair shops and dealerships were required to have a different network accessible testing tools for each particular make of vehicle. Standardization changed that and today the diagnostic test connector (DTC) has a standard pin definition and industry-wide standard diagnostic test codes so that a single hand-held "STAR" tester can access the vehicle's electronic systems and determine where a fault lies regardless of the vehicle's manufacturer.

4.9 References

[1] R. H. Johnston, "A history of automobile electrical systems," *SAE Automotive Engineering, Historical Series,* pp. 53-66, Sep. 1996.

[2] T. Heiter, R. Kallenbach, C. Kramer, and S. Schustek, "Recent developments in starting systems," *SAE International Congress and Exposition,* Paper No. 97113, Detroit, MI, Feb. 1997.

[3] J. M. Miller, A. Emadi, A. V. Rajarathnam, and M. Ehsani, "Current status and future trends in more electric car power systems," in *Proc. of 1999 IEEE Vehicular Technology Conference,* Houston, TX, May 1999.

[4] A. Emadi, M. Ehsani, and J. M. Miller, "Advanced silicon rich automotive electrical power systems," 18th *Digital Avionics Systems Conference on Air, Space, and Ground Vehicle Electronic Systems,* St. Louise, Missouri, Oct. 1999.

[5] J. G. Kassakian, H. C. Wolf, J. M. Miller, and C. J. Hurton, "Automotive electrical systems-circa 2005," *IEEE Spectrum,* pp. 22-27, Aug. 1996.

[6] J. M. Miller, D. Goel, D. Kaminski, H. P. Schoner, and T. Jahns, "Making the case for a next generation automotive electrical system," *Convergence Transportation Elect. Association Congress,* Dearborn, MI, Oct. 1998.

[7] J. M. Miller and P. R. Nicastri, "The next generation automotive electrical power system architecture: Issues and challenges," 17th *Digital Avionics Systems Conference on Air, Space, and Ground Vehicle Electronic Systems,* Bellevue, WA, Oct./Nov. 1998.

[8] L. A. Khan, "Automotive electrical systems: architecture and components," 18th *Digital Avionics Systems Conference on Air, Space, and Ground Vehicle Electronic Systems,* St. Louise, Missouri, Oct. 1999.

[9] S. Muller and X. Pfab, "Considerations implementing a dual voltage power network," *Convergence Transportation Electronics Association Congress*, Dearborn, MI, Oct. 1998.

[10] J. G. Kassakian, "The future of power electronics in advanced automotive electrical systems," *Proc. of IEEE Power Electronics Specialist Conf.*, 1996, pp. 7-14.

[11] N. Traub, "Dual/higher voltage – A global opportunity," *SAE Strategic Alliance Report*, pp. 1-4.

[12] A. G. Williams, "The system approach and the impact of the design of higher voltage electrical power system," *DIACS Colloquia Abstract*, May 2001, pp. 1-4.

[13] D. J. Perreualt and V. Caliskan, "Automotive power generation and control," *LEES Technical Report TR-00-003*, MIT, May 2000.

[14] C. P. Cho and D. R. Crecelius, "Vehicle alternator/generator trends towards next millennium," *Proceedings of the IEEE International Vehicle Electronics Conference*, vol. 1, Sep. 1999, pp. 433-438.

[15] E. C. Lovelace, T. M. Jahns, J. L. Kirtley Jr., and J. H. Lang, "An interior PM starter/alternator for automotive application," *Lab. for Electromagnetic and Electronic systems*, MIT.

[16] R. R. Henry, B. Lequesne, S. Chen, J. J. Ronning, and Y. Xue, "Belt driven starter-generator for future 42-V systems," *SAE Paper No. 2001-01-0728*, March 2001.

[17] W. X. Li and Y. X. Yanliang, "The study of permanent magnet starter motor with auxiliary poles," 5th *International Conference on Electric Machines and Systems*, vol. 2, Aug. 2001, pp. 838-841.

[18] *Starting Device for an Internal Combustion Engine and Method for Starting the Internal Combustion Engine*, Daimler Chrysler AG, US Patent No. 6,240,890, June 2001.

[19] *Starter/Generator for an Internal Combustion Engine, Especially an Engine of a Motor Vehicle*, ISAD Electronics Systems GmbH and Co. KG, Grundl and Hoffman GmbH, US Patent No. 6,177,734, Jan. 2001.

[20] L. M. C. Mhango, "An experimental study of alternator performance using two-drive ratios and a novel method on speed boundary operation," *IEEE Trans. on Industry Application*, vol. 28, no. 3, pp. 625-631, May-June 1992.

[21] J. R. Quesada, V. Lemarquand, J. F. Charpentier, and G. Lemarquand, "Study of triangular winding alternators," *IEEE Trans. on Magnets*, vol. 38, no. 2, pp. 1361-1364, March 2002.

[22] F. B. Reiter Jr., K. Rajashekara, and R. J. Krefta, "Salient pole generator for belt-driven automotive alternator application," 36th *Industry Application Conference*, vol. 1, Sep.-Oct. 2001, pp. 437-442.

[23] M. Naidu, N. Boules, and R. Henry, "A high-efficiency high-power-generation system for automobiles," *IEEE Trans. on Industry Applications*, vol. 33, no. 6, pp. 1535-1543, Nov.-Dec. 1997.

[24] B. Fahimi, "A switched reluctance machine based starter/generator for more electric cars," *IEEE Electric Machines and Drives Conference*, June 2001, pp. 73-78.
[25] C. P. Cho and R.H. Johnston, "Electric motors in vehicle applications," in *Proc. IEEE Vehicle Electronics Conference*, Sep. 1999, pp. 193-198.
[26] R. F. Wall and H. L. Hess, "Induction machines as an alternative for automotive electrical generation and starting systems," in *Proc. International Conference on Electric Machines and Drives*, pp. 499-501, May 1999.
[27] J. M. Miller, A. R. Gale, P. J. McCleer, F. Leonardi, and J.H. Lang, "Starter-alternator for hybrid electric vehicle: comparison of induction and variable reluctance machines and drives," in *Proc. 33rd IAS Annual Meeting Industry Application Conference*, vol. 1, Oct.1998, pp. 513-523.
[28] J. Wai and T. M. Jahns, "A new control technique for achieving wide constant power speed operation with an interior PM alternator machine," in *Proc. 36th IEEE Industry Application Conference*, vol. 2, Oct. 2001, pp. 807-814.
[29] F. Caricchi, F. Crescimbini, F. G. Capponi, and L. Solero, "Permanent-magnet, direct-drive, starter/alternator machine with weakened flux linkage for constant-power operation over extremely wide speed range," in *Proc. 36th IAS Annual Meeting Industry Application Conference*, vol. 3, Sep.-Oct. 2001, pp. 1626-1633.
[30] F. Caricchi, F. Crescimbini, E. Santini, and L. Solero, "High-efficiency low-volume starter/alternator for automotive applications," *IEEE Trans. on Industry application*, vol. 1, pp. 215-222, Oct. 2000.
[31] B. Fahimi and A. Emadi, "Robust position sensorless control of switched reluctance motor drives over the entire speed range," in *Proc. IEEE 33rd 2002 Power Electronics Specialist Conference*, Cairns, Queensland, Australia, June 2002.
[32] M. E. Elbuluk and M. D. Kankam, "Potential starter/generator technologies for future aerospace applications," *IEEE Aerospace and Electronics Systems Magazine*, vol. 12, no. 5, pp. 24-31, May 1997.
[33] H. Rehman, X. Xu, N. Liu, G. S. Kahlon, and R. J. Mohan, "Induction motor drive system for the Visteon integrated starter-alternator," *25th IEEE Industry Electronic Conference*, vol. 2, 1999, pp. 636-641.
[34] M. Naidu and J. Walter, "A 4kW, 42V induction machine based automotive power generation system with a diode bridge rectifier and a PWM inverter," *IEEE Transactions*, pp. 449-456, 2001
[35] Y. Dou and H. Chen, "A design research for hybrid excitation rare earth permanent magnet," in *Proc. 5th International Conference of Electrical Machines and Systems*, vol. 2, Aug. 2001, pp. 856-859.
[36] C. Zhang and F. Tian, "Research on improving permanent magnetic generator output characteristic," in *Proc. 5th International Conference on Electrical Machines and Systems*, vol. 2, Aug. 2001, pp. 850-852.

[37] A. de Vries, Y. Bonnassieux, M. Gabsi, F. d'Oliveira, and C. Plasse, "A switched reluctance machine for a car starter-alternator system," in *Proc. IEEE Electric Machines and Drives Conference*, June 2001, pp. 323-328.
[38] D. Holt, "42V update," *Service Tech Magazine*, pp. 14-16, Sep. 2001.
[39] M. A. Masrur, R. J. Hampo, and J. M. Miller, "Rotational sensorless scalar control of three-phase induction motors and its application to automotive electric power assist steering," *Proceedings of the Institution of Mechanical Engineers*, vol. 214, Part D, pp. 33-44, Paper No. D00799, June 1999.
[40] H. K. Khalil, E. G. Strangas, and J. M. Miller, "A torque controller for induction motors without rotor position sensors," *IEEE International Conference on Electric Machines*, Spain, Sep. 1996.
[41] R. Frank, R. Meyer, and J. Reiter, "The role of power electronics in future automotive systems," *Congress on Transportation Electronics, Convergence 1998*, Hyatt Regency Hotel, Dearborn, MI, Oct. 1998.
[42] A. Graf, "Semiconductors in the 42V PowerNet," *Intertech, 42V Automotive Systems Conference*, Chicago, Sep. 2001.
[43] J. G. Kassakian, H. C. Wolf, J. M. Miller, and C. J. Hurton, "Automotive electrical systems circa 2005," *IEEE Spectrum*, Aug. 1996.
[44] L. Aschliman and J. M. Miller, "Digital output AC current sensor for automotive applications," *IEEE WPET92*, Hyatt-Regency, Dearborn, MI, Oct. 1992.
[45] F. L. Rees and J. M. Miller, "50 to 100 volt DC inverters for vehicular motor drive systems - The use of MOSFET's & IGBT's compared," *IEEE WPET92*, Hyatt-Regency, Dearborn, MI, Oct. 1992.
[46] J. M. Miller, "Integrated power module requirements for automotive applications: Achieving higher power density," *Power Conversion-Intelligent Motion, Power Quality Conference*, Hyatt-Regency, Irvine, CA, Sep. 1992.
[47] G. Leen and D. Heffernan, "Expanding automotive electronic systems," *Computer*, vol. 35, no. 1, pp. 88-93, Jan. 2002.
[48] G. Leen, D. Heffernan, and A. Dunne, "Digital networks in automotive vehicle," *Computing and Control Engineering Journal*, vol. 10, no. 6, pp. 257-266, Dec. 1999.
[49] CAN 2.0 Specification, downloaded from http://www.can-cia.de/.
[50] M. Fars, K. Ratcliff, and M. Barbosa, "An introduction to CAN open," *Computing and Control Engineering Journal*, vol. 10, no. 4, pp. 161-168, Aug. 1999.
[51] W. Xing, H. Chen, and H. Ding, "The application of controller area network on vehicle," in *Proc. IEEE International Vehicle Electronics Conference*, vol. 1, 1999, pp. 455-458.
[52] G. Leen and D. Heffernan, "Time triggered controller area network," *Computing and Control Engineering Journal*, vol. 12, no. 6, pp. 245-256, Dec. 2001.

[53] T. Führer, B. Müller, W. Dieterle, F. Hartwich, R. Hugel, and M. Walther, "Time triggered communication on CAN," *Robert Bosch GmbH*, downloaded from http://www.can-cia.de/.
[54] F. Hartwich, B. Müller, T. Führer, and R. Hugel, "CAN network with time triggered communication," *7th International CAN Conference*, Amsterdam, Netherlands.
[55] B. Wiegand, "SAE J1850 class B data communication networks interface," *Communication Standards for European On-Board-Diagnostics Seminar (Ref. No. 1998/294), IEE*, pp. 5/1-5/9, 1998.
[56] C. A. Lupini, T. J. Haggerty, and T. A. Braun, "Class 2: General Motor's version of SAE J1850," *Delco Electronics Corporation*, U.S.A.
[57] J. Oliver, "Implementing the J1850 protocol," *Intel Corporation*, http:/www/intel.com/design/intarch/papers/j1850_wp.htm.
[58] M. Rabel, A. Schmeiser, and H. P. Grossman, "Integrating IEEE 1394 as infotainment backbone into the automotive environment," in *Proc. 2001 IEEE 53rd Vehicular Technology Conference*, vol. 3, 2001, pp. 2026-2031.
[59] M. Scholles, K. Frommhagen, L. Kleinmann, P. Nauber, and U. Schelinski, "IEEE 1394 firewire system for industrial and factory automation applications," in *Proc. 8th IEEE Conference On Emerging Technologies and Factory Automation*, vol. 2, 2001, pp. 627-630.
[60] G. Baltazar and G. P. Chapelle, "Firewire in modern integrated military avionics," *IEEE Aerospace and Electronics Systems Magazine*, vol. 16, no. 11, pp. 12-16, Nov. 2001.
[61] D. Steinberg and Y. Birk, "An empirical analysis of the IEEE 1394 serial bus protocol," *IEEE Micro*, vol. 20, no. 1, pp. 58-65, Jan./Feb. 2000.
[62] *MOST Specification Framework,* downloaded from http://www.mostnet.de/.
[63] H. Kopetz, G. McCall, P. Mortana, E. Dilger, L. A. Johansson, M. Krug, P. Lideñ, B. Müller, U. Panizza, S. Poledna, A. Schedl, J. Södenberg, M. Strömber, and T. Thurner, downloaded from the website of Dr. Hermann Kopetz, University of Vienna.
[64] *LIN Specification 1.1,* downloaded from http://www.lin-subbus.org, March 2000.
[65] J. W. Specks and A. Rájnak, "LIN protocol development tools & software interfaces for LIN in vehicles," in *Proc. 9th International Conference on Electronic Systems for Vehicles*, Oct. 2000.
[66] Material available on http://www.palowireless.com/.
[67] J. C. Haarsten, "The Bluetooth radio system," *IEEE Personal Communications*, vol. 7, no. 1, pp. 28-36, Feb. 2000.
[68] Zhang Pei, Li Weidong, Wang Jing, and Wang Youzhen, "Bluetooth-the fastest developing wireless technology," in *Proc. 2000 International Conference on Communication Technology*, vol. 2, 2000, pp. 1657-1664.
[69] Bluetooth Whitepaper published by AU Systems.

[70] R. Jordan and C. T. Abdallah, "Wireless communication and networking: an overview," *IEEE Antennas and Propagation Magazine,* vol. 44, no. 1, pp. 185-193, Feb. 2002.

[71] S. Avancha, A. Joshi, and T. Finn, "Enhanced service discovery in Bluetooth," *Computer,* vol. 35, no. 6, pp. 96-99, Jan. 2002.

[72] A. Kansal and U. B. Desai, "Mobility support for Bluetooth public access," in *Proc. 2002 International Symposium on Circuits and Systems,* vol. 5, 2002, pp. 725-728.

[73] P. Murphy, E. Welsh, and J. P. Frantz, "Using Bluetooth for short term ad-hoc connections between moving vehicles," in *Proc. IEEE 5th Vehicular Technology Conference,* vol. 1, 2002, pp. 414-418.

[74] Myoung-Cheol Shin, Seong-Il Park, Sung-Won Lee, Se-Hyeon Kang, and In-Cheol Park, "Area efficient digital baseband module for Bluetooth wireless communication," Division of Electrical Engineering, KAIST, Taejon, KOREA.

[75] E. A. Bretz, "The car: just a Web browser with tires," *IEEE Spectrum,* vol. 38, no. 1, pp. 92-94, Jan. 2001.

[76] W. Kellerer, C. Bettstetter, C. Schwingenschlogl, and P. Sties, "(Auto) Mobile communication," *IEEE Personal Communications,* vol. 8, no. 6, pp. 41-47, Dec. 2001.

[77] G. Leen and D. Heffernan, "Vehicles without wires," *Computing and Control Engineering Journal,* pp. 205-211, Oct. 2001.

[78] R. Nüsser and R. M. Pelz, "Bluetooth based wireless connectivity in an automotive environment," in *Proc. 52nd IEEE Vehicular Technology Conference,* vol. 4, 2000, pp. 1395-1942.

[79] L. B. Fredriksson, "Bluetooth in automotive applications," *Bluetooth Technology Conference,* 2001.

[80] L. B. Fredriksson, "Bluetooth in automotive diagnostics," *Bluetooth Technology Conference,* 2001.

[81] H. Wunderlich and M. Schwab of Daimler-Chrysler R & D, Germany and L. B. Fredriksson of KVASER, Sweden, "Potential of Bluetooth in automotive applications," *Bluetooth Technology Conference,* 2000.

[82] H. Wunderlich, M. Schwab, L. B. Fredriksson, M. Nikola, "Bluetooth in automotive industry: enabling effortless connectivity between devices," *Bluetooth Technology Conference,* 2000.

[83] J. Kardach of Intel, "Bluetooth technology overview," March 1999.

[84] D. J. Goodman, "Wireless Internet: promises and challenge," *IEEE Comp. Magazine,* 2000.

[85] R. Shorey and B. A. Miller, "The Bluetooth technology: merits and limitations," in *Proc. IEEE Conference on Personal Wireless Communication,* 2000, pp. 80-84.

[86] J. M. Miller, "Barriers and opportunity for power train integrated power electronics," *Center for Power Electronic Systems, CPES,* Invited Paper, VT Seminar, Blacksburg, VA, April 2002.

5

Electric and Hybrid Electric Vehicles

Automakers are under immense pressure today to improve fuel economy and also reduce vehicle emissions. However, to retain profitability, automakers must also ensure that the vehicles being manufactured are affordable, profitable, and desirable to prospective customers. It has been well established that full hybrid vehicle (HV) technology is the most cost effective way to improve fuel economy, reduce emissions, and still maintain customer demands of performance, comfort, and cost. The most implementation-amenable and perhaps the lowest cost hybrid technology for meeting these goals is sometimes referred to as a "soft hybrid." A soft hybrid utilizes a small 6-10 kW integrated starter generator (ISG), which is sometimes also referred to as a combined starter/alternator (CSA) or as an integrated starter-alternator (ISA). A small ISG, or what has been called a belt-ISG, can be mounted on the front of the engine and utilizes the FEAD (front engine accessory drive) belt to start the engine. An ISG can also be integrated into the transmission bell housing and coupled to the rear of the engine crankshaft, in what is known as a crankshaft-ISG. The operational strategy resulting from shutting the engine "off" at stoplights and other conditions when the engine is not needed to provide mobility is referred to as "idle-stop". Because the ISG is typically rated from 3 to 8 times the power of today's engine starter motor, the engine can be re-started in a much shorter time than in a vehicle equipped with only a conventional flywheel mounted starter. The impact is starting performance that is brisk, silent, and transparent to the customer.

Today, hybrid vehicles (HVs) are synonymous with vehicles that offer a greater fuel economy and lower emissions when compared to their conventional production platforms. The development of an affordable hybrid technology faces challenges on several fronts. Challenges include, but are not limited to, their technical content and development, corporate challenges, government regulations, program challenges, selecting technology partners, and producing a hybrid vehicle that customers find fun to drive. HVs are attractive and fun to drive because they offer multiple attributes and features that consumers want. Hybrids may soon offer all-wheel drive capability, traction control, regenerative braking, zero emissions, active air conditioning at vehicle stops, and silent re-starts. Moreover, hybrids are charge-sustaining, meaning they don't have to be plugged in, need fewer re-fueling stops, and are on par with their conventional production counterparts in 0 to 60 mph performance, passing maneuvers, and top speed.

5.1 Principles of Hybrid Electric Drivetrains

Integrating an electric machine drive system into the power train of a hybrid electric vehicle represents a challenging exercise in packaging complex electromechanical and power electronic subsystems. Typically, the ISG or HV traction motor and its attendant power and control electronics are physically partitioned because power electronics has not yet evolved to the stage in which fully packaged drives can be realized. A similar situation exists for the control and sensor subsystems necessary for a fully functional high performance drive. Hardware partitioning requires that more attention be given to installation issues and to mitigating system interactions. The vehicle power supply and electrical infrastructure must be capable of supporting the power and energy levels demanded. We visualize the full hybrid electric vehicle, particularly what is known in the industry as a low storage requirement hybrid, as employing only the amount of electrical capacity needed to crank the ICE from stop to idle speed in less than 0.3s and to off-load the ICE during transient operation such as quick acceleration and deceleration. A power-assist HV is another name for low storage HV. During normal driving, the hybrid traction system electric machine is designed to operate as a high efficiency alternator. Alternator efficiency exceeding 80% is required if the cost and complexity of hybridization is to meet industry targets of twice fuel economy of a conventional five passenger car. The corporate average fuel economy (CAFÉ) specified by EPA for this class of vehicle is 27.5 mpg. The cost of providing electricity on conventional vehicles, passenger cars, and light trucks can be calculated based on an ICE's 40% marginal efficiency. Using gasoline feedstock having a density of 740 g/l, an energy density of 8,835 Wh/l (32 MJ/l) and today's alternator efficiency of 45% on average, we can make some cost of electricity comparisons. Today, the cost of on-board electricity is ~$0.21/kWh when the fuel price is $1.25 per U.S. gallon, significantly higher than residential electricity cost and far too expensive for a hybrid electric vehicle. If the mechanical to electrical energy conversion efficiency is increased from 45% to 85%, the cost of on-board electricity drops to ~$0.11/kWh. Another way of looking at the difference in efficiency is to evaluate its impact on vehicle fuel efficiency for an 80 mpg (3 liter/100km) car. When calculated for the average speed over the Federal Urban Drive Cycle (23 mph) and for an average vehicle electrical load of 800W, today's alternator lowers the fuel economy by 5.86 mpg, whereas the HV system only lowers fuel economy by 3.1 mpg. Hence, the higher efficiency of a HV system is worth 2.76 mpg in an 80 mpg vehicle, or 0.35 mph fuel economy reduction per 100W of electrical load.

Figure 5.1 illustrates from the standpoint of electrical system voltage (traction battery potential) the various classifications and hybrid functional capability of these vehicles. At the low end, a 14V belt ISA system, and at the high end, a high voltage full hybrid such as the Toyota Prius, generally offer the best value because in the first instance the vehicle modifications necessary to realize idle-stop are minimal and in the latter case the benefits of adding hybridization are high, leading to high value.

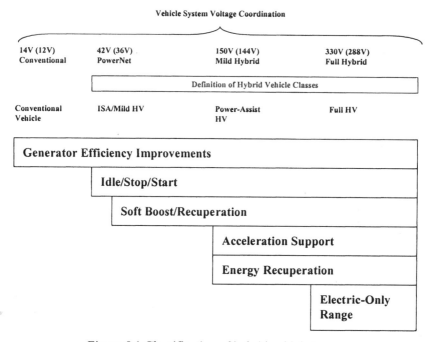

Figure 5.1 Classification of hybrid vehicle types.

We summarize Figure 5.1 as follows. Low end systems such as 14V/42V B-ISA/ISG should target small displacement engines (<2.8 liter). Medium voltage systems 42V/150V are most suited to crankshaft mounted systems for light duty vehicles including small trucks. The high voltage systems (>200V, typically 288 – 320V) are most suited to full hybrid functionality that includes:

- Stop/start, non-idling, or idle stop systems.
- Launch and boost (acceleration assist). Power assist hybrid.
- Regenerative braking. Energy recuperation is desired of all hybrid implementations.
- Electric-only range (may incur additional tax incentives). Possible on only the high end HV due to a need for energy storage system capacity of >2 kWh assuming a passenger car consumption of about 0.5 kWh/mile and 50% SOC.
- Relative power capacity of these systems ranges from 10 kW to >80kW.

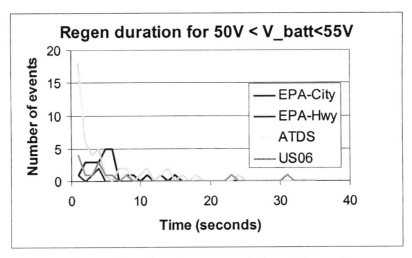

Figure 5.2 Duration of stop events during a drive cycle.

Energy recuperation is highly transient and fleeting. Figure 5.2 illustrates the duration of braking events for the more popular drive cycles. The real world drive cycle or customer drive cycle, ATDS, has most braking events of 5 seconds or less. In fact, there are few braking events having a duration of greater than 10 seconds. HVs replenish their energy storage systems during braking events by capturing as much of the available vehicle kinetic energy as possible during these events. For electric system power ratings of 20 kW or higher, the HV is capable of capturing virtually all of the available braking energy, except for emergency stops or extremely aggressive braking events.

Hybridization leads to downsized internal combustion engine power plants for equal vehicle performance while delivering improved fuel economy. This is most evident by comparing existing hybrid vehicles in the market to conventional vehicles in terms of engine peak power normalized to vehicle mass, or kW/100kg. Figure 5.3 illustrates this finding graphically and shows that hybrids on the road today exhibit a peak power (engine plus electric system) of 6.74 kW/100 kg compared to some comparator conventional vehicles (CV)

Electric and Hybrid Electric Vehicles

having 8.1 kW/100 kg specific power for the same performance. The difference translates to fuel economy benefit. The five hybrids mapped in Figure 5.3 are the Toyota Prius, Honda Insight, and the three US Partnership for a New Generation of Vehicle (PNGV) hybrids from GM (Precept), Ford (Prodigy), and Daimler-Chrysler.

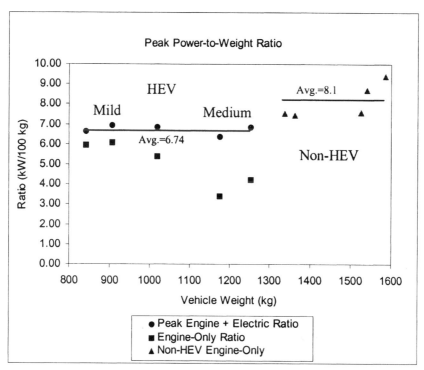

Figure 5.3 Comparison of HV and CV peak power to weight metric (Courtesy of Delphi Automotive).

We noted in the previous chapter that engine cranking is one of the more challenging aspects of the vehicle electrical system. In a hybrid vehicle, it is desirable that the hybrid traction motor, or in mild hybrid cases, the ISA, provide engine start during both key start during cold conditions and strategy start, also known as warm restart. Inline 4 cylinder and V-engine 6 cylinder engines have been characterized in climate chambers to assess their cold breakaway torque and cranking torque. Figure 5.4 illustrates the level of torque that the HV electric machine must deliver, under cold battery conditions, to spin the engine to >450 rpm.

It is interesting to note in Figure 5.4 that cold cranking effort does not really change the requirements for starting torque until the temperature drops below approximately −10°C and steadily increases as temperature is further decreased. The cold engine motoring torque illustrated represents the ISA torque needed to sustain engine cold motoring at approximately 450 rpm. A V8 engine has similar cranking to the V6 engine. The distinction between cranking torque extends from three characteristics of the engine in question. Cranking torque is proportional to the number of crankshaft journal bearings (5 in an I4 but only 4 in V6 and V8 engines), the total piston ring circumference, and, to a limited extent, the compression ratio. Piston ring circumference accounts for the scrubbing friction, which is displacement dependent. Compression ratio is a minor variable since it remains relatively constant among various engines. The FEAD loads and inertia also represent part of the cranking load depicted in Figure 5.4. A manual transmission with the clutch disengaged is also represented by this figure. However, an automatic transmission places a significant burden on engine cranking because the torque converter impeller is fixed to the engine crankshaft. The impeller not only has significant polar moment of inertia, typically 0.1 kg-m^2/rad, but viscous friction as well. An automatic transmission can significantly increase the cranking torque and inertia.

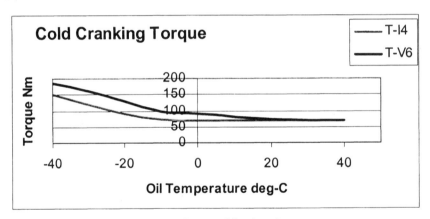

Figure 5.4 Engine cranking requirements.

5.2 Architectures of Hybrid Electric Drivetrains

Due to the environmental concerns, there is a definite development towards new propulsion systems for future cars in the form of electric and hybrid electric vehicles (EV and HEV). Electric vehicles are known as zero emission vehicles. They use batteries as electrical energy storage devices and electric motors to propel the automobile. Hybrid vehicles combine more than one energy source for propulsion. In heat engine/battery hybrid systems, the mechanical power available from the heat engine is combined with the electrical energy stored in a battery to propel the vehicle. These systems also require an electric drivetrain to convert electrical energy into mechanical energy, just like in an EV.

Electric and Hybrid Electric Vehicles 195

Hybrid electric systems can be broadly classified as series or parallel hybrid systems [1]-[4]. The series and parallel hybrid architectures are shown in Figures 5.5 and 5.6, respectively. In series hybrid systems, all the torque required to propel the vehicle is provided by an electric motor. On the other hand, in parallel hybrid systems, the torque obtained from the heat engine is mechanically coupled to the torque produced by an electric motor. In an EV, the electric motor behaves exactly in the same manner as in a series hybrid. Therefore, the torque and power requirements of the electric motor are roughly equal for an EV and series hybrid, while they are lower for a parallel hybrid.

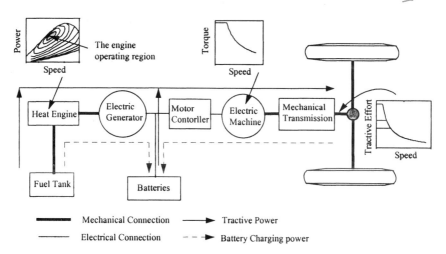

Figure 5.5 Series HEV architecture.

5.3 Electrical Distribution System Architectures

Figure 5.7 depicts the conventional electrical power distribution system architecture for hybrid electric vehicles. It is a DC system with main high voltage bus, e.g., 300V or 140V. A high voltage storage system is connected to the main bus via the battery charge/discharge unit. This unit discharges and charges the batteries in motoring and generating modes of the electric machine operation, respectively. There are also two other charging systems, which are on-board and off-board. The off-board charger has three phase or single phase AC/DC rectifiers to charge the batteries when the vehicle is parked at a charging station. The on-board charger, as is shown in Figure 5.7, consists of a starter/generator and a bi-directional power converter. In the generating mode, an internal combustion engine provides mechanical input power to the electric generator. Therefore, the electric generator supplies electric power to the bi-directional power converter, providing high voltage DC to the main bus. Moreover, in the motoring mode, i.e., cranking the engine, a high voltage DC system via the bi-directional power converter provides input electric power to the electric machine, which is a starter to the vehicle engine.

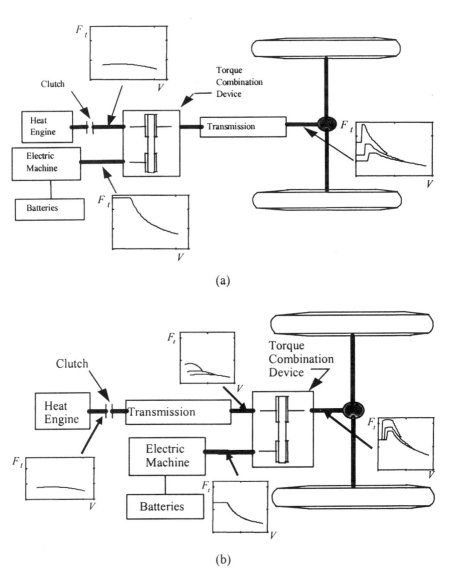

Figure 5.6 Parallel HEV architectures: (a) engine-motor-transmission configuration, (b) engine-transmission-motor configuration.

Electric and Hybrid Electric Vehicles

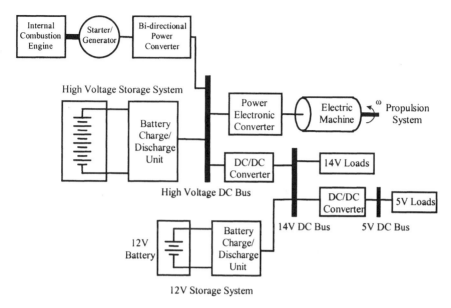

Figure 5.7 Conventional electrical power distribution system architecture for hybrid electric vehicles.

In Figure 5.7, electric propulsion system is feeding from the main high voltage bus. Furthermore, conventional low power 14V and 5V DC loads are connected to the 14V bus. A low voltage 14V bus is connected to the main bus with a step-down DC/DC converter. A 12V storage system via the battery charge/discharge unit is also connected to the low voltage bus. It should be mentioned that Figure 5.7, without internal combustion engine, starter/generator, and bi-directional power converter, shows the electrical power distribution system architecture of electric vehicles.

5.4 More Electric Hybrid Vehicles

As described, demand for higher fuel economy, performance, and reliability as well as reduced emissions will push the automotive industry to seek electrification of ancillaries and engine augmentations. This is the concept of more electric vehicles (MEV). Expansion of the MEV concept to HEV leads to more electric hybrid vehicles (MEHV). In future MEV and MEHV, throttle actuation, power steering, anti-lock braking, rear-wheel steering, air-conditioning, ride-height adjustment, active suspension, and electrically heated catalyst will all benefit from electrical power systems.

Figure 5.8 shows the architecture of the MEHV electrical power system. It is a multi-voltage hybrid (DC and AC) electrical power distribution system with a main high voltage (e.g., 300V or 140V) DC bus providing power for all loads. Conventional loads as well as new electrical ancillary and luxury loads associated with the more electric environment are feeding from the main bus via different DC/DC and DC/AC power electronic converters.

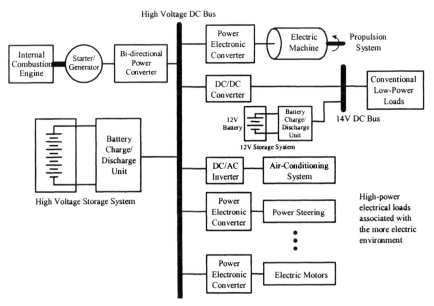

Figure 5.8 MEHV electrical power system architecture.

5.5 Hybrid Control Strategies

When designing a hybrid vehicle, the first challenge is choosing the component sizes (engine and motor) and the mechanical setup of the vehicle (parallel, series, or series-parallel). These decisions are made based on the target cost and performance of the vehicle. At the same time, the designer must think of the ways in which these components will interact. The vehicle must operate as a conventional vehicle where the performance of the vehicle corresponds to the actions of the driver.

A hybrid control strategy controls the flow of energy between the components of the HEV. Energy flow is designed so that the demands of the driver are met while reaching the objectives set for the control strategy. These objectives may be to:

- Improve the fuel economy of the vehicle.
- Reduce emissions.
- Maintain or surpass the objectives set for the performance of the vehicle.

The objectives listed above are usually competing; therefore, improving one means hindering another objective. For example, for the Mercedes CI 1.7L engine, the region where the engine is most efficient does not coincide with the region where NOx emissions are low (Figures 5.9 and 5.10). This is why most control strategies use a weighting function to define what objectives are more important than others. Usually, performance is the most important constraint.

Control strategy would differ for different types of vehicles. The types of vehicles can be broadly grouped by their mechanical setup (series, parallel, or series-parallel), intended use (sports car, SUV, city or inter-city driving), and by state-of-charge (SoC) control (charge depleting or sustaining).

The issue with designing a control strategy is that the problem is non-casual. This means that we cannot incorporate the driver's future action since there is no information on what the driver will do next. Therefore, a good control strategy is not optimized for only one drive cycle. For example, the driver can accelerate at any time for any period of time and the control strategy has to be defined so to be able to withstand this requirement. This is why charge sustaining control strategies are preferred. They provide for more power to be available at any given time since the control strategy keeps the battery state-of-charge (SoC) high. Again, the performance of the vehicle cannot be compromised.

Control strategies determine the operating points of an engine via the efficiency and emission maps of the engine, while maintaining the battery SoC.

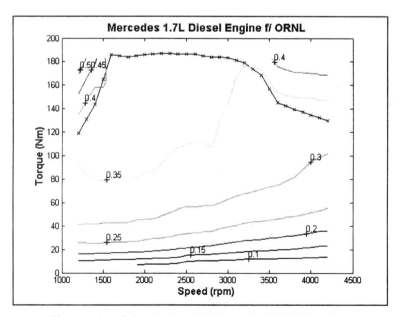

Figure 5.9 Efficiency map of a Mercedes CI 1.7L engine.

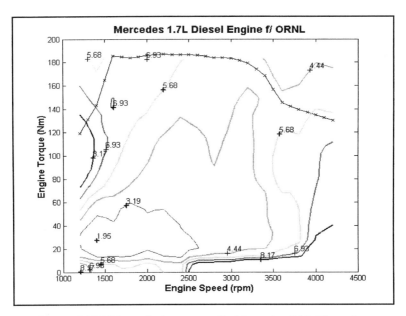

Figure 5.10 NOx emissions map of a Mercedes CI 1.7L engine.

In this section, we look at the differences between parallel and series control strategies as well as the differences between charge sustaining and charge depleting control strategies. We also look at two practical examples, namely the control strategies of the Toyota Prius and Honda Insight.

5.5.1 Parallel Hybrids

For parallel hybrids, the control strategy decides on the torque split between the engine and the motor. The speed is defined by the speed of the shaft.

5.5.1.1 Charge Sustaining Parallel Hybrids

For this control strategy, operation can be defined by the following constraints:

- When the battery SoC is above a certain threshold, the motor is used to assist with the propulsion and to keep the motor out of the low efficiency region.
- Once the battery SoC goes below a certain threshold, the control strategy changes to bring the SoC above the minimum by exerting extra torque from the ICE.
- Battery recharging strategy can be overridden if the ICE alone cannot meet the power demand of the drivetrain.
- Regenerative breaking is always used.

Electric and Hybrid Electric Vehicles

The advantage is that the vehicle has unlimited range since the batteries are never depleted. Also, the vehicle performance is not compromised since the batteries always have enough power to supply the motor. Figure 5.11 shows a typical SoC history over five urban dynamometer driving schedule (UDDS) cycles for this control strategy.

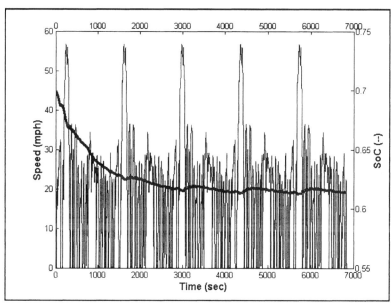

Figure 5.11 SoC history over five UDDS cycles for charge sustaining parallel hybrids.

5.5.1.2 Charge Depleting Parallel Hybrids

For this control strategy, operation can be defined by the following constraints:

- The motor is used for all propulsion below a certain threshold speed.
- When the threshold speed is reached, the control strategy switches into a mode where the ICE is used for propulsion, while the motor is used for propulsion assistance.
- As the battery gets depleted, the ICE kicks in at lower speeds.

This control strategy avoids using the ICE at low efficiency regions; it only allows the engine to operate at high power demands, where the engine is more efficient. However, the battery may get depleted to very low levels and the performance of the vehicle can be compromised if high accelerations are needed when the SoC is at its low. Figure 5.12 shows a typical SoC history over five UDDS cycles for this control strategy.

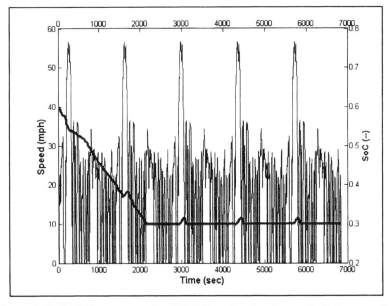

Figure 5.12 SoC history over five UDDS cycles for charge depleting parallel hybrids.

5.5.2 Series Hybrids

In parallel hybrids, all the torque that propels the vehicle comes from the electric motor. The ICE is only used to replenish the batteries via the generator. Series setups are usually used in larger vehicles that justify the use of two electric motors and that can fully utilize the capabilities of the small ICE that is on-board for energy generation.

5.5.2.1 Charge Sustaining Series Hybrids

In this control strategy, the load of the ICE would try to follow the load of the motor. Figure 5.13 shows a typical SoC history over five UDDS cycles for charge sustaining series hybrids.

5.5.2.2 Charge Depleting Series Hybrids

In this control strategy, the load of the ICE would be off until the battery reaches a low SoC. Once the low SoC is reached, it will be on until a high state of charge is reached. The benefit of this control strategy is that the engine is allowed to operate at one single most efficient operating point. This greatly improves the efficiency of the ICE. However, the battery losses are increased dramatically due to high charge and discharge rates. In addition, battery life is shortened due to this aggressive use. Figure 5.14 shows a typical SoC history over five UDDS cycles for charge depleting series hybrids.

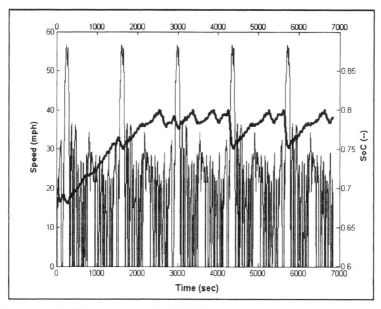

Figure 5.13 SoC history over five UDDS cycles for charge sustaining series hybrids.

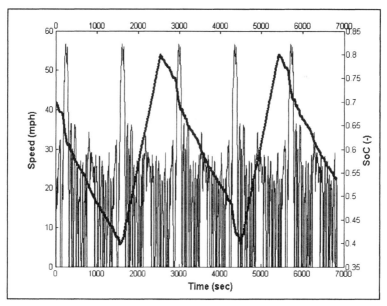

Figure 5.14 SoC history over five UDDS cycles for charge depleting series hybrids.

5.5.3 Practical Models

Most practical models try to combine the benefits of the control strategies defined above. This is similar to the manner in which a series-parallel hybrid uses the advantages of both the series and parallel setups. Therefore, the most advanced practical HEVs are series-parallel charge sustaining-depleting models. Below, we explain two of practical control strategies that have been implemented in actual vehicles.

5.5.3.1 Toyota Prius

The Toyota Prius is a full size car similar to a Toyota Corolla. The engine uses a strategy similar to a parallel charge depleting strategy in the sense that the engine turns on only when the car is traveling above 12 mph. However, the control strategy is charge sustaining, as can be seen from Figure 5.15. In this sense, it is similar to the charge sustaining control strategy. If the battery SoC is below 50%, the engine is loaded so that the battery is replenished to this threshold value. Also, if the SoC is way above 50%, the electric motor is loaded more to bring this value down.

The Toyota Prius is optimized for low fuel consumption and extremely low emissions in the "around town" driving scenario. Still, the highway fuel economy is about 38 mpg. Below are the characteristics of the vehicle:

- 1254 kg curb weight
- 52 kW gasoline engine
- 33 kW electric motor
- 6.5 Ah Ni-MH battery pack
  - 38 modules for total voltage of 237.6 V
  - Total power 1778 W-h
  - Mass 53.3 kg

5.5.3.2 Honda Insight

The Honda Insight is a 2-seat coupe that uses its 1L engine as the principal power source, with additional power provided by a 10kW electric motor. The control strategy does not allow the engine to idle, but the engine works at all speeds.

The Insight is capable of 100 mph speeds while averaging between 60 and 70 mpg overall. Such high fuel economy values are also attributed to the fact that the vehicle is very aerodynamic.

The control strategy is similar to a charge sustaining parallel vehicle in the sense that the electric motor is only used to start the engine and to assist the engine with propulsion. The ICE is only switched off when the vehicle is at a stop. The motor is designed so that it provides about 10 N-m of torque at any vehicle speed. This keeps the state of charge of the battery at a near constant value. Figure 5.16 shows the SoC history over one UDDS cycle for Honda Insight. Below are the characteristics of the vehicle:

- 856 kg curb weight
- 50 kW gasoline engine
- 10 kW electric motor
- 6.5 Ah Ni-MH battery pack
  - 20 modules for total voltage of 144 V
  - Total power 936 W-h
  - Mass 35.2 kg

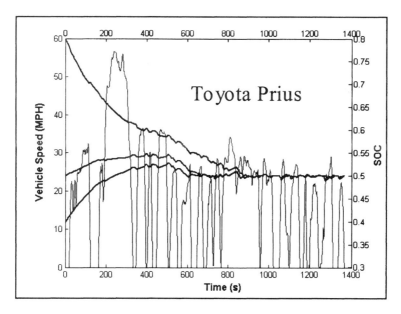

Figure 5.15 SoC history over one UDDS cycle for Toyota Prius.

In summary, the choice of control strategy greatly influences the performance, fuel economy, and emissions of a vehicle. In addition, as can be seen from the graphs, series control strategies allow for larger variations in the battery state of charge. This is intuitive since, in series hybrids, all torque is provided by the electric motor. Therefore, it is expected that the battery SoC varies more since the batteries are used more. It is important to notice that all control strategies that have been implemented in actual vehicles are charge sustaining. The reason is that the performance of the vehicle should never be compromised at the cost of better fuel economy.

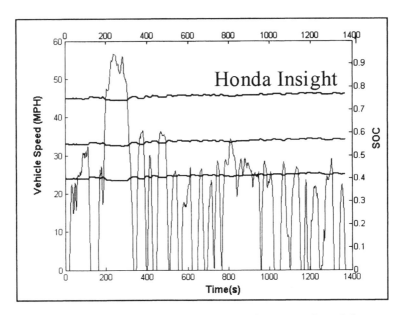

Figure 5.16 SoC history over UDDS cycle for Honda Insight.

5.6 Hybridization Effects

It is not just electrification of existing functions in the vehicle that consume power; several components are needed in addition to the HV motor/generator (M/G) and battery to implement hybridization. Among these are the motor controller, high voltage wiring overlay, power electronics cooling system, climate control, and regenerative brakes. The implementation of these components along with the HV traction components is a major challenge. In this section, we focus on the effects of hybridization on supporting subsystems. These supporting subsystems include electric drive air conditioning (A/C) if used, electro-hydraulic or electro-mechanical brakes, electric assist power steering, energy storage system climate control, and the human-machine interface.

The A/C system can be directly driven by the hybrid power train or independently as a standalone system or by both the engine and electric drive. The latter system is referred to as a hybrid A/C because it still retains the engine driven belt but also has a built in electric motor for engine off operation. Some of the hybrid A/C compressors are the two-stage, variable displacement rotary vane type. The displacement is higher when driven with the engine belt and lower when driven with the electric motor to avoid over-rating the electric system. A DC brushless motor drives the hybrid A/C compressor in the electric mode when the ICE is off and the operator requests cabin cooling.

Fuel economy aggressive HVs are equipped with a series regenerative brake system or RBS. The HV M/G unit captures energy from the RBS to charge the high voltage battery. An electro-hydraulic brake system (EHB) replaces the hydraulic power assist ABS unit. The EHB system consists of a hydraulic electronic control unit (HECU) to replace the ABS, and an actuator control unit (ACU) replaces the conventional booster and master cylinder assembly. The ACU assembly consists of a conventional master cylinder, a reservoir, and brake pedal pressure and speed sensors. The HECU consists of a motor-pump assembly to pressurize brake fluid, an accumulator to store highly pressurized brake fluid, valves to regulate brake fluid flow to brake calipers, and electronics for valve controls.

A computer resolves the brake pedal pressure and its application speed to determine how much braking effort is needed to stop the vehicle. It also computes how much mechanical brake application is required after utilizing M/G regenerative method of braking (energy recuperation). The electronically controlled valves release brake fluid to the calipers based on the driver's brake pedal application and the regenerative braking capabilities of the M/G. In the case of electrical system power failure, the electronic valves open and the conventional master cylinder sends pressurized brake fluid directly to the brake calipers, bypassing the RBS system.

An RBS system also maintains proper balance between front and rear axle brakes, accommodates anti-lock brake (ABS) functionality, and is compatible with vehicle stability programs such as the electronic stability program (ESP) or interactive vehicle dynamics (IVD) or, as some refer to it, vehicle stability control (VSC).

The EHB system offers many advantages to customers over conventional hydraulic power assist ABS. It utilizes series regeneration capability to improve fuel economy by permitting the M/G to recuperate vehicle dynamic energy before application of the service brakes (friction or foundation brakes). It reduces the stopping distance in comparison to the conventional brakes. Also, it isolates the brake pedal from the ABS pressure pulses. ABS pedal pulsation during brake application is believed to cause some drivers to release their foot pressure on the brake pedal during emergency brake applications. Finally, in the EHB system, the pedal has a shorter range of travel and requires less force for stopping. For these reasons, the pedal does not give a feel of braking effort being excessive. Furthermore, the knowledge that all that energy is being routed to the battery and saved instead of being wasted as heat in the brake pads is reassuring to the environmentally conscious operator.

The instrument cluster, or human-machine-interface, as in use by most HVs, has two unique gauges (an energy available gauge and an electric power charge/assist gauge) as well as warning lamps to alert the operator of hybrid functional failure. Their instrument display or the message center's provide the status and deliver warning messages about various system functions. Gear status in a production vehicle is displayed using mechanical linkages, whereas it is an electronic display in the HV cluster.

Hybrid gauges and warning lamps rely on CAN communication. The HV cluster is linked to the vehicle system for the standard gauge displays and warning lamp functions. Additional circuits are added to connect the cluster to the message center control switch, panel dimmer switch, park brake switch, and power supply in ignition key on mode.

Packaging of all the hybrid components is difficult in any type of vehicle. When it comes to the battery and its power distribution system, this fact becomes painfully apparent. As with a full hybrid, finding a package location for the battery that does not intrude into cabin or some other already used space can be very challenging. The location is generally not near the M/G and control electronics but somewhere further away such as beneath the rear passenger seat or beneath the rear floor pan. The routing of high power and high voltage cables is typically done beneath the vehicle's floor pan. This location provides a convenient path to route such heavy cables and the metal floor pan acts to shield any AC fields from cabin contents and passengers.

In any hybrid power train, the fact that two independent power sources have a role in propulsion means that certain precautions are in order. Electrical distribution of high power from the traction battery to the M/G(s) or traction motor must be adequately safeguarded from short circuits to the vehicle chassis or other components. It is also necessary that galvanic isolation be included so that during extended stand times the high voltage battery does not become discharged. Typically, disconnect means are provided in or near the battery compartment in the form of high current contactors.

Packaging the HV battery in the HEV is generally a considerable challenge since it is desirable to modify the original vehicle as little as possible, as well as to maintain customer comfort and safety, and retain the maximum cargo space. If the battery package location is beneath the cargo floor, it is likely the package space for the spare tire would be lost.

With the full complement of hybrid technologies, the HV is equipped to offer customers idle stop, launch assist, boosting for passing and lane change maneuvers, regenerative braking to replenish the battery and accessory power for a wide variety of consumer electronics features and functions. Notable offerings are 120V/220V 50/60Hz AC power-points, heated seats and heated glass and steering wheel, and a host of other features.

5.7 42V System for Traction Applications

Generally, a high-voltage bus such as 300V or 140V is required for the electric propulsion system of a hybrid electric car. In fact, fully hybridized vehicles require expensive high-voltage insulation for the electric propulsion system. Furthermore, there is a drastic change required in conventional cars to obtain hybrid electric vehicles. Therefore, more electric cars have received more

Electric and Hybrid Electric Vehicles

attention than hybrid electric cars from both the automotive industry and the academy in the recent years.

In order to have the advantages of both more electric and hybrid electric concepts, [5]-[7] show that lightly hybridized cars are feasible with switched reluctance machine (SRM) based traction systems. The main advantage of the proposed system is having a common low-voltage (42) electrical power system for traction as well as other automotive loads.

5.7.1 Lightly Hybridized Vehicles

As is depicted in Figure 5.17, in a more electric hybrid car, other than the high-voltage system, which is required for the traction load, not only the traditional 14V system is required for the low-power loads, but also a higher voltage (42V) is necessary for powering the new introduced loads in the more electric environment.

In [5], we have shown that it is possible to drive the traction load (up to 10kW) from the low-voltage (42V) system in order to avoid the high-voltage system. The main advantage of the proposed low-voltage system is establishment of a standard system voltage (42V) for manufacturing of more electric and hybrid electric cars. Avoiding expensive and complex high-voltage insulation and integration of a common electrical power system for traction as well as other automotive loads are other advantages of the low-voltage power system.

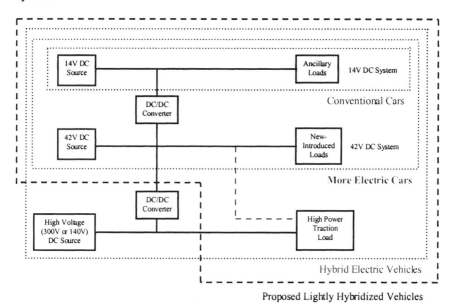

Figure 5.17 Automotive electrical power systems.

In [6] and [7], the suitability of SRM drives for the current intensive environment of the proposed low-voltage system has also been presented. In fact, there is not much change in weight, volume, and cost of the SRM for the low-voltage traction, compared to the high-voltage traction. References [6] and [7] also elaborate on different power electronic converter options to drive the high-current traction motor. It is concluded that the proposed low-voltage system is feasible not only from the power electronic point of view, but also from the electric machine point of view to drive current intensive traction SRMs in lightly hybridized electric cars.

A low-voltage (42V DC) electrical power system for traction applications in lightly hybrid electric cars is shown in Figure 5.18. In the storage system, 36V batteries are used. They provide power to the main 42V DC bus via battery charge/discharge unit. An electric machine for traction is feeding from the main bus with a bi-directional power electronic converter. In Figure 5.18, high power new-introduced and luxury (hotel) loads are directly connected to the main bus. However, low power conventional loads are feeding power from a low-voltage 14V DC bus. A step-down PWM DC/DC converter provides power from the main bus to the 14V bus. It is possible to have a 12V storage system connected to the low-voltage bus with a battery charge/discharge unit.

In this low-voltage system, the electric machine is current intensive. Consequently, its power electronic driver must have the high-current capabilities. In addition, the electrical power system must be able to deliver the required high current to the electric propulsion system.

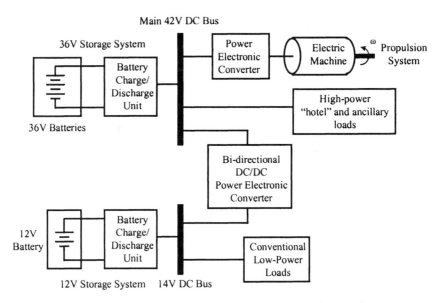

Figure 5.18 Low-voltage electrical power system for traction.

Electric and Hybrid Electric Vehicles

5.7.2 Low-Voltage Storage System

As was mentioned, conventional HEVs have high-voltage (i.e., about 300V or 140V) storage systems. Generally, these storage systems utilize high-voltage batteries consisting of many cells stacked in series in order to achieve the required high voltage. The high number of cells is a disadvantage of the high-voltage storage systems mainly due to the maintenance problems. However, the number of cells in a low-voltage storage system, such as the proposed 42V, is much less than the number of cells in a high-voltage system. Therefore, low-voltage systems have considerably fewer cell failures. In addition, locating and repairing of the cell failures are easier.

Furthermore, the operation of the low-voltage systems, especially under environmental conditions and wet weather, is much safer and less sensitive compared to the conventional high-voltage systems. In addition, in a low-voltage storage system, there is no need for expensive electrical insulation required for the high-voltage batteries. However, larger connectors and thicker cables are required. Moreover, the high currents may cause high losses.

With the same power rating, the energy storage capacity of a low-voltage storage system is exactly as in a high-voltage storage system. In fact, instead of the series connection of the battery cells in a high-voltage system, the same number of battery cells would be connected in a series-parallel configuration to achieve the required low voltage. Therefore, the number of battery cells that are connected in series in a low-voltage system is considerably less than the number of battery cells that are connected in series in a high-voltage system. This is also another advantage of the proposed 42V propulsion system because of the lower battery failure rate.

5.7.3 Low-Voltage Main System with High-Voltage Bus for Propulsion

A low-voltage (42V) main electrical power system with high-voltage bus for propulsion is shown in Figure 5.19. In this system, a step-up DC/DC converter boosts the main 42V to the high-voltage, i.e., 300V or 140V, to drive the traction machine. Providing high current associated with low-voltage for the step-up DC/DC converter is still the main problem. Besides, increasing 42V to 300V or 140V needs expensive step-up DC/DC conversion, which may not be economically feasible. In addition, there might be dynamic problems such as destabilizing effects of the power electronic converters to the low-voltage DC power system of the automobile. Another approach is using a transformer. However, since the rating is high, using a transformer may not be possible due to the weight and volume problems besides the efficiency and economic issues.

5.8 Heavy Duty Vehicles

Since oil resources are limited, we are becoming concerned about our dependence on oil for transportation. We are also concerned about air pollution, which is associated with many human health problems. Cars and trucks we drive, mostly powered by petroleum-based fossil fuels like gasoline and diesel, are a major source of these problems. Heavy-duty vehicles like buses and trucks form a considerable portion of the land vehicles and cause most of the air pollution. The aim of this section is to point out the possible and available electrical technology today that could help reduce these factors and improve performance and reliability.

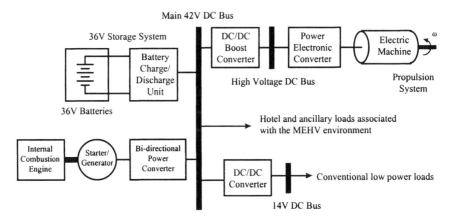

Figure 5.19 Low-voltage (42 V) main electrical power system with high-voltage bus for propulsion.

5.8.1 Conventional Heavy Duty Vehicles

Similar to those of other land vehicles, the electrical system of a heavy duty vehicle was originally comprised of three elements: the battery system, the charging system, and the cranking system. In order to improve the vehicle's fuel economy, emissions, performance, and reliability, the mechanical and hydraulic loads need to be changed to electrical loads. Almost all the conventional and new introduced electric and electronic auxiliary and hotel loads present in a light duty vehicle are also present in heavy duty vehicles, but at higher power levels. Some of the most common electric loads are presented in Table 5.1 and Figure 5.20.

The Electric Vehicle Association of the Americas (EVAA) advocates the commercialization of electric transportation technologies as an effective means of justifying air and noise pollution, greenhouse gases, and over-reliance on petroleum by the transportation sector. The organization is headquartered in Washington, DC. The EVAA has many buses and trucks available for sale and leasing. Table 5.2 shows some specifications for an electric bus that is available at the EVAA [8].

Table 5.1 Typical electric loads in a heavy duty vehicle.

Electric Load	Power (W)
Front Defrost	250
Radio	60
Front Wipers	40
Brake Lights	500
Turn Signals	200
External Lights	800
Radiator Cooling Fan	400
Catalyst Heater	3500
Starter	3500
Oil Pump	600
Water Pump	600

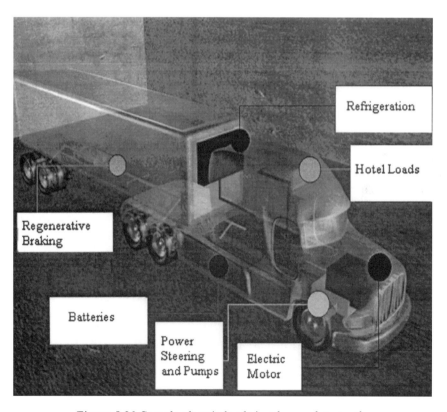

Figure 5.20 Sample electric loads in a heavy duty truck.

Table 5.2 EVAA electric bus specifications.

Length	22 ft
Height	112 in
Curb Weight	14,500 lb
Gross Vehicle Weight	20,000 lb
Battery	Lead acid
Power	140 kW
Acceleration: 0-25	12 sec
Maximum Speed	42 mph
Regenerative Braking	Yes

Electric Vehicles International (EVI) [9] is another leading manufacturer of electric buses, trolleys, and trucks. EVI's products have logged over 1 million miles of quiet, reliable, and pollution-free service worldwide for the past 10 years. Their products are designed with an advanced electric drive system utilizing AC technology, which results in increased range and gradeability. As an example, Table 5.3 shows the specifications for a typical EVI electric heavy-duty delivery truck.

Table 5.3 Typical EVI electric heavy duty delivery truck specifications.

Voltage	320 V
Capacity	3000-5000 lb
Maximum Speed	32 mph
Batteries	Lead acid, 12 V
Controller	AC controller, 390 amp
Motor	56 kW
Charger	220 V
Regenerative Braking	Yes

5.8.2 Hybrid Electric Heavy Duty Vehicles

As we have explained, a hybrid electric vehicle combines a supplementary power unit, such as an internal combustion engine, with batteries and an electric motor; it results in lower emissions and higher fuel economy than in conventional vehicles. Nowadays, there are many hybrid electric buses and trucks that are being driven and available for sale. Hybrid-electric drive systems offer high power and efficiency. Hybrids are not zero-emission vehicles, because they have an internal combustion engine; however, they considerably reduce emissions compared with emission from the internal combustion engine alone. HEVs can be designed with engines that use conventional or alternative fuels, or fuel cells. The fuel efficiency is also increased because hybrids consume considerably less fuel than conventional propulsion systems, regardless of the type of fuel used. Again, another great advantage with heavy duty HEVs is the ability to use regenerative braking. Regenerative braking allows the propulsion system to put a load on the drive axle during braking and convert kinetic energy into electrical energy. The vehicle stores that energy on-board to be used to drive the wheels at another time. This regenerative braking can serve brake wear and increase the overall fuel economy of the vehicle by recycling energy that is normally lost during braking.

Although there are advantages to using HEVs, there are some disadvantages as well. The energy storage device for electric and hybrid electric vehicles today is the lead-acid battery. These energy storage devices add major weight to the bus or truck. Current battery technology does not have the energy storage capacity per unit weight that would optimize the use of HEV technology for the range and load capacity of buses. The biggest obstacles for the new energy storage devices are cost and reliability.

Figure 5.21 shows a hybrid electric bus operating at New York City Metropolitan Transportation Authority (MTA). The department has committed itself to establishing the cleanest bus fleet in the world and significantly reducing the air pollution in New York City. Table 5.4 presents specifications of the hybrid bus [10].

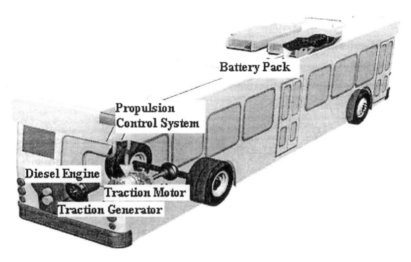

Figure 5.21 New York City Metropolitan Transportation Authority (MTA) hybrid bus.

In Figure 5.21, the diesel engine powers a traction generator that provides primary power through the propulsion control system to the traction motor and recharges the batteries. The traction motor drives the wheels and regenerates power during braking. Batteries provide supplemental power to a traction motor during acceleration and grade climbing. The propulsion control system manages the flow of power to make the bus move as the driver commands and uses regenerative braking to slow the bus and simultaneously recharge the batteries. The system is integrated. During acceleration, power flows from the traction generator and battery pack to the traction motor; during cruise mode, power flows from the traction generator to drive the traction motor and recharge the batteries as needed; and, during braking, the traction motor acts as a generator sending power to the batteries for recharging. The smaller diesel engine, operating at a more constant speed and with better overall fuel economy, can significantly reduce overall bus emission [10].

Table 5.4 Hybrid bus fleet facts.

Length	40 ft
Width	102 in
Height	125 in
Curb Weight	31,840 pounds
Seats	31
Engine	DDC series 30 diesel
Rating	230 hp at 2300 rpm 605 ft-lb at 1500 rpm
Diesel Fuel Storage	100 gallons
Traction Generator	170 kW at 2000 rpm
Traction Motor	187 kW, 346 V (rms) at 500 Hz
Traction Batteries	Sealed lead acid 2 roof mounted battery tubs 23 (12V) batteries in each tub 580V total
Regenerative Braking	Yes

ISE Research-ThunderVolt [11], based in San Diego, is a privately held firm specializing in development of technologies and products for electric, hybrid electric, and fuel cell vehicles, with a focus on heavy duty vehicles such as buses, trucks, and tractors. ThunderVolt^{TM} hybrid electric system offers the high power and torque needed to drive trucks up to 80,000 lb. Therefore, drivers in class 8 trucks can enjoy the benefits of electric vehicle technology. Table 5.5 shows class 8 truck specifications [11].

5.8.3 Fuel Cell Heavy Duty Vehicles

A fuel cell operates like a battery, although they have some distinct physical differences. A fuel cell converts chemical energy directly into electricity by combining oxygen from the air with hydrogen gas. Unlike conventional batteries, a fuel cell is supplied with reactants externally. Therefore, while a battery is discharged, a fuel cell, as long as the supply of fuel is provided, never faces the discharging problem.

Table 5.5 Class 8 truck specifications.

Performance	
Acceleration 0-30 mph 0-60 mph	15 sec 50 sec
All Electric Range	5-10 miles
Hybrid Range	300-500 miles
Maximum Seed	75 mph
General Specifications	
Auxiliary Power Unit	Various engine models, combined with 200 kW air cooled permanent magnet generator
Drive Motor	400 kW (536 hp) peak, 325 kW (436 hp) continuous AC induction motor
Motor Controller	500 KVA modular, variable frequency controller
Gear Ratio	4.4:1 planetary reduction
Steering	Electro-hydraulic system
Braking	Regenerative braking
Charger	Modular 6.1 kW-20 kW charger onboard or offboard
Generator	80 kW continuous

Fuel cell powered buses are already running in several cities. The bus was one of the first applications of the fuel cells because fuel cells need to be large to produce enough power to drive a vehicle. Most buses use either direct hydrogen fuel cells (DHFC) or direct methanol fuel cells (DMFC). Furthermore, as a major step towards reducing emissions from heavy duty vehicles, most transit authorities have opted to use compact natural gas (CNG) fueled buses.

Argonne National Laboratory (ANL) has developed a comprehensive model of a polymer electrolyte fuel cell (PEFC) power system for transportation. The major objective of their analysis is to examine system efficiency, performance, and component sizes [12]. The power generation and power requirements of the various components for such a 50 kW electric system are shown in Table 5.6.

In 1998, Georgetown University, Washington, introduced the first commercially viable, liquid-fueled, fuel cell powered transit bus. This 40-foot electric bus uses a 100 kW phosphoric acid fuel cell (PAFC) as the primary energy source. They later upgraded this bus to a proton exchange membrane fuel cell (PEMFC) bus in 2001. Some useful technical data for this 40-foot PEMFC transit bus is summarized in Table 5.7 [13]. Figure 5.22 shows Georgetown's 40-foot PEMF bus.

Table 5.6 Power requirements for a typical 50 kW fuel cell system.

Component	Power (kW)
Fuel-Cell Stack	52.5
Expander	9.2
Compressor	9.3
Radiator Fan	1.5
Condenser Fan	0.6
Water Pump	0.1
Fuel Pump	<5 W

Table 5.7 Characteristics of the Georgetown University fuel cell bus.

Characteristic	Specification
Weight	39,500 lb
Propulsion System	PEMFC/NiCd hybrid
Fuel	Methanol
Motor Drive	AC induction
Motor Power	186.5 kW (250 hp)
Acceleration (0-30 mph)	14.5 sec
Top Speed	66 mph
Driving Range	350 miles
Noise Level	10 dB below ICE

Another association that works with fuel cell buses is the Electric Vehicle Association of the Americas (EVAA) [8]. This industry association works to advance electric vehicle transportation technologies. Direct hydrogen fuel cells (DHFC) are used to provide the entire power to the propulsion system. XCELLSiS is a fuel cell engine that is used for transportation applications. A brief summary of the technical data of the DHFC bus is shown in Table 5.8 [8].

In summary, over the past few years, several electric, hybrid electric, and fuel cell buses and trucks have been demonstrated in the United States, Canada,

Europe, and Japan. Thus, it is clear that the electrical power systems of heavy duty vehicles are growing rapidly because of their high efficiency, comfortness, and low emission. Chapter 9 will explain fuel cell based vehicles, including heavy duty vehicles, in detail.

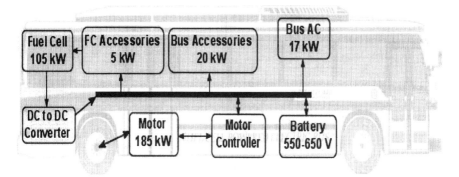

Figure 5.22 Georgetown University fuel cell bus.

Table 5.8 Characteristics of the EVAA DHFC fuel cell bus.

Characteristic	Specification
Fuel	Direct hydrogen
Fuel Cell Engine Weight	Nearly 5000 lb
Power	205 kW
Horsepower	250 hp
Voltage	600-900 V DC
Net Efficiency	~ 40%
Driving Range	225 miles
Fuel Cell Operating Temperature	70-80 °C
Power Conditioning	Liquid cooled IGBT inverter
Motor Drive	Liquid cooled BLDC motor

5.9 Electric Dragsters

Dragsters are cars that are used in special drag races. The main aim of such races is to test the speed and maximum torque that can increase the speed of the car. In the case of dragsters, we are not concerned so much with the cost, as we are with the efficiency and performance. In fact, the vehicle should be able to achieve ground-breaking speeds within a few seconds.

However, a natural question arises about the need for electric dragsters. Since these cars evolve constantly, we need trained personnel to build and maintain them. If a drag racer starts with an electric dragster and builds it on his/her own, a huge cost will be saved since the cost of gas engines are quite prohibitive as compared to motors. We do not consider the pollution factor in this case, since drag cars race for a short while, usually less than 10-15 minutes.

Internal structure of an electric dragster is similar to that in an all-electric vehicle. The series structure consists of battery, controller (driver), and electric motor. A transmission is usually used to control the speed of the car, as in a conventional dragster.

5.9.1 Components

The battery is one of the main components. Most major carmakers today go either for well researched lead acid batteries or for nickel metal hydride (NiMh) batteries. The latter batteries are increasingly used. Most electric dragsters can use a minimum battery voltage of 350V. Batteries as high as 550V are also used for faster electric dragsters.

The number of batteries in an electric dragster depends on the battery voltage, type, size, desired nominal system voltage, and size of the vehicle. In general, a higher system voltage results in increased performance.

Suppose that we use standard NiMh batteries of 36V. We have to use at least 10 of these batteries in series to obtain the desired system voltage. If we use Li-ion batteries, we will have far better performance; but, the cost will be much higher.

The controller is also an important part of the dragster since it actually controls the amount of voltage and current that goes through the system, thus increasing or decreasing the speed and torque of the dragster. The controller transforms the battery's direct current into alternating current (for AC motors only) and regulates the energy flow from the battery. Unlike the carburetor, the controller can also reverse the motor rotation so the vehicle can go in reverse and convert the motor to a generator so that the kinetic energy of motion can be used to recharge the battery when the brake is applied.

These controllers adjust speed and acceleration by solid-state switches. Switching devices such as IGBTs rapidly interrupt (turn on and turn off) the electricity flow to the motor. High power (high speed and/or acceleration) is achieved when the intervals that the current is put off are short. Low power occurs when the intervals are longer. Some of the controllers also have regenerative braking capabilities. Regenerative braking is a process by which the motor is used as a generator to recharge the batteries when the vehicle is slowing down. During regenerative braking, kinetic energy normally absorbed by the brakes and turned into heat is converted to electricity by the motor/controller and is used to re-charge the batteries. Regenerative braking not only increases the range of an electric vehicle, it also decreases brake wear and reduces maintenance cost.

Electric and Hybrid Electric Vehicles

The electric motor is the heart of any electric dragster. Electric motors convert electrical energy into mechanical energy. The size of the motor varies from a few hp for small electric dragsters to tens of hp for larger dragsters. Table 5.9 shows typical 1 hp motor characteristics for a very small electric dragster.

Table 5.9 Typical 1 hp motor characteristics.

	DC Brush	Brushless DC	AC Induction
Peak efficiency (%)	84-90	93-96	94-95
Efficiency at 10% load (%)	78-85	75-83	93.5-95
Maximum speed (rpm)	4,000-5,500	4,000-9,000	9,000-15,500
Cost per shaft hp	$100-150	$100-130	$50-75
Relative cost of the controller to DC brush type	1	3-5	6-8

5.9.2 External Structure and General Factors

Now that we have looked at the internal structure of the dragsters, we come to the external design of the dragsters. There are several factors that influence the efficiency and performance of a dragster once it is built. Some of the important factors are wind, humidity, barometric pressure, altitude, outside temperature, road conditions, availability of desired components, and cost of the components. Of these, some factors we can control and some we cannot. Hereafter, we take into consideration factors that we can control, such as weight of the dragster, before adding the batteries.

If we choose a material that is light yet durable, it is to our advantage since this can actually benefit the efficiency and the torque. Use of a light durable alloy or hard resin is definitely a good option since it complements the increase in weight due to the batteries and motors. Composite materials can also be used. It is the term used for the materials made of reinforced fibers, continuous or otherwise, which are added to a material known as the matrix, which has a much lower mechanical strength. The matrix is responsible for maintaining the geometrical arrangement of the part and transmitting external stresses to the fiber.

When the dragster is racing at a high speed, the most influential force on it besides the force due to friction is air resistance. Aerodynamic resistance can be divided into several forces: pressure resistance, which basically depends on the shape of the vehicle; air friction resistance over the surface of the vehicle; density resistance, which is produced by the parts standing away from the vehicle like wing mirrors; and resistance to the air flow through internal vehicle ducts (radiator and air ducts).

The greatest effect (70%) is caused by pressure resistance which depends on many vehicle design factors: front spoiler, the shape of the vehicle front, the angle of hood slope, the angle of front windscreen inclination, the external shape of the roof, the angle of inclination of the rear window, and the shape of the vehicle rear.

Transmission is also of great importance. Automatic transmissions reduce efficiency by about 20% and, since the torque factor is different, they do not shift at the right time. We should not use the clutch or flywheel during the dragster construction. Direct drive is not a good option, as the gears provide better motor efficiency on acceleration. Manual 4-speed and 5-speed are the best options available.

Rolling resistance on the dragster is determined by how much energy is required for it to move over the road. Even fresh pavement is riddled with surface imperfections that slow a vehicle down. Without suspension, the vehicle's weight is unsprung and must be lifted up and over these imperfections for the vehicle to move forward. With suspension, the majority of the weight is sprung and imperfections are absorbed by the suspension. Only a small portion of the weight of the vehicle and driver needs to be lifted. It takes far less energy to lift that weight than the whole weight of the dragster. Thus, to the road, a suspended vehicle feels significantly lighter than an unsuspended vehicle and will have less rolling resistance.

Suspension also directly reduces tire rolling resistance. Tire rolling resistance is not as much about tire width or tire pressure as it is about consistency of tire contact patch. The more consistent the tire's contact patch is with the road, the less rolling resistance the vehicle will have. By redirecting the load into the suspension system, the tires are kept from having to deflect as much. Further rolling resistance also involves certain other factors like using one wheel or two to turn, chamber, center of gravity and weight distribution, and brake drag.

There are many designs that exist today for electric dragsters that enable the geometry of the system to maintain proper angles while turning. The known ones are:

- One wheel in front and two in the back, or the tricycle design.
- Two wheels in front and one in the back.

For electric dragsters, the preference is the tricycle design. This is because it allows for a uni-body design. This design of dragster can be constructed using light weight Kevlar honeycomb fiber board for a lighter and more aerodynamic vehicle. The single rear wheel design requires a much heavier structure to keep the body from flexing because of the added weight and turning forces on the rear wheel.

Front wheel drive is preferable. The front wheel is pulling the vehicle in the direction that the driver wants to go. This reduces friction in cornering. When the drive wheel turns in the direction of travel, there is less friction between the road and tires. When using a rear wheel drive, the drive wheel is trying to push in a straight line instead of turning the vehicle.

If the rear of the vehicle starts to swing out in a corner, it applies power to the front wheel to pull the vehicle out instead of having to use the brakes in front, which will only slow the speed down.

Better system balance is the other reason. This is because weight is more evenly distributed over each wheel, which increases stability. There is less stress on drive components, wheel bearing friction, and tire wear. In a single back wheel design, more weight is over the back wheel than over the front wheels. There is a myth that two wheels in front are more stable. That may be true when the center of gravity (COG) is above the center of the wheel axles, but is not when the COG is below the center of the wheels. The tricycle would be very difficult to flip in any situation and can turn just as tight a corner.

In addition, rolling resistance is greater with two wheels that are trying to steer around a turn. The geometry of the wheels is very difficult to maintain because the inside wheel wants to turn at a different speed than the outside wheel. This causes an increase in tire-to-road friction and the wheel bearing friction. These two wheels must maintain a parallel configuration, and this is very difficult to do when used for turning. The two wheels in back configuration always stays true to parallel.

The motor is much more efficient at the coolest temperature possible. The batteries are much more efficient at the warmest temperature possible. When the motor is mounted in the front, fresh air can be better directed around and through the motor, keeping it cool. With the batteries in the rear of a two rear wheel dragster, it is easier to keep them out of the air flow and, thus, they can be better insulated to keep them warm.

5.9.3 Conventional Dragsters

Conventional fuel dragsters are fuelled by conventional sources like petrol, methanol, nitro-methane, or such other organic sources. Top fuel dragsters must weigh a minimum of 1900 lbs (860 kg) and may not have a wheelbase exceeding 300 inches (7.6m) or less than 140 inches (3.5m). The rear brake rotors measure 11 inches in diameter and are made from either steel of carbon fiber activated via a hand lever in the cockpit and utilized only on the rear tires. The car's primary braking system is a pair of parachutes that can produce up to five negative G's of stopping power, enough to throw even a well belted driver forward in the cockpit.

Keeping such a dragster (6000 hp) glued to the racing surface at more than 300 mph (482 kph) is no easy task; the driver's best friends are the huge rear wing and the front canard wings that keep the dragster from being an airborne missile. The carbon fiber rear wings produce 2750 to 2900 kg of down force on the rear tires. The amount of down force depends on many factors: the number of elements (horizontal slats) on the rear wing, the angle of attack of the wing (is setting relative to level) on the wing, the type of air, and how the car is set up. The canards, which are usually made from magnesium, can produce as much as 720 kg of down force, though the front tires are subjected to only 120 kg because of the rear wings leverage.

The powerful elixir that helps coax more than 10 hp from each of the engines 500 cubic inches is the fuel nitro methane. Nitro methane is produced by the nitration of propane, and the end result is CH_3NO_2. It is fed by fuel pumps that deliver 225 liters per minute - the equivalent of 8 shower heads - a typical system will use 55 liters of fuel during the burnout procedure and quarter mile run.

The engine of choice for the majority of teams is an aluminum replica of the 426 Chrysler hemi. At least four manufacturers sell the engine blocks, which cost around $5800, and each engine normally displaces around 500ci (8.1 liters) once the crankshaft and rotation piston assemblies are in place. When equipped with cylinder heads and connecting rods carved from chunks of aluminum, and when the intake manifold, supercharger, and fuel pumps are bolted in place, a nitro engine producing 6000 hp costs approximately $65,000.

A typical Chassis is fabricated from over 300 feet (90 m) of 4130 chrome-moly tubing and costs between $40,000 and $60,000. For the price, the customer also receives the front spindles, a steering box, seat, floor pan, motor plates, rear end plates, wing struts, a body, fuel tank, brake handle and master cylinder, clutch pedal, fuel shut-off and parachute release levers, control cables, oil overflow tank, and, in most cases, front wing and nose assembly. The chassis alone weigh approximately 600 lb (270 kg).

To prevent a loss of traction, power is transferred from the engine to the rear tires via a complex timer controlled clutch system. The centrifugal pressure that squeezes together the four discs and three steel floater plates is applied gradually in a series of infinitesimal stages controlled by hydraulic fluid-powered throw-out bearing and ram until complete one-to-one lock up with the engine and drive train is achieved, approximately three seconds into the run. Clutch temperatures can soar to an excess of 1000 °F. Most top fuel cars run a standard gear end ratio of 3.20:1. Two types of front tires are used – small airplane style tires for quicker reaction times and larger bicycle-size units for better elapsed times.

5.10 Modeling and Simulation of Automotive Power Systems

Due to the environmental concerns, the auto industry is developing new propulsion systems in the form of electric, hybrid electric, more electric, and fuel cell vehicles. As a result, the National Renewable Energy Laboratory (NREL) has developed a tool that is meant to accelerate the development of these new, fuel-efficient vehicles. In 1994, the first version of ADvanced VehIcle SimulatOR (ADVISOR) software was first made available [23], [24]. Since then, the software has been updated and new features have been added. Major car manufacturers and universities have been involved in developing this software.

Electric and Hybrid Electric Vehicles

ADVISOR is a set of model, data, and script text files for use with MATLAB and Simulink, available from MathWorks Inc. It is designed for analysis of the performance, fuel economy, and emissions of conventional, electric, hybrid electric, and fuel cell vehicle. The power of ADVISOR is that it predicts the performance of a vehicle without the need to actually assemble a test vehicle. It helps in choosing and sizing the components of a vehicle to give the optimal performance, fuel economy, and emission characteristics.

The backbone of ADVISOR models is the Simulink block diagram shown in Figure 5.23 (for a parallel HEV). Each subsystem of the block diagram has a MATLAB file (m-file) associated with it that defines the parameters of the subsystem. The user can alter both the block diagram and the m-files to suit his/her needs.

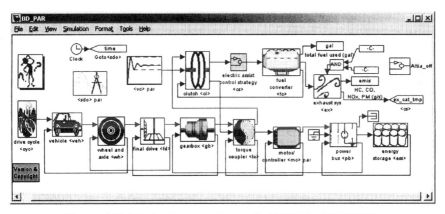

Figure 5.23 Block diagram of a parallel HEV in ADVISOR.

As shown in Figure 5.23, the vehicle model in ADVISOR is modular. This means that each subsystem of the vehicle is separated and is only connected to the rest of the model via input and output ports. Such a setup is convenient for vehicle development since each physical subsystem can be easily interchanged in order to be able to measure the effect of using different subsystems in the vehicle. The parameters for each subsystem are obtained from experimental data.

In general, the efficiency and limiting performances define the operation of each component. For example, the ICE is modeled using an efficiency map that is obtained by experiments. The efficiency map of a Geo 1.0L (43 kW) engine is shown in Figure 5.24. The maximum torque curve is also shown in this map. Maximum torque curve is another constraint to the engine subsystem. Also, the emissions and fuel consumption maps of the engine must be available to be able to simulate the engine.

There are no capabilities in ADVISOR for component modeling. On the other hand, ADVISOR is modular so one can replace a certain subsystem with a

more detailed model as long as the same inputs and outputs are provided. In general, the user takes two steps:
- Defining a vehicle using measured or estimated component and overall vehicle data.
- Prescribing a speed-versus-time trace along with road grade that the vehicle must follow.

ADVISOR then puts the vehicle through its paces, making sure it meets the cycle to the best of its ability and measuring (or offering the opportunity to measure) just about every torque, speed, voltage, current, and power passed from one component to another.

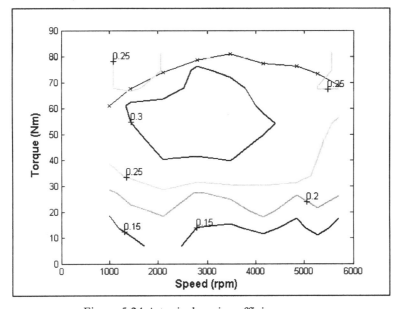

Figure 5.24 A typical engine efficiency map.

ADVISOR has a graphical user interface (GUI) that makes the program easier to use. There are three screens involved in the simulation – two input screens and one results screen. The first screen (Vehicle Input) allows the user to choose the vehicle components and the type of vehicle that he/she wants to simulate. First, the user chooses the type of vehicle (series HEV, parallel HEV, conventional, or fuel cell) and then all components of the vehicle can be defined. The screen is shown in Figure 5.25.

Next, the user chooses what kind of tests he/she wants to perform on the vehicle (Figure 5.26). There are many kinds of specialized tests and drive cycles that the user can chose from. A drive cycle shows how the vehicle speed changes with time. For example, the urban dynamometer driving schedule (UDDS) drive cycle is shown in the upper left corner of Figure 5.26.

Electric and Hybrid Electric Vehicles 227

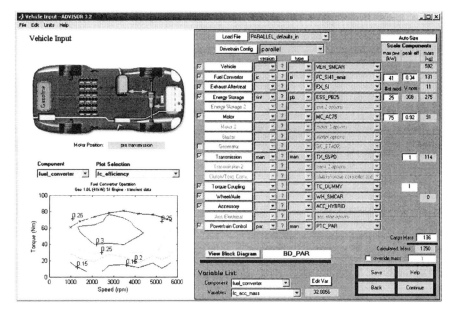

Figure 5.25 Vehicle input screen in ADVISOR.

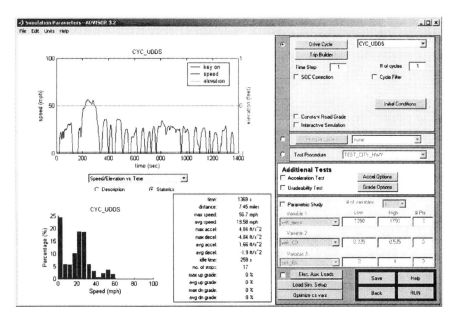

Figure 5.26 Simulation parameters screen.

The last screen shows the results of the simulation (Figure 5.27). On the right side, we see the fuel economy, emissions, gradeability, and acceleration performance of the simulated vehicle. We can also plot parameters such as battery state of charge or motor output power versus time. Simulation times can extend to 1 hour for complex optimization routines. The average simulation time is about 5-10 minutes.

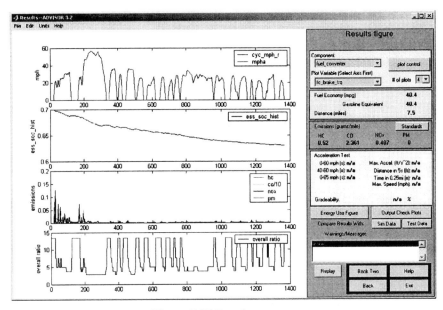

Figure 5.27 Results screen.

For the latest version of ADVISOR, the functionality of the software was improved by allowing links to other software packages like Ansoft's Simplorer and Avanti's Saber. This powerful software packages allows for a more detailed look at the electric systems of the vehicle. Saber is especially functional for more electric vehicle studies since it allows for a detailed analysis of the electric subsystems in a vehicle. In summary, ADVISOR allows us to answer questions like:

- Was the vehicle able to follow the trace (drive cycle)?
- How much fuel and/or electric energy were required in the attempt?
- What were the peak powers delivered by the components?
- What was the efficiency of the transmission?

By iteratively changing the vehicle definition and/or driving cycle, the user can go on to answer questions such as:

- At what road grade can the vehicle maintain 55 mph indefinitely?

- What's the smallest engine we can put into this vehicle to accelerate from 0 to 60 mph in 12 seconds?
- What's the final drive ratio that minimizes fuel use while keeping the 40 to 60 mph time below 3 seconds?

5.11 References

[1] C. C. Chan, "An overview of electric vehicle technology," *Proc. of the IEEE*, vol. 81, no. 9, pp. 1202-1213, Sep. 1993.

[2] C. C. Chan and K. T. Chau, "An overview of power electronics in electric vehicles," *IEEE Trans. on Industrial Electronics*, vol. 44, no. 1, pp. 3-13, Feb. 1997.

[3] A. Emadi, B. Fahimi, M. Ehsani, and J. M. Miller, "On the suitability of low-voltage (42 V) electrical power system for traction applications in the parallel hybrid electric vehicles," *Society of Automotive Engineers (SAE) Journal*, Paper No. 2000-01-1558, 2000.

[4] M. Ehsani, K. M. Rahman, and H. A. Toliyat, "Propulsion system design of electric and hybrid vehicles," *IEEE Trans. on Industrial Electronics*, vol. 44, no. 1, pp. 19-27, Feb. 1997.

[5] A. Emadi, B. Fahimi, M. Ehsani, and J. M. Miller, "On the suitability of low-voltage (42 V) electrical power system for traction applications in the parallel hybrid electric vehicles," *Society of Automotive Engineers (SAE) Journal*, Paper No. 2000-01-1558, 2000; and, in *Proc. SAE 2000 Future Car Congress*, Arlington, Virginia, April 2000.

[6] A. Emadi, "Low-voltage switched reluctance machine based traction systems for lightly hybridized vehicles," *Society of Automotive Engineers (SAE) Journal*, SP-1633, Paper Number 2001-01-2507, pp. 41-47, 2001; and, in *Proc. SAE 2001 Future Transportation Technology Conference*, Costa Mesa, CA, Aug. 2001.

[7] A. Emadi, "Feasibility of power electronic converters for low-voltage (42V) SRM drives in mildly hybrid electric traction systems," in *Proc. IEEE 2001 International Electric Machines and Drives Conference*, Cambridge, MA, June 2001.

[8] Electric Vehicle Association of the Americas, "Electric drive technologies and products," http://www.evaa.org/.

[9] Electric Vehicles International LLC, http://www.evi-usa.com/.

[10] U.S. Department of Energy, "Diesel hybrid electric buses," *New York City Transit*.

[11] ThunderVolt, "The bus drive system for the 21st century," http://www.isecorp.com/.

[12] R. Kumar, R. Ahluwalia, E. D. Doss, H. K. Geyer, and M. Krumpelt, "Design, integration, and trade-off analysis of gasoline-fueled polymer electrolyte fuel cell systems for transportation," Argonne National Laboratory, Argonne, IL.

[13] A Federal Transit Administration Project, "Clean, quit transit buses are here today," http://fuelcellbus.georgetown.edu/.
[14] C. W. Ellers, "Electric transportation the challenge is yours," Electronic Transportation Design, Yachats, OR.
[15] R. J. Kevala, "Development of a liquid-cooled phosphoric acid fuel cell/battery power plant for transit bus applications," Booz, Allen & Hamilton Inc, Bethesda, Maryland.
[16] California Environmental Protection Agency, "Fuel cell electric vehicles," Sacramento, CA, http://www.arb.ca.gov/.
[17] Environmental and Energy Study Institute, "The most important environmental technology," Washington DC.
[18] Howstuffworks, "How fuel cells work," http://www.howstuffworks.com/.
[19] Institute of Transportation Studies, "Fuel efficiency comparison of advanced transit buses using fuel cell and engine hybrid electric drivelines," University of California.
[20] Office of Transportation Technology, "Fuel cells for transportation," http://www.cartech.doe.gov/.
[21] T. Rehg, R. Loda, and N. Minh, "Development of a 50 kW, high efficiency, high power density, CO-tolerant PEM fuel cell stack system," Honeywell Engines & Systems, Torrance, CA.
[22] Y. Baghzouz, J. Fiene, J. Van Dam, L. Shi, E. Wilkinson, and R. Boehm, "Modifications to a hydrogen/electric hybrid bus," Center for Energy Research, University of Nevada, Las Vegas.
[23] National Renewable Energy Laboratory, "ADVISOR documentation," See http://www.ctts.nrel.gov/analysis/.
[24] K. B. Wipke, M. R. Cuddy, and S. D. Burch, "ADVISOR 2.1: A user-friendly advanced powertrain simulation using a combined backward/forward approach," *IEEE Trans. on Vehicular Technology,* vol. 48, no. 6, pp. 1751-1761, Nov. 1999.
[25] S. Onoda, S. M. Lukic, A. Nasiri, and A. Emadi, "A PSIM-based modeling tool for conventional, electric, and hybrid electric vehicle studies," in *Proc. 2002 IEEE Vehicular Technology Conference,* Vancouver, BC, Canada, Sep. 2002.

6
Aircraft Power Systems

Increasing use of electric power to drive aircraft subsystems is the opportunity that exists to significantly improve the aircraft power system performance, reliability, and maintainability. The extent of these changes will certainly depend on the cost effective production of power electronic and control electronic devices, as well as fault tolerant electrical distribution systems, electric drives, and microprocessors.

This chapter briefly discusses the conventional and advanced aircraft power system architectures, components, electrical power generation technologies, and stability. In addition, disadvantages, opportunities for improvement, and current trends in the aircraft power systems are explained. The future aircraft power systems will employ hybrid multi-voltage level DC and AC systems, separate buses for power and control, and an intelligent power management center.

6.1 Conventional Electrical Systems

Increasing use of electric power to drive aircraft subsystems that, in the conventional aircraft, have been driven by a combination of mechanical, electrical, hydraulic, and pneumatic systems, is seen as a dominant trend in advanced aircraft power systems. This is the concept of more electric aircraft (MEA) [1]-[5]. Recent advances in the areas of power electronics, electric drives, control electronics, and microprocessors are already providing the impetus to improve the performance of aircraft electrical systems and their reliability [6]-[14]. As a result, the MEA concept is seen as the future direction of aircraft power system technology.

In the aircraft electrical system, different types of loads require power supplies that are different from those provided by the main generators. For example, in an advanced aircraft power system having a 270V DC primary power supply, certain components are employed which require 28V DC or 115V AC supplies for their operation. Therefore, future aircraft power systems will employ multi-voltage level hybrid DC and AC systems. It, consequently, becomes necessary to employ not only components which convert electrical power from one form to another, but also components which convert the supply to a higher or lower voltage level. As a result, in a modern aircraft, different kinds of power electronic converters, such as AC/DC rectifiers, DC/AC inverters, and DC/DC choppers, are required. In addition, in the variable speed constant frequency (VSCF) systems, solid-state bi-directional converters are used to condition variable-frequency power into a fixed frequency and voltage [15]-[20]. Moreover, bi-directional DC/DC converters are used in battery charge/discharge units. Therefore, MEA electrical distribution systems are mainly in the form of multi-converter power electronic systems. Due to extensive interconnection of components, a large variety of dynamic interactions is possible.

6.1.1 Vehicle Operating Loads

Feiner in [5], [6] divides the power utilization subsystems of an aircraft into two general classes. Based on [5], [6], the first group are subsystems dedicated to the vehicle operation which are needed in the air transportation system. On the other hand, the second group are subsystems that are used for the well-being and comfort of the passenger and cargo.

The power needed for the subsystems in an aircraft is currently derived from mechanical, electrical, hydraulic, and pneumatic sources or a combination of these. Figure 6.1 shows the conventional subsystems which are driven from electrical sources [8], [9]. This distribution network is a point-to-point topology in which all the electrical wiring is distributed from the main bus to different loads through relays and switches. This kind of distribution network leads to expensive, complicated, and heavy wiring circuits.

6.1.2 Advanced Electrical Loads

There is a trend in MEA towards replacement of more engine driven mechanical, hydraulic, and pneumatic loads with electrical loads due to performance and reliability issues. Some of the loads considered are flight control systems, electric anti-icing, environmental systems, electric actuated brakes, electromechanical valve control, air-conditioning system, utility actuators, fuel pumping, and weapon systems. In fact electrical subsystems may require a lower engine power with higher efficiency. Also, they can be used only when needed. Therefore, MEA can have better fuel economy and performance. Figure 6.2 shows the main electrical power subsystems in the MEA power systems.

Aircraft Power Systems 233

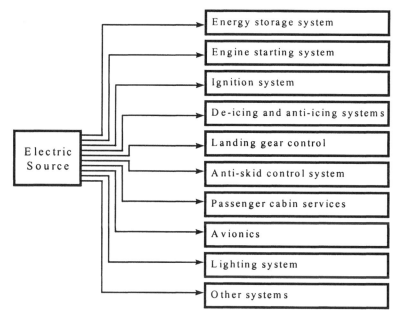

Figure 6.1 Conventional aircraft electrical subsystems.

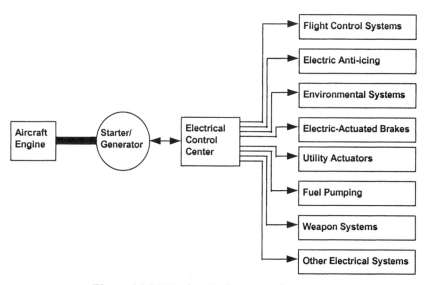

Figure 6.2 MEA electrical power subsystems.

Most of the future electric loads require power electronic controls. In future aircraft power systems, power electronics will be used to perform three different tasks. The first task is simple on/off switching of loads, which is performed by mechanical switches and relays in conventional aircraft. The second task is the control of electric machines. The third task, which will be explained later, is not only changing the system voltage to a higher or lower level, but also converting electrical power from one form to another using DC/DC, DC/AC, and AC/DC converters. Similar to power electronic converters, motor drives are essential elements of the MEA. The permanent magnet synchronous motor (PMSM), the switched reluctance motor (SRM), and the induction motor (IM) are the main electric drive systems for aircraft applications. In [11], the PMSM, IM, and SRM in the ranges of 10 to 300hp were compared in terms of efficiency and power density for a given dimensional envelope.

6.2 Power Generation Systems

In conventional aircraft, the wound-field synchronous machine has been used to generate AC electrical power with constant frequency of 400Hz. This machine/drive system is known as a constant speed drive (CSD) system [15]-[16]. Figure 6.3 shows a typical constant speed drive system. In Figure 6.3, a synchronous generator supplies an AC constant frequency voltage to the AC loads in the aircraft. Then, AC/DC rectifiers are used to convert the AC voltage with fixed frequency at the main AC bus to multi-level DC voltages at the secondary buses, which supply electrical power to the DC loads. Excitation voltage of the synchronous generator and firing angles of the bridge rectifiers are controlled via the control system of the CSD system. Recent advancements in the areas of power electronics, control electronics, electric motor drives, and electric machines have introduced a new technology, the variable speed constant frequency (VSCF) system. The main advantage of the VSCF system is providing better starter/generator systems. Other advantages are higher reliability, lower recurring costs, and shorter mission cycle times [15]. There are many other advantages that are explained in [11], [15]-[18].

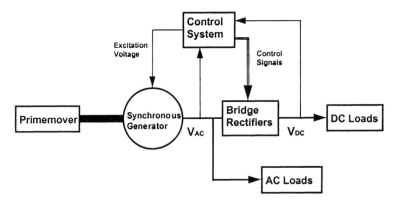

Figure 6.3 Typical constant speed drive (CSD) system.

Aircraft Power Systems

Figure 6.4 shows the block diagram of a typical variable speed constant frequency starter/generator system. In the generating mode, the aircraft engine, which has variable speed, provides mechanical input power to the electric generator. Therefore, the electric generator supplies variable frequency AC power to the bi-directional power converter, which provides AC constant frequency voltage to the main bus. In the motoring mode, the constant frequency AC system via the bi-directional power converter provides input electric power to the electric machine, which is a starter to the aircraft engine.

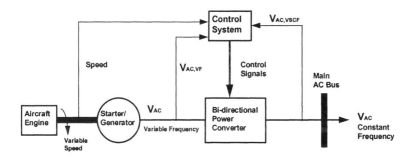

Figure 6.4 Typical variable-speed constant-frequency starter/generator system.

Synchronous, induction, and switched reluctance machines are three candidates for VSCF starter/generator systems. The comparison of these machines for aircraft applications is based on the electromagnetic weight, power density, efficiency, control complexity and features, complexity of design and fabrication, reliability, and thermal robustness [11], [14]-[16].

6.3 Aircraft Electrical Distribution Systems

In order for the electric power available at the generating sources to be made available at the terminals of the power-consuming equipment, some organized form of electrical distribution throughout an aircraft is essential. On the other hand, in order to optimize aircraft performance, reliability, and life cycle cost, the MEA emphasizes the utilization of electrical power systems instead of mechanical, hydraulic, and pneumatic power transfer systems. As a result, electrical power systems of MEA have a larger capacity and more complex configuration. In fact, in the MEA electrical power systems, a number of different types of loads are used which require power supplies different from those standard supplies provided by the main generator. Therefore, future aircraft electrical power systems will employ multi-voltage level hybrid DC and AC systems. For example, in an advanced aircraft power system having a 270V DC primary power supply, certain instruments and electronic equipment are employed which require 28V DC and 115V AC supplies for their operation, and, as we have already seen, DC cannot be entirely eliminated even in aircraft which are primarily AC in concept. Furthermore, we may also note that even within the pieces of consumer equipment themselves, certain sections of their circuits require different types of power supply and/or different levels of the same kind of the supply. It, therefore, becomes necessary to employ not only equipment which will convert electrical power from one form to another, but also equipment which will convert one form of supply to a higher or lower value. As a result, in a modern aircraft, different kinds of power electronic converters such as AC/DC rectifiers, DC/AC inverters, and DC/DC choppers are required. In addition, in the VSCF systems, solid-state bi-directional converters are used to condition variable-frequency power into a fixed frequency and voltage. Moreover, bi-directional DC/DC converters are used in battery charge/discharge units. Therefore, the future aircraft electrical power systems will employ multi-converter power electronics systems. Figure 6.5 shows an advanced aircraft power system architecture in which we have several power electronic converters.

In Figure 6.5, the loads are controlled by intelligent remote modules. Therefore, the number and length of wires in the harness are reduced. Furthermore, by interconnection between remote modules via communication/control buses, it is possible to have a power management system (PMS). The primary function of the PMS is time phasing of the duty-cycle of loads in order to reduce the peak power demand [21]. Other functions of the PMS are battery management and charging strategy in a multiple battery system, load management, management of the starter/generator system including the regulator, and provision and control of a high integrity supply system.

The distribution control network of Figure 6.5 simplifies the vehicle's physical design and assembly process and offers additional benefits from the amalgamation with intelligent power management control:

- They put all loads under intelligent control; therefore, power management feature can be readily integrated into the existing control with minimal cost.
- Power management strategy can help optimize the size of the generators and batteries.
- Communications inherent with a networked vehicle system can give improved performance with minimal increase in complexity and cost.
- Vehicle economy can be improved using the knowledge of the battery state in a networked system.

Furthermore, the main distribution system can also be changed from DC to AC. The main advantage of AC distribution systems is easy conversion to different voltage levels by transformers. Furthermore, AC machines can simply be used.

Aircraft Power Systems

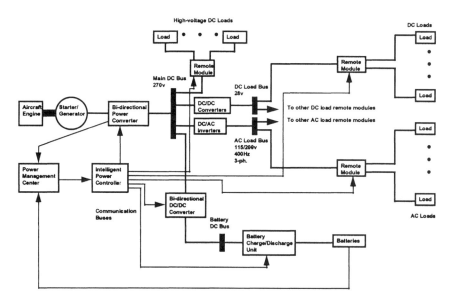

Figure 6.5 The concept of an advanced aircraft power system architecture of the future.

6.4 Stability Analysis

For studying the stability, the common approach is to linearize the system around its operating point and obtain the small-signal linearized model of the system. Then the stability of the system can be determined using linear system stability analysis methods such as Routh-Hurwitz, root locus, Bode plot, and Nyquist criterion. Although small-signal stability can be governed using this linearized approach, large-signal stability of the system cannot be determined. In the other words, the absence of small-signal instability do not guarantee large-signal stability of the system.

Large-signal stability refers to the ability of the system to move from one steady state operating point following a disturbance to another steady state operating point. In the large-signal stability assessment, the way that the aircraft power system responds to the various types of disturbance, such as change in power demand, loss of generations, short circuits, and open circuits, is important. In addition, the dynamics of the system are affected by interconnection between the components.

Stability analysis of the aircraft power systems is best performed via time domain simulation using large-signal models. It depends on the actual control and protection circuit dynamics which may include but is not limited to undervoltage, overvoltage and overcurrent protections, switching effects of power electronic converters, nonlinearities due to magnetic saturation, leakage, semiconductor operation, temperature variations, and aging.

Dynamic behavior of an aircraft power system can be improved and its stability margins can be increased by considering the following issues:

- Using suitable fast-response protection devices.
- Avoiding heavily loaded connections.
- Overrating the sources to ensure an appropriate reserve in power generation.
- Overrating the power distribution system as an appropriate reserve in the ability of delivering power to the loads.
- Avoiding long links.
- Managing the loads properly according to the operating conditions of sources and the distribution system to avoid overloading.
- Controlling the loads via the control commands of the management center in such a way that the system always operate around its nominal power.
- Avoiding overvoltage and undervoltage.

A conservative approach to designing aircraft power systems may be to over-design the ratings of the sources and loads by conservatively defining their margins of stability and enhancing the protection mechanisms to react to any unforeseen instability during the operation. However, there is a trade-off between operating a system close to its stability limit and operating a system with an excessive reserve in power generation and distribution capability. In fact, financial considerations always determine the extent to which any of the mentioned issues can be implemented.

6.5 References

[1] A. Emadi and M. Ehsani, "Electrical system architectures for future aircraft," in *Proc. 34th Intersociety Energy Conversion Engineering Conference,* Vancouver, British Columbia, Canada, Aug. 1999.

[2] A. Emadi and M. Ehsani, Chapter 21 titled "More electric vehicles: 21.1 aircraft and 21.2 terrestrial vehicles", *CRC Handbook of Power Electronics*, CRC Press, Nov. 2001.

[3] R. E. Quigley, "More electric aircraft," in *Proc. 1993 IEEE Applied Power Electronics Conf.*, San Diego, March 1993, pp. 609-911.

[4] J. A. Weimer, "Electrical power technology for the more electric aircraft," in *Proc. IEEE 12th Digital Avionics Systems Conf.*, Fort Worth, Oct. 1993, pp. 445-450.

[5] L. J. Feiner, "Power-by-wire aircraft secondary power systems," in *Proc. IEEE 12th Digital Avionics Systems Conf.*, Fort Worth, Oct. 1993, pp. 439-444.

[6] L. J. Feiner, "Power Electronics transforms aircraft systems," in *Proc. WESCON'94*, Anaheim, Sep. 1994, pp. 166-171.

[7] J. S. Cloyd, "Status of the United States Air Force's more electric aircraft initiative," *IEEE AES Systems Magazine*, pp. 17-22, April 1998.
[8] E. H. J. Pallett, *Aircraft Electrical Systems*, Pitman Publishing Ltd., 1988.
[9] T. K. Eismin, R. D. Bent, and J. L. McKinley, *Aircraft Electricity and Electronics*, McGraw-Hill Book Company, 1994.
[10] T. L. Skvareniana, S. Pekarek, O. Wasynczuk, P. C. Krause, R. J. Thibodeaux, and J. Weimer, "Simulation of a more electric aircraft power system using an automated state model approach," in *Proc. 31st Intersociety Energy Conversion Engineering Conf.*, Washington, Aug. 1996, pp. 133-136.
[11] R. Krishnan and A. S. Bharadwaj, "A comparative study of various motor drive systems for aircraft applications," in *Proc. 1991 IEEE Industry Application Conf.*, Dearborn, Oct. 1991, pp. 252-258.
[12] W. G. Homeyer, E. E. Bowles, S. P. Lupan, P. S. Walia, and M. A. Maldonado, "Advanced power converters for more electric aircraft applications," in *Proc. the 32st Intersociety Energy Conversion Engineering Conf.*, Aug. 1997, pp. 591-596.
[13] K. C. Reinhardt and M. A. Marciniak, "Wide-bandgap power electronics for the more electric aircraft," in *Proc. 31st Intersociety Energy Conversion Engineering Conf.*, Washington, Aug. 1996, pp. 127-132.
[14] E. Richter, "High temperature, lightweight, switched reluctance motors and generators for future aircraft engine applications," in *Proc. American Control Conference*, 1988, pp. 1846-1851.
[15] M. E. Elbuluk and M. D. Kankam, "Potential starter/generator technologies for future aerospace application," *IEEE Aerospace and Electronics Systems Magazine*, vol. 11, no. 10, pp. 17-24, Oct. 1996.
[16] J. G. Vaidya, "Electrical machines technology for aerospace power generators," in *Proc. 1991 Intersociety Energy Conversion Engineering Conf.*, vol. 1, 1991, pp. 7-12.
[17] T. L. Skvarenina, O. Wasynczuk, P. C. Krause, W. Z. Chen, R. J. Thibodeaux, and J. Weimer, "Simulation and analysis of a switched reluctance generator/more electric aircraft power system," in *Proc. 31st Intersociety Energy Conversion Engineering Conf.*, Washington, Aug. 1996, pp. 143-147.
[18] E. Richter and C. Ferreira, "Performance evaluation of a 250kW switched reluctance starter/generator," in *Proc. 1995 IEEE Industry Applications Conf.*, Orlando, Oct. 1995, pp. 434-440.
[19] E. Richter, C. Ferreira, J. P. Lyons, A.V. Radun, and E. Ruckstadter, "Initial testing of a 250kW starter/generator for aircraft applications," *SAE Aerospace Atlantic Conf. Rec.*, Dayton, April 1994.
[20] T. M. Jahns and M. A. Maldonado, "A new resonant link aircraft power generation system," *IEEE Trans. on Aerospace and Electronic Systems*, vol. 29, no. 1, pp. 206-214, Jan. 1993.

[21] M. A. Maldonado, N. M. Shah, K. J. Cleek, P. S. Walia, and G. J. Korba, "Power management and distribution system for a more electric aircraft (MADMEL)- program status," in *Proc. 32st Intersociety Energy Conversion Engineering Conf.*, Aug. 1997, pp. 274-279.

[22] R. E. Kalman and J. E. Bertram, "Control system analysis and design via the second method of Lyapunov: I. continuous time systems," *American Society of Mechanical Engineering*, 1960.

7
Space Power Systems

Electrical power is the most critical ingredient for the international space station (ISS). This is because it is the primary resource available to the crew in order to live comfortably aboard the ISS. Apart from that, the availability of electricity aboard the ISS also facilitates safe operation of the station and helps aerospace scientists to perform further research and development work while in outer space. Supplying electricity to such a huge system indeed poses electrical engineers with an enormous challenge. This chapter basically describes the entire process of overcoming this challenge through the detailed discussion of the electrical power system (EPS) of the ISS. Firstly, photovoltaic (PV) systems, as primary power producing units, shall be discussed in detail. In addition, battery and flywheel technologies, as secondary energy storage devices, will be presented. Furthermore, this chapter aims to introduce spacecraft power systems for small satellites and interplanetary missions. Various alternate power sources for space missions are also discussed. Finally, the Earth Orbiting System (EOS) TERRA power system and the EPS for Space Based Radar (SBR) satellites, designed by NASA, are explained as examples.

The power system architectures of spacecrafts will be explained in detail, with much emphasis devoted to PV power conversion techniques and the solar array technology. Battery technology and energy management issues will be discussed wherein battery design issues and various degradation factors will be highlighted. The role of power electronics with regard to the power conversion systems within the EPS and the battery charge/discharge units (BCDU) for the battery packs will also be discussed. In general, DC/DC bi-directional power electronic converters are widely used in aerospace power system applications. Furthermore, various other smaller DC loads are fed from low power DC/DC converter units (DDCU).

7.1 Introduction

As mentioned before, the presence of electric power on the ISS is extremely critical to the station crew. This is because the ISS depends on electricity for the purpose of comfortable existence in space, along with a high degree of safety. Furthermore, scientific experimentations would not be possible in the absence of electric power. Thus, without a doubt, the design of the EPS for the ISS is indeed a critical issue.

Firstly, it must be well understood that the EPS of the ISS is the biggest power system in outer space. Also, it is a well-known fact that the entire ISS power system is DC based and not AC. Furthermore, it is also obvious that the only readily available power source to provide energy to the ISS is the solar rays from the sun. Hence, it is understandable that the constant R&D conducted by NASA is directed towards the PV array technology. The EPS of the ISS uses PV cells, assembled together to form arrays, in order to produce high levels of power [1]. In addition, the ISS relies on rechargeable Ni-H_2 (nickel-hydrogen) batteries to provide a secondary energy storage facility. This is because the PV cells are not always under the influence of sunlight, which is incidentally known as the "eclipse" part of the orbit [1]. Thus, the secondary batteries ensure the continuity of power aboard the ISS. Obviously, during the "sunlit" part of the orbit, the batteries get recharged.

Although the above processes seem simple and straightforward, the entire process of collecting solar energy, converting it to usable electricity, and distributing it to various loads leads to the building up of excess heat, which could hamper the operation and reliability of the spacecraft. In order to dissipate this excess heat, the ISS employs special radiators, which shall be described later, while discussing the EPS for the ISS in detail.

7.2 International Space Station

As was mentioned earlier, probably the only feasible power source to be used in outer space by the ISS is solar rays from the sun. However, since the sun's rays are not always incident on the solar panels of the ISS, there becomes a necessity to provide secondary/back-up power. This power, initially, was to be met by Ni-Cd (nickel-cadmium) batteries; however, due to a short life span, these were not found to be satisfactory. The next best choice was found to be rechargeable Ni-H_2 batteries, which are much heavier but provide longer operating time. A pictorial view of the ISS, with its distinct solar array structure, is shown in Figure 7.1.

Basically, the power management and distribution (PMAD) system of the ISS distributes power at a voltage of 160V DC [1], [2]. This is essentially done by a series of power electronic switches. Furthermore, in order to meet the operational conditions, DC/DC power electronic Buck converters units step down the voltage from 160V to 120V DC. This 120V system is called the secondary power system, which basically serves the ISS electric loads. The

Space Power Systems 243

DC/DC converter units (DDCUs) also satisfy the isolation demands of the secondary system from the primary system. Also, at the same time, they maintain a good level of power quality throughout the station [1]. A schematic representation of the ISS EPS is shown in Figure 7.2.

Figure 7.1 Pictorial view of the International Space Station (photo courtesy of the National Aeronautics and Space Administration [NASA] Glenn Research Center, Cleveland, Ohio, U.S.A).

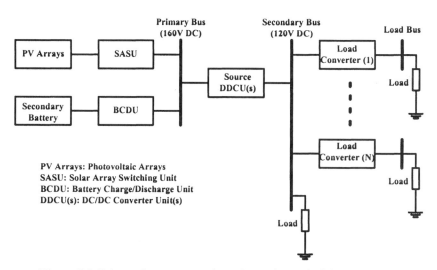

Figure 7.2 Schematic representation of one channel of the ISS EPS.

Basically, there exist two main sections of the ISS EPS, namely, the Russian segment and the US segment, which will be explained in detail later.

The Russian segment basically provides power to the core loads and payloads on the Russian segment of the ISS. The entire power system, including both the U.S. and Russian segments, produces about 110kW of power, which is basically sufficient to serve approximately 50 average sized homes. Apart from that, an additional 50kW of power is reserved for R&D purposes [1].

The ISS EPS basically consists of numerous power channels, but the schematic representation in Figure 7.2 shows only one channel for the sake of simplicity [1], [2]. In fact, the EPS uses eight PV arrays to convert sunlight to electricity. Due to the fact that the space station requires extremely high power levels, the solar arrays consist of nearly 300,000 silicon solar cells. Also, as mentioned earlier, in order to avoid any interruptions in power supply, NASA have developed rechargeable $Ni-H_2$ batteries, which are connected in parallel with the solar arrays, as is depicted in Figure 7.2. Both of these systems (viz., the solar array power system and the secondary battery system) are connected to the primary bus, which, in turn, delivers power at a voltage of 160V DC. As mentioned earlier, the secondary batteries get recharged during the "sunlit" part of the orbit and they are put in use during the "eclipse" part of the orbit. Basically, about 40 such battery cells are stacked in series, with their temperature and pressure monitored at all times. They are placed inside an enclosure called the orbital replacement unit (ORU), which allows easy removal and replacement of batteries in orbit [1]. The $Ni-H_2$ batteries are generally expected to last for about 4-6 years' duration.

It is interesting to note at this point that, when the solar arrays produce power, the station structure tends to attain a voltage about the same as the array voltage [1]. Hence, under such conditions, there exists a potential problem due to the fact that the space station could be exposed to arcing from the surface to the nearby surroundings. In light of this problem, engineers at NASA have designed a special hollow cathode assembly, which is the most critical component of a device called the "plasma contactor", which essentially grounds the entire structure [1]. Connecting a small supply of gas into ions and electrons and discharging this stream into space achieves the process of grounding. Therefore, this stream carries along with it the excess electrons that create the surface charge in the first place.

Another problem that may occur during the design of the EPS is that storing and distributing electricity with batteries builds up excess heat that can damage critical equipments. To eliminate this problem, engineers at NASA have opted to use liquid ammonia radiators, which dissipate the heat away from the spacecraft [1].

Having studied the working and design of the ISS EPS in brief thus far, it becomes imperative to study each section of the power system in detail. Primarily, space based power systems are excellent examples of multi-converter power electronic systems [2]. Referring back to Figure 7.2, it can be seen that the power at the primary source bus is produced at 160V DC as pointed out earlier. The primary bus voltage either feeds power to the $Ni-H_2$ batteries or

Space Power Systems

extracts power from it via a battery charge/discharge unit (BCDU). The BCDU is essentially a bi-directional DC/DC power electronic converter. Also, a DC/DC converter unit (DDCU) provides power at a stepped down voltage of 120V DC to the secondary bus [2]. In addition, all the smaller loads are also fed from low-power DC/DC converters, which reduce the voltage level even further in accordance with the load-bus requirements. Hence, as is fairly clear from the description of the ISS EPS, the entire system is DC based and is power electronics intensive.

The primary aim now is to take a look at the various functions of the ISS EPS. The EPS functions mainly consist of a primary power system, a secondary power system, and support systems [3]. Each one of these functions is described in the sub-sections below.

7.2.1 Primary Power System

The basic component of the primary power system of the EPS is known as the "power channel", which is basically a collection of hardware components comprising a solar array, which acts as a primary power source [3]. Apart from the solar arrays, the power channel also includes sequential shunt units (SSU), beta gimbal assembly (BGA), electronics control units (ECU), BCDUs, etc. A basic power flow diagram of the PV module is shown in Figure 7.3.

As is clear from Figure 7.3, the EPS comprises of a primary power system and a secondary power system. In general, the primary power system consists of the following 3 functions:

(a) Primary power generation
(b) Primary power storage
(c) Primary power distribution

As is obvious by now, primary power generation involves conversion of solar energy to electrical energy as well as suitable regulation of electrical energy. As is shown in Figure 7.3, the primary job of the SSU is to receive power from the PV array and maintain the voltage of the primary bus at 160V DC. But, at the same time, the equipments that use this primary power are designed for even higher operating voltages, in the range of 130-173V DC [3]. Also, as seen in Figure 7.3, the beta and alpha gimbals are used to rotate the solar arrays in the direction of the sun in order to provide maximum power to the space station.

The primary power storage function is achieved by using rechargeable $Ni-H_2$ batteries as was explained earlier. Each of the batteries has their own BCDU in order to control their respective state of charge (SOC).

Finally, as depicted in Figure 7.3, during the sunlit/isolation period of the orbit, the DC switching unit (DCSU) directs power from the PV arrays to the ISS via the primary bus. Furthermore, the DCSU also transfers power from the PV array to the $Ni-H_2$ batteries for charging purposes.

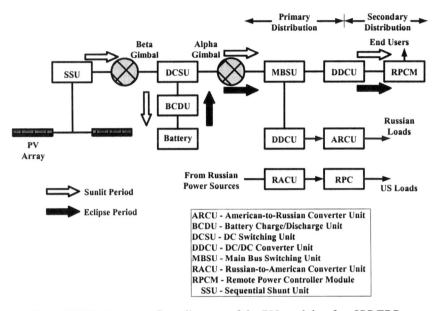

Figure 7.3 Basic power flow diagram of the PV module of an ISS EPS.

In the primary power generation section of Figure 7.3, the solar cells are basically made up of silicon and have nearly 15% conversion efficiency from sunlight to electricity [4]. The I-V curve of the PV array and a superimposed representative load line are depicted in Figure 7.4.

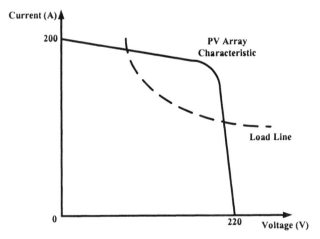

Figure 7.4 Typical PV array curve with representative load line.

As is clear from Figure 7.4, the dotted load line intersects the PV array characteristic at 2 distinct points, which are basically the operating points. Referring back to Figure 7.3, the SSU uses solid state switches operating at high frequencies on the order of 20 kHz to dynamically shunt power from individual array strings in order to follow load demand and, at the same time, maintain the response of the array at a low value operating point [4]. Also, from Figure 7.3, the 6.25 kW DDCUs provide about 20dB isolation between primary and secondary systems. In addition, they step the voltage down to 120V, as was highlighted earlier.

7.2.2 Secondary Power System

In a broad sense, the secondary power system consists of 2 main functions:

(a) Secondary power conversion
(b) Secondary power distribution

The secondary power conversion system uses one type of orbital replacement Unit (ORU), which is essentially the DDCU, which, in turn, converts primary power to secondary power. The output of the DDCUs provides 120V-regulated power with approximately 150% current-limiting capability for co-coordinated fault protection [4]. Again, as mentioned previously, each DDCU feeds the secondary power system, consisting of remote power controllers (RPCs) and cables, in order to supply power to major aerospace loads and in-flight experimental set-ups.

It is also important to note that the DDCUs and RPCs are used to feed all loads of the US and other international partners, except for the Russian segment, which uses power electronic converters of different ratings for their 28V networks. The RPCs basically form a critical part of the remote power control module (RPCM), which, in turn, forms the base of the secondary power distribution system [4]. Secondary power primarily originates from the DDCU (DC/DC Buck converters) and is then distributed through a network of ORUs called secondary power distribution assemblies (SPDAs) and remote power distribution assemblies (RPDAs).

7.2.3 Support Systems

The last of the EPS functions includes support systems, which possess the following sub-functions:

(a) Thermal control
(b) Grounding
(c) Command and control

Thermal control and grounding issues have already been highlighted earlier in this Chapter. An additional point to be noted about grounding is that the grounding function is incorporated in a single point ground (SPG) topology, which basically maintains all equipment on the ISS at a common potential.

In order to monitor and control the operation of the EPS, there exists a command and control function, which primarily governs over all the above-mentioned functions. These functions are realized either by software applications or hardware, either on-board the station or from the ground. Control and monitoring done from the ground fulfills the requirements of EPS operations, analysis, and planning.

Having discussed the primary functions of the ISS EPS, it is necessary to point out where exactly there exists a scope for future development work. Such R&D work may offer significant improvement in the operation of the ISS later on in its existence. One such alternate could be to consider the usage of gallium arsenide (GaAs) solar arrays. This is because an important difference between silicon and GaAs arrays is the fact that the GaAs array makes optimum use of its concentrators to focus sunlight on the PV cells, thus leading to a considerable increase in the array output efficiency [4]. Apart from this, another area of future development work is opting for alternate energy storage devices.

If the $Ni-H_2$ batteries that are now being utilized are improved significantly, they could improve their energy density characteristics, which would ultimately lead to significant increase in average available power on the ISS. Also, in this regard, fibrous nickel-cadmium (Ni-Cd) (FNC) batteries, flywheels, and fuel cell technologies present reasonable alternatives to provide the necessary secondary power [4]. FNCs operate over a wider temperature range than $Ni-H_2$ batteries and are also less expensive.

Flywheels show large storage capacities, recovering efficiencies, and operational life compared to batteries. Furthermore, another up-and-coming technology, with regard to secondary power source provision, is the fuel cell. Fuel cells, as explained in Chapter 9, possess enormously high energy densities (about 4000 Wh/kg) compared to that of batteries (about 100 Wh/kg).

7.2.4 Power Exchange Between the U.S. and Russian Segments

As mentioned earlier, the ISS EPS consists of a hybrid system comprising 2 major segments, namely, a 120V U.S.-built segment and a 28V as well as 120V Russian-built segment [5]. The 2 systems are independent of each other, but they are interconnected via DC/DC converters. This arrangement allows appropriate power transfer depending on system demands.

A major portion of the above discussion involves the US segment of the ISS EPS. In order to present a brief idea about the Russian segment, this section will, apart from the power exchange issues, also describe the Russian segment EPS overview.

The Russian power system can be broadly divided into 2 basic categories: namely, a 28V network and a 120V network. There exist two 28.5V systems, each rated at 3.5 kW and 4.4 kW, respectively [5]. As for the 120V system, it is

Space Power Systems 249

still in its design stages and, hence, it will not be discussed in detail here. The fundamental Russian system components include:

(a) Solar arrays
(b) Solar array regulators
(c) American-to-Russian converter units (ARCUs)
(d) Ni-Cd batteries
(e) Battery charge/discharge units (BCDUs)
(f) Main bus-bar and bus filters

At many times during the operation, there exists a need to transfer some amount of power from the American segment to the Russian segment and vice versa. Referring back to Figure 7.3, it can be seen that power is generally transferred via DDCUs known as Russian-to-American converter units (RACUs) or American-to-Russian converter units (ARCUs). Typically, the RACU converts $28.5V \pm 0.5V$, from the Russian power system to an output of $123V \pm 2.0V$ for usage by the U.S. segment [5]. The ARCUs, installed in the U.S. segment of the EPS can, likewise, aid in power transfer from the U.S. segment to the Russian segment. The ARCUs are rated at 1.3 kW and 1.5 kW, respectively. The 1.3 kW ARCUs operate at voltages between 110V and 130V DC, whereas the 1.5 kW ARCUs are capable of operating at voltages ranging between 110V and 180V DC [5].

7.2.5 Future Flywheel Technology Development

Flywheel technology is being encouraged as a replacement for the existing Ni-H$_2$ batteries used as secondary energy storage devices. This section takes a look how this technology would actually operate if it indeed substituted the rechargeable batteries as a secondary energy storage device.

It is a well-known fact that rechargeable Ni-H$_2$ batteries possess a design life of approximately 6-7 years, whereas the entire life of the ISS itself spans about 15 years [5]. Therefore, this will require periodic replacement of the battery packs, which means that repeated costs need to be incurred. NASA basically intends to successfully demonstrate the usage of flywheels to store energy aboard the ISS. Thus, significant savings in terms of launch-mass and cost can be experienced by eliminating the need to replace the batteries from time to time.

The flywheel energy storage system (FESS), wherein one battery and its accessories are replaced by a flywheel system, is shown in Figure 7.5. In addition, major requirements of power, in accordance with battery and flywheel characteristics, are summarized in Figures 7.6 and 7.7. Each flywheel energy storage unit (FESU) consists of 2 flywheel modules and associated control electronics [5]. Each flywheel module primarily consists of:

(a) Composite rotor
(b) Motor/generator

250   Chapter 7

(c)   Magnetic actuators
(d)   Bearing systems
(e)   Rotor caging systems
(f)   Sensors
(g)   Housing

During the "sunlit" part of the orbit, the counter-rotating flywheels are spun up to their maximum speeds by drawing energy from the PV system. These speeds could range between 60,000 and 70,000 rpm. This is known as the charging cycle of the flywheels [5], [6]. During the "eclipse" part of the orbit, the flywheels drive the generator, thereby decelerating the flywheels from their maximum speed down to about 20,000 rpm in order to feed power as output to the primary bus. This cycle is known as the discharge cycle of the flywheels [5], [6].

Having discussed the basic modes of operation of the FESS, it is necessary to point out that the replacement of the BCDU and battery module with an FESU, as discussed here, poses some degree of uncertainty about the system stability and interaction. Hence, it becomes critical that major issues such as fault protection, current sharing, power quality, and overall system stability are exhaustively investigated before any replacement procedure is indeed implemented [6].

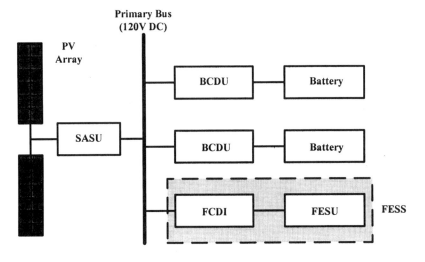

FCDI - Flywheel Control and Data Interface
FESU - Flywheel Energy Storage Unit
FESS - Flywheel Energy Storage System

Figure 7.5 Hybridized representation of a flywheel and battery energy storage system.

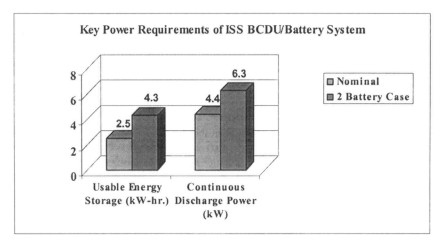

Figure 7.6 Key power requirements of the ISS BCDU/Battery system.

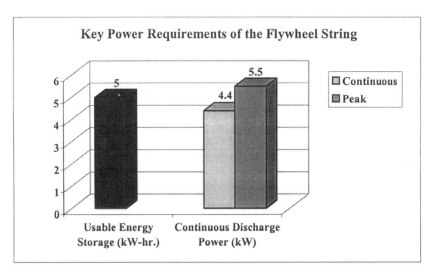

Figure 7.7 Key power requirements of the flywheel energy storage system (FESS).

7.3 Spacecraft Power Systems

Space vehicles generally travel millions of miles away from the earth, more often than not encountering harsh environments on the way. Power for these vehicles must, hence, be extremely efficient and reliable. As a rough estimate, the spacecraft loads require power in the range of about 0.5-2.5 kW.

Basically, PV arrays and radioisotopes are considered the best options to power spacecraft loads.

A spacecraft's electrical power system primarily consists of a primary generating system (PV array or radioisotope) and a power management and distribution system. Furthermore, the EPS may also comprise a secondary energy storage system (batteries) for feeding other spacecraft loads. The remainder of this section will look at the various power producing techniques aboard a typical spacecraft that is orbiting the earth or traveling to distant planets.

The main focus here is to discuss the conceptual design and performance analysis of a typical spacecraft electrical power system (EPS). Traditionally, spacecraft based power systems are solar electric propulsion (SEP) intensive [7]. The spacecraft EPS usually uses a channelized 500V DC power management and distribution (PMAD) architecture. A typical block diagram representation of a spacecraft power system is shown in Figure 7.8. Each channel consists of a PV array, a solar pointing gimbal, and an array regulation unit (ARU), which, in turn, feeds a control power distribution unit (PDU) [7]. This is essentially done through power distribution cables.

As is clear from Figure 7.8, the PDU distributes power from the primary PV arrays to the power processing units (PPUs) of the thruster. The PDU basically contains payload power supplies, BCDUs for lithium-ion batteries, controllers, etc.

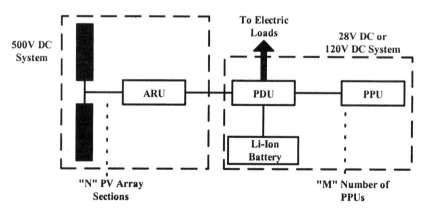

Figure 7.8 Block diagram representation of a typical spacecraft EPS.

Only a small amount of energy is stored in the Li-ion batteries (about 15 kWh) since the thrusters are idle during the "eclipse" period. This small amount of energy is enough to feed payload and SEP stage loads.

A specific voltage level of 500V DC was selected for the PMAD operation since it greatly reduces the PPU size, complexity, and power losses

[7]. Other obvious advantages include lower current operation and, hence, smaller conductor sizes as well as consequent cost savings. At the same time, it must be kept in mind that the 500V DC voltage level is still low enough to permit the use of standard aerospace power cables. In general, the SEP architecture proves to be an attractive solution for interplanetary missions in outer space.

7.3.1 Alternate Power Sources for Spacecraft Power Systems

It must be well understood by now that all earth-orbiting satellites rely on batteries to deliver electric power during the "eclipse" period. Such eclipses occur on numerous occasions in a day and, hence, each eclipse period demands a recharge of the batteries [8]. In fact, the major criteria in selecting batteries for aerospace power systems are charge/discharge cycles and depth of discharge. Initially, Ni-Cd batteries were preferred for the purpose of providing spacecraft power. However, it was soon discovered that these only possessed a maximum life span of 8 years. This fact holds true provided at least 60% of the battery's capacity remains after each discharge [8]. Also, Li-ion batteries could be good choice in the near future since they can deliver up to 200 Wh/kg. But again, their demonstrated lifetime in a satellite environment is only a few thousand charge/discharge cycles.

Hence, more recently, much heavier rechargeable $Ni-H_2$ batteries have gained popularity for spacecraft power systems. Typically, $Ni-H_2$ batteries have a proven life of about 20,000-30,000 charge/discharge cycles, which ultimately results in a low life cycle cost and, at the same time, fulfills the 10 year life expectancy of the satellite. Another attractive alternate power system for manned space missions since the early 1960s has been the "fuel cell". For long space missions, it has been identified that energy storage and conversion based on hydrogen technology is of critical importance. The alkaline fuel cell (AFC), with an immobile electrolyte, is considered one of the best options for powering space shuttles [9]. In recent years, the proton exchange membrane fuel cell (PEMFC) has gained favor over the AFC for space shuttle applications. For example, NASA is considering upgrading its fuel cell program by replacing the existing AFC units with PEMFC units [9].

In the further few topics of this Section, some examples of spacecraft power systems designed for real space missions shall be explained. These sets of descriptions will, hence, provide a wider knowledge of spacecraft power systems and their finer details.

7.3.2 Earth Observing System (EOS) TERRA Spacecraft

The Earth Observing System (EOS) TERRA spacecraft represents the first high voltage spacecraft EPS implemented by NASA. The average power requirement of the spacecraft is about 2.5 kW, with a minimum lifetime of 5 years [10]. To meet this power requirement, the spacecraft primarily makes use of gallium arsenide (GaAs) on germanium solar cells. The power unit has about 30,000+ such PV cells, which are capable of delivering about 5 kW of power. In order to provide power during the "eclipse" part of the orbit, the TERRA spacecraft uses two 54-cell, series-connected Ni-H$_2$ batteries [10]. Furthermore, the required control is achieved from a PDU, which is specifically designed to provide 120V DC (± 4%) under any load conditions. This regulated voltage, in turn, is achieved via the sequential shunt unit (SSU) and the 2 BCDUs. A pictorial view of the EOS TERRA spacecraft is shown in Figure 7.9.

Figure 7.9 Pictorial view of the EOS TERRA spacecraft (photo courtesy of the National Aeronautics and Space Administration [NASA] EOS TERRA Spacecraft, http://terra.nasa.gov/).

7.3.3 Electrical Power Systems for Space Based Radar (SBR) Satellites

Typically, a space based radar (SBR) satellite is utilized to improve surveillance capabilities in the 21st century. Some of the major applications of SBR satellites include control of air traffic and supervision of sovereign territory [11]. This section covers some of the critical design issues with regard to the EPS as well as problems associated with generating and distributing power on the SBR spacecraft.

Fundamentally, the SBR power system differs from existing power systems in a number of ways. Firstly, the radar payload requires tens of kilowatts of DC input power. Secondly, it must be kept in mind that the secondary loads, namely, transmitter/receiver modules (T/R modules), are distributed over a huge area. Lastly, the radar also demands a critical transient response from the DC power supply [11].

It is absolutely essential to also discuss here the various payloads that a spacecraft may carry along on a flight. As mentioned earlier, the dominant loads are basically the T/R modules. These primary payloads, as they are called, consume a huge share of the spacecraft's total power as is depicted in Figure 7.10.

In a recent study conducted on SBR satellite payloads, it was concluded that the secondary payloads, although substantial, actually do not pose any direct influence on the EPS design [11]. In addition, further studies led to the conclusion that, whenever in "ON" state, the radar requires a DC input of about 20-25 kW. In order to summarize the various types of primary and secondary payloads, refer to Figure 7.11, which gives a brief overview of the same.

A basic model of the SBR power system is represented in Figure 7.12. In this regard, there exist a few design issues and tradeoffs, which shall be discussed here. One of the major design factors is based on the radar system's load characteristics. Once the overall power and energy levels are decided for the design purposes, a number of other factors need to be considered. These include voltage types and levels within the various areas of design, power distribution architecture, location of various elements and modules, control system architecture, and component technologies [11].

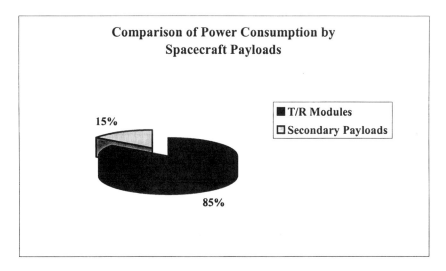

Figure 7.10 Comparison of power consumption by spacecraft payloads.

As for future developments of the SBR spacecraft power system, the usage of foldable PV arrays is widely being considered. Moreover, with regards to the power distribution system, a major concern is the power conversion and conditioning system. It is yet not clear whether localized or centralized power conditioning units should be used [11]. In the central approach, power converters in the range of 5-10 kW and 85-90% efficiency would be required. On the other hand, the local conditioners demand lower power of about 1-5 kW but need to operate at higher efficiencies, in the range of 90-95%. Hence, such design requirements pose stiff challenges that ultimately need to be overcome in order to make the design of the SBR power system a successful one. In this regard, further investigations of critical technologies are being planned so they can be implemented in the near future.

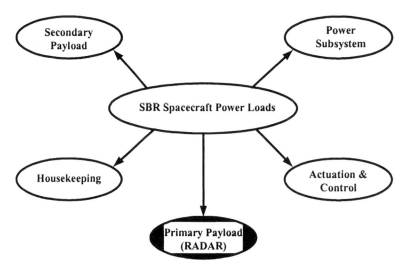

Figure 7.11 Overview of SBR spacecraft power loads.

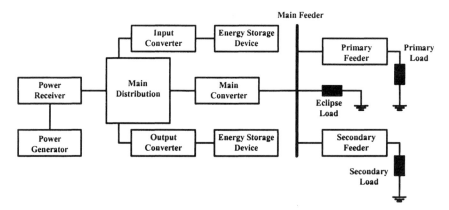

Figure 7.12 Basic model of an SBR power system.

Space Power Systems 257

Electrical power is the most critical resource for space vehicular power systems because it allows the crew to live comfortably, to safely operate the station, and to perform useful scientific experiments. As was explained, PV arrays provide the obvious source of power to the ISS. However, the sun's rays are not always available and, hence, during the "eclipse" period, the ISS relies on banks of nickel-hydrogen ($Ni-H_2$) rechargeable batteries to provide continuous power. This basic operation of the ISS EPS was exhaustively covered in previous section and the finer details of design and control were also studied.

Furthermore, the future scope for R&D to be carried out on the ISS EPS was discussed. Various alternate power sources along with their operational characteristics and trade-offs were studied. As people from around the world continue their exploration and development of space aboard the ISS, they will rely upon this well-designed power system to harness the sun's energy. Quite simply, the ISS constitutes the biggest power system in outer space and presents engineers in the field of power systems and power electronics with huge windows of opportunity.

7.4 Modeling and Analysis

This section describes a novel modular approach for the modeling and simulation of multi-converter power electronic space systems. These systems may consist of many individual converters connected together to form large, complex systems. Due to this heavy interconnection of converters, a large variety of dynamic interactions is possible. Conventionally, averaging techniques are used for the modeling of these systems [12]-[23]. However, rapid and large-signal dynamics cannot be followed by the averaging methods. Therefore, we use a generalized method in which we consider the average of the state variables as well as the harmonics. The main advantage of this large-signal approach is that there is no restriction on the size of the signal variations. Use of this method, in addition to simplifying the analysis procedure, allows increasing the time step for analysis of the system and, therefore, reduces the required computation time and computer memory considerably. Furthermore, these models are flexible so that future modifications are possible; also, they can be used in different systems.

7.4.1 Modeling and Simulation Considerations

Of primary concern in a multi-converter power electronic system is the way that the system responds to dynamics caused by interactions between power electronic converters, and also to both changing power demands and various disturbances. A changing power demand introduces a wide spectrum of dynamic changes into the system; each of them occurs on a different time scale. In this context, the fastest dynamics are due to sudden changes in demand and are associated with the transfer of energy between the power sources and the loads.

Slightly slower dynamics are the voltage control actions needed to maintain system operating conditions, until finally the very slow dynamics corresponding to the way in which the generation is adjusted to meet the slow demand variations take effect. Similarly, the way the system responds to disturbances also covers a wide spectrum of dynamics and associated time frames.

Testing of space power systems consisting of many different converters is extremely expensive due to the size and complexity of these systems. Therefore, it is very important to develop reliable large-signal models to analyze these systems. Modeling and simulation of these systems before and during the design has many advantages: allowing the evaluation of converters and interactions among them, reducing costly redesigns, supporting trade studies and parametric studies, and supporting test case definitions and explanation of test anomalies.

If the system is defined as a large-signal model, as is required for the system level studies, then clearly linearized averaged models are invalid. Time domain simulation of non-linear, time-varying system modules, which includes the protection circuitry, system control dynamics, and limitations, must be used for accuracy of overall system performance simulation. Failure to accurately include the non-linear protection circuitry and actual feedback control dynamics and limits, i.e., dynamic range, of each individual source and load converters, could render the large-signal study results. However, transient simulations using switching models of power converters require vast computer resources and long simulation times, and often cause simulation convergence errors. The averaged model runs much faster than the comparable switching model and does not require excessive computer resources [12]-[16]. The main limitation is the small variations condition for state variables [13], [14]. Therefore, the possibility of using averaging methods for multi-converter power electronic systems is nil, since these systems have state variables that generally show large-signal variations and oscillating behavior.

By using the generalized state space averaging method, which employs the Fourier series with time-dependent coefficients, various kinds of converters having different waveforms of state variables can be large-signally modeled. Indeed, simplifying approximations are generally used in this method, during which terms of little importance are omitted from the mentioned series [24]-[27]. Application of this more general method with maintaining the DC and first harmonic coefficients to DC/DC symmetrical resonant converters, in which a domain harmonic exists, gives good results [24]. Also, the results in other types of single-ended DC/DC converters are good [25], [26]. In this section, this more general method and its application in multi-converter power electronic systems are described. The main advantages of this modular approach are: (1) increasing the time step for analysis of the system and therefore reducing the required computation time and computer memory, (2) simplifying the analysis procedure, (3) simulation of rapid and large-signal dynamics, (4) simulation of internal state variables of each converter in the system, (5) being flexible such that future modifications are possible, (6) reducing the complexity of modeling large systems by modeling a less complex subsystem, (7) the models can be used in different systems, i.e., in a package for simulation of multi-converter power systems with different topologies.

7.4.2 Generalized State Space Averaging Method

The generalized state space averaging method is based on the fact that the waveform $x(t)$ can be approximated with arbitrary precision in the *(t-T, t]* range by the Fourier series:

$$x(t) = \sum_{k=-n}^{n} <x>_k (t) e^{jk\omega t} \tag{7.1}$$

where

$$\omega = \frac{2\pi}{T}$$

$$<x>_k (t) = \frac{1}{T} \int_{t-T}^{t} x(\tau) e^{-jk\omega\tau} d\tau \tag{7.2}$$

In relation (7.1), the value of n depends on the required degree of accuracy, and if n approaches infinity, the approximation error approaches zero. If we only consider the term $K=0$, we have the same state space averaging method [12]. If a state variable does not have an oscillating form and is almost constant, we only use the term *(K=0)*. Also, if a state variable only has an oscillating form similar to a sine wave, we use the terms $K=-1,1$. This method is named first harmonic approximation. In addition, if a state variable has a DC coordinate and also has an oscillating form, we use the terms $K=-1,0,1$. However, the more terms we consider, more accuracy we have.

Selection of T for modeling of each converter is very important and it should be considered carefully. For instance, it is the switching period in DC/DC converters and the main wave period of the output voltage in DC/AC inverters. The $<x>_k (t)$ is the complex Fourier coefficient. These Fourier coefficients are functions of time since the interval under consideration slides as a function of time. The analysis computes the time-evolution of these Fourier coefficients as the window of length T slides over the actual waveform. Our approach is to determine an appropriate state space model in which the coefficients (7.2) are the state variables. Certain properties of the Fourier coefficients (7.2) are key for the analysis and are detailed below.

(a) Differentiation with respect to time: The time derivative of the kth coefficient is computed to be

$$\frac{d}{dt}<x>_k (t) = -jk\omega <x>_k (t) + <\frac{d}{dt}x>_k (t) \qquad (7.3)$$

This formula is very important to form an averaged model involving the Fourier coefficients. The case where ω is time varying will also need to be considered for the analysis of the systems where the drive frequency is not constant. In this case, formula (7.3) is only an approximation, but, for slowly varying $\omega(t)$, it is a good approximation [24].

(b) Transforms of functions of variables:

$$<f>_k = <f(x_1, x_2, ..., x_n)>_k \qquad (7.4)$$

In most cases, it is impossible to obtain an explicit form for (7.4) in terms of a finite number of the coefficients $<x>_i$. A procedure for exactly computing (7.4) is available in the case where f is polynomial. This procedure is based on the following convolutional relationship.

$$<x.y>_k = \sum_i <x>_i <y>_{k-i} \qquad (7.5)$$

where the sum is taken over all integers i. In many cases, we shall rely on many of the terms in (7.5) which are negligibly small. The quantity in (7.4) can be computed in the case where f is a polynomial by considering each homogeneous term separately. The constant and linear terms are trivial to transform. The transforms of the quadratic terms are computed using (7.5). Homogeneous terms of higher order are dealt with factoring each such term into the products of two lower order terms. Then the procedure can be applied to each of the factors.

7.4.3 Modeling and Analysis of Multi-Converter Power Electronic Systems

In this section, through the application of the generalized state space averaging method, we present an approach for the modeling of multi-converter power electronic systems. As an example, we model the representative system of Figure 12.1, which is presented in Chapter 12. We also compare the resulted responses with the system time precise solution in the presence of different dynamics in the system. A modular approach is used in modeling and analyzing of multi-converter power electronic systems. The converters and subsystems of the system are modularized and subsequently interconnected to form the complete system. Modularizing the system into converters and subsystems has several advantages: (1) converters and subsystems models can be used in different systems, (2) it reduces the complexity of modeling large systems by modeling a less complex subsystem, (3) the proposed models can be verified with manageable test conditions.

Figure 7.13 shows the interconnecting converters in a multi-converter power electronic system. Each converter receives its voltage source from the preceding converter along with other converters in the stage, and supplies outputs to the following load converters no. *1, ..., N*, in addition to resistive loads, all of which is represented by resistance *R*.

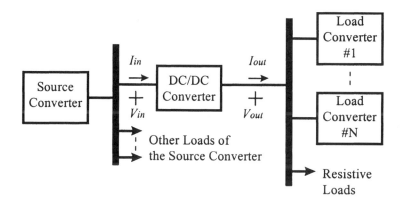

Figure 7.13 Interconnecting converters in the multi-converter power electronic systems.

7.4.3.1 DC/DC Converters in the Multi-Converter Power Electronic Systems

In this section, we suppose that the converter of Figure 7.13 is a PWM DC/DC Buck, Boost, Buck-Boost, or Cuk converter operating in continuous conduction mode, with switching period T and duty cycle d. To apply the generalized state space averaging method, first a commutation function $u(t)$ is defined as

$$u(t) = \begin{cases} 1, 0 < t < dT \\ 0, dT < t < T \end{cases} \quad (7.6)$$

This commutation function depends on the circuit switching control, which determines when the circuit topology changes according to time.

7.4.3.1.1 Buck Converter

Figure 7.14 shows a DC/DC PWM Buck converter. The unified set of circuit state variable equations is obtained by applying (7.6) to the two sets of topological circuit state space equations.

$$\begin{cases} \dfrac{di_L}{dt} = \dfrac{1}{L}[v_{in}u(t) - v_o] \\ \dfrac{dv_o}{dt} = \dfrac{1}{C}[i_L - i_{out}] \\ i_{in} = i_L u(t) \end{cases} \qquad (7.7)$$

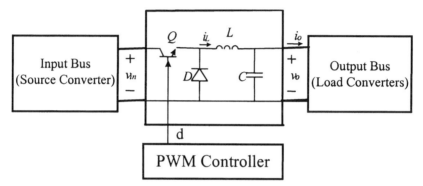

Figure 7.14 DC/DC PWM Buck converter.

Nevertheless, in the set of equations of the generalized state space averaged model, the actual state space variables are the Fourier coefficients of the circuit state variables, which are, in this case, i_L and v_o. Using the first-order approximation to obtain i_L and v_o, we have six real state variables, as follows:

$$\begin{aligned} &<i_L>_1 = x_1 + jx_2, <i_L>_0 = x_5 \\ &<v_o>_1 = x_3 + jx_4, <v_o>_0 = x_6 \end{aligned} \qquad (7.8)$$

Since i_L and v_o are real,

$$<i_L>_{-1} = <i_L>_1^*, <v_o>_{-1} = <v_o>_1^* \qquad (7.9)$$

where the operator * means the conjugate of a complex number. By applying the time derivative property of the Fourier coefficients in (7.7), and further substituting the Fourier coefficients of the communication function $u(t)$,

$$<u(t)>_0 = d, <u(t)>_1 = \dfrac{j}{2\pi}(e^{-j2\pi d} - 1) \qquad (7.10)$$

one comes to

$$\begin{bmatrix} \dot{x}_1 \\ \dot{x}_2 \\ \dot{x}_3 \\ \dot{x}_4 \\ \dot{x}_5 \\ \dot{x}_6 \end{bmatrix} = \begin{bmatrix} 0 & \omega & -\frac{1}{L} & 0 & 0 & 0 \\ -\omega & 0 & 0 & -\frac{1}{L} & 0 & 0 \\ \frac{1}{C} & 0 & -\frac{1}{RC} & \omega & 0 & 0 \\ 0 & \frac{1}{C} & -\omega & -\frac{1}{RC} & 0 & 0 \\ 0 & 0 & 0 & 0 & 0 & -\frac{1}{L} \\ 0 & 0 & 0 & 0 & \frac{1}{C} & -\frac{1}{RC} \end{bmatrix} * \begin{bmatrix} x_1 \\ x_2 \\ x_3 \\ x_4 \\ x_5 \\ x_6 \end{bmatrix}$$

$$+ \begin{bmatrix} \frac{d}{L} & 0 & \frac{1}{2\pi L}Sin 2\pi d \\ 0 & \frac{d}{L} & -\frac{1}{2\pi L}(1-Cos 2\pi d) \\ 0 & 0 & 0 \\ 0 & 0 & 0 \\ \frac{1}{\pi L}Sin 2\pi d & -\frac{1}{\pi L}(1-Cos 2\pi d) & \frac{d}{L} \\ 0 & 0 & 0 \end{bmatrix} * \begin{bmatrix} Re\{<v_{in}>_1\} \\ Im\{<v_{in}>_1\} \\ <v_{in}>_0 \end{bmatrix}$$

$$+ \sum_{j=1}^{N} \begin{bmatrix} 0 & 0 & 0 \\ 0 & 0 & 0 \\ \frac{1}{C} & 0 & 0 \\ 0 & \frac{1}{C} & 0 \\ 0 & 0 & 0 \\ 0 & 0 & -\frac{1}{C} \end{bmatrix} * \begin{bmatrix} Re\{<i_{in_j}>_1\} \\ Im\{<i_{in_j}>_1\} \\ <i_{in_j}>_0 \end{bmatrix} \quad (7.11)$$

$$<i_{in}>_1 = \left[dx_1 + \frac{1}{2\pi}Sin(2\pi d)x_5 \right] + j\left[dx_2 - \frac{1}{2\pi}(1-Cos(2\pi d))x_5 \right] \quad (7.12)$$

$$<i_{in}>_0 = \frac{1}{\pi}Sin(2\pi d)x_1 - \frac{1}{\pi}(1-Cos(2\pi d))x_2 + dx_5 \quad (7.13)$$

where i_{in_j} is the input current of the load converter j and R is the resistive load of the converter. Equations (7.11)-(7.13) represent the generalized state space averaged model of the Buck converter. The circuit state variables are calculated and given by

$$\begin{aligned} i_L &= x_5 + 2x_1 Cos\omega t - 2x_2 Sin\omega t \\ v_o &= x_6 + 2x_3 Cos\omega t - 2x_4 Sin\omega t \end{aligned} \quad (7.14)$$

7.4.3.1.2 Boost Converter

Figure 7.15 shows a DC/DC PWM Boost converter. The unified set of circuit state variable equations is obtained by applying (7.6) to the two sets of topological circuit state space equations.

$$\begin{cases} \dfrac{di_L}{dt} = \dfrac{1}{L}\left[v_{in} - (1-u(t))v_o\right] \\ \dfrac{dv_o}{dt} = \dfrac{1}{C}\left[(1-u(t))i_L - i_{out}\right] \\ i_{in} = i_L \end{cases} \qquad (7.15)$$

By applying the generalized state space averaging method, as was done for the Buck converter, the state space equations of the converter can be written as

$$\begin{bmatrix} \dot{x}_1 \\ \dot{x}_2 \\ \dot{x}_3 \\ \dot{x}_4 \\ \dot{x}_5 \\ \dot{x}_6 \end{bmatrix} = \begin{bmatrix} 0 & \omega & -\dfrac{1-d}{L} & 0 & 0 & \dfrac{Sin2\pi d}{2\pi L} \\ -\omega & 0 & 0 & -\dfrac{1-d}{L} & 0 & -\dfrac{Sin^2\pi d}{\pi L} \\ \dfrac{1-d}{C} & 0 & 0 & -\dfrac{1}{RC} & \omega & -\dfrac{Sin2\pi d}{2\pi C} & 0 \\ 0 & \dfrac{1-d}{C} & -\omega & -\dfrac{1}{RC} & \dfrac{2Sin^2\pi d}{\pi C} & 0 \\ 0 & 0 & \dfrac{Sin2\pi d}{\pi L} & -\dfrac{2Sin^2\pi d}{\pi L} & 0 & -\dfrac{1-d}{L} \\ -\dfrac{Sin2\pi d}{\pi C} & \dfrac{2Sin^2\pi d}{\pi C} & 0 & 0 & \dfrac{1-d}{C} & -\dfrac{1}{RC} \end{bmatrix} * \begin{bmatrix} x_1 \\ x_2 \\ x_3 \\ x_4 \\ x_5 \\ x_6 \end{bmatrix}$$

$$+ \begin{bmatrix} \dfrac{1}{L} & 0 & 0 \\ 0 & \dfrac{1}{L} & 0 \\ 0 & 0 & 0 \\ 0 & 0 & 0 \\ 0 & 0 & \dfrac{1}{L} \\ 0 & 0 & 0 \end{bmatrix} * \begin{bmatrix} Re\{<v_{in}>_1\} \\ Im\{<v_{in}>_1\} \\ <v_{in}>_0 \end{bmatrix}$$

$$+ \sum_{j=1}^{N} \begin{bmatrix} 0 & 0 & 0 \\ 0 & 0 & 0 \\ \dfrac{1}{C} & 0 & 0 \\ 0 & \dfrac{1}{C} & 0 \\ 0 & 0 & 0 \\ 0 & 0 & -\dfrac{1}{C} \end{bmatrix} * \begin{bmatrix} Re\{<i_{in_j}>_1\} \\ Im\{<i_{in_j}>_1\} \\ <i_{in_j}>_0 \end{bmatrix} \qquad (7.16)$$

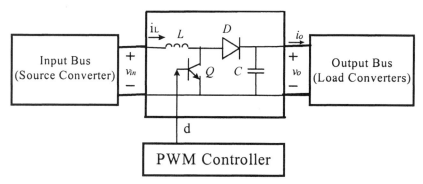

Figure 7.15 DC/DC PWM Boost converter.

7.4.3.1.3 Buck-Boost Converter

Figure 7.16 shows a DC/DC PWM Buck-Boost converter. The unified set of circuit state variable equations is obtained by applying (7.6) to the two sets of topological circuit state space equations.

$$\begin{cases} \dfrac{di_L}{dt} = \dfrac{1}{L}\left[v_{in}u(t) - (1-u(t))v_o\right] \\ \dfrac{dv_o}{dt} = \dfrac{1}{C}\left[(1-u(t))i_L - i_{out}\right] \\ i_{in} = i_L u(t) \end{cases} \qquad (7.17)$$

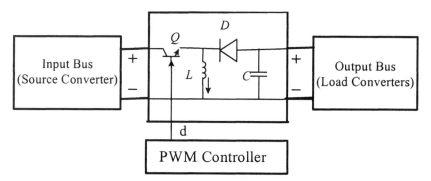

Figure 7.16 DC/DC PWM Buck-Boost converter.

By applying the generalized state space averaging method, as was done for the Buck converter, the state space equations can be written as

$$\begin{bmatrix} x_1 \\ x_2 \\ x_3 \\ x_4 \\ x_5 \\ x_6 \end{bmatrix} = \begin{bmatrix} 0 & \omega & \frac{1-d}{L} & 0 & 0 & -\frac{Sin2\pi d 0}{2\pi L} \\ -\omega & 0 & 0 & \frac{1-d}{L} & 0 & \frac{Sin^2\pi d}{\pi L} \\ -\frac{1-d}{C} & 0 & -\frac{1}{RC} & \omega & \frac{Sin2\pi d}{2\pi C} & 0 \\ 0 & -\frac{1-d}{C} & -\omega & -\frac{1}{RC} & -\frac{Sin^2\pi d}{\pi C} & 0 \\ 0 & 0 & \frac{Sin2\pi d 0}{\pi L} & \frac{2Sin^2\pi d}{\pi L} & 0 & \frac{1-d}{L} \\ \frac{Sin2\pi d}{\pi C} & -\frac{2Sin^2\pi d}{\pi C} & 0 & 0 & -\frac{1-d}{C} & -\frac{1}{RC} \end{bmatrix} * \begin{bmatrix} x_1 \\ x_2 \\ x_3 \\ x_4 \\ x_5 \\ x_6 \end{bmatrix}$$

$$+ \begin{bmatrix} \frac{d}{L} & 0 & \frac{1}{2\pi L}Sin2\pi d \\ 0 & \frac{d}{L} & -\frac{1}{2\pi L}(1-Cos2\pi d) \\ 0 & 0 & 0 \\ 0 & 0 & 0 \\ \frac{1}{\pi L}Sin2\pi d & -\frac{1}{\pi L}(1-Cos2\pi d) & \frac{d}{L} \\ 0 & 0 & 0 \end{bmatrix} * \begin{bmatrix} Re\{<v_{in}>_1\} \\ Im\{<v_{in}>_1\} \\ <v_{in}>_0 \end{bmatrix}$$

$$+ \sum_{j=1}^{N} \begin{bmatrix} 0 & 0 & 0 \\ 0 & 0 & 0 \\ \frac{1}{C} & 0 & 0 \\ 0 & \frac{1}{C} & 0 \\ 0 & 0 & 0 \\ 0 & 0 & -\frac{1}{C} \end{bmatrix} * \begin{bmatrix} Re\{<i_{in_j}>_1\} \\ Im\{<i_{in_j}>_1\} \\ <i_{in_j}>_0 \end{bmatrix} \quad (7.18)$$

7.4.3.2 DC/AC Inverters in the Multi-Converter Power Electronic Systems

Figure 7.17 shows a single-phase voltage source DC/AC inverter, which connects bus #3 to the AC bus in the representative system of Figure 12.1. We suppose that the inverter at the output side is connected to a series RLC load.

By the use of commutation function $u(t)$, which is determined by the inverter control, the state equations of the AC side of the inverter can be written as

$$\begin{cases} \frac{di_L}{dt} = \frac{1}{L}[v_{in}u(t) - Ri_L - v_c] \\ \frac{dv_c}{dt} = \frac{1}{C}[i_L] \end{cases} \quad (7.19)$$

For the modeling of the inverter, we use the first order approximation of the generalized state space averaging method. Therefore, we have four real state variables, as follows:

$$<i_L>_1 = x_1 + jx_2$$
$$<v_c>_1 = x_3 + jx_4 \qquad (7.20)$$

For the square wave inverter, the commutation function $u(t)$ is defined as

$$u(t) = \begin{cases} 1, & 0 < t < \dfrac{T}{2} \\ -1, & \dfrac{T}{2} < t < T \end{cases} \qquad (7.21)$$

The first harmonic of the commutation function is $<u>_1 = -j\dfrac{2}{\pi}$. After mathematical calculations, the final state space equations are obtained as

$$\dot{X} = \begin{bmatrix} -\dfrac{R}{L} & \omega & -\dfrac{1}{L} & 0 \\ -\omega & -\dfrac{R}{L} & 0 & -\dfrac{1}{L} \\ \dfrac{1}{C} & 0 & 0 & \omega \\ 0 & \dfrac{1}{C} & -\omega & 0 \end{bmatrix} X + \begin{bmatrix} 0 \\ -\dfrac{2}{\pi L}<v_{in}>_0 \\ 0 \\ 0 \end{bmatrix} \qquad (7.22)$$

where ω is the angular frequency of the square wave inverter. We assume that ω_0 is the natural frequency of the RLC load. The method of first harmonic approximation well describes the circuit behavior for frequencies $\omega \geq \omega_0$, but for frequencies $\omega < \omega_0$, because of the existence of high order harmonics, such as third harmonic for $\omega = \dfrac{1}{3}\omega_0$, the acquired error is prominent. The reason is that the RLC circuit bandwidth is not effective in filtering of $\omega = \dfrac{1}{3}\omega_0$.

7.4.3.3 Effects of Different Parameters on the Generalized State Space Averaging Method

As was mentioned earlier, one of the approximations of this method is neglecting high order harmonics, an option that might depend on the required degree of accuracy. If a K order approximation is used, higher order complex Fourier coefficients are neglected. On the other hand, rapid changes in the waveform of a state variable depend on the existence of high order Fourier coefficients. Thus, it is clear that a first or second order approximation is not able to completely follow rapid changes. As is shown in the waveforms, the inductors currents have sharp peaks, and, therefore, greater errors have occurred in following the current waveforms by the proposed method. The output voltage has slower changes, and, as a result, less error is observed in this regard.

From the simulation of individual converters based on their stand alone operation, it becomes evident that when the switching frequency is not much higher than the converter natural frequencies, the approximation order is an important factor in improving the model accuracy, though at the expense of increasing the calculations. In addition, it can be observed that the results of the generalized state space averaged model with first order approximation are better approximations than the corresponding ones of the topological model when the duty cycle is around 0.5 (absence of even harmonics) and for specific state variables. It can also be seen that the topology complexity, in terms of number of components, does not determine the approximation order for a satisfactory model.

As converter dynamic complexity is increased, the converter reaches the stable position slower. Non-zero terms in the state space matrix show the dynamic complexity of the converters. Another point worth mentioning is that in addition to the converter dynamics, the kind of state variable has also an effect in the speed of damping of the transient state. For example, the damping speed is much higher for the capacitor voltage than for the inductor current.

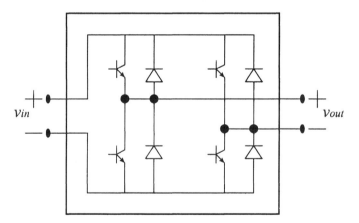

Figure 7.17 A single phase DC/AC inverter.

Furthermore, as the complexity of the considered converter increases, assuming the converter parameters to be the same, higher errors occur in the following of actual curves. For example, the observed error for the Buck converter is less than that of the Buck-Boost converter. In other words, converter dynamic has an effect on the accuracy of the model. In a manner that the nearer a Jordan form square matrix in its ideal state is to a diagonal form, its

Space Power Systems

consideration regarding the theory of control systems is easier. This is because it can be disintegrated into a series of lower order systems. More non-diagonal terms of a square matrix (i.e., terms outside the Jordan blocks) result in increased system complexity with regard to the theory of control systems. This is the same issue that was referred to as the dynamic complexity of the converters.

7.4.4 Analysis of the Representative System

The representative multi-converter power electronic system of Figure 12.1 is analyzed by the above-mentioned method. The acquired outcomes of this method, along with the results of the exact topological models in the presence of different dynamics in the system, are illuminated in Figure 7.18-7.22. If the second-order approximation is used, the corresponding model matrix has ten real state space variables for Buck, Boost, and Buck-Boost converters. Indeed, the use of higher order approximations improves the model accuracy, though at the expense of more complexity in the calculations.

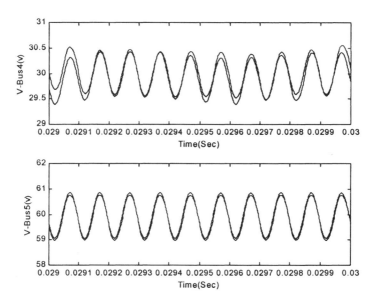

Figure 7.18 Voltage simulation of the steady state operation by the exact topological models (solid line) and generalized state space averaging method (dotted line).

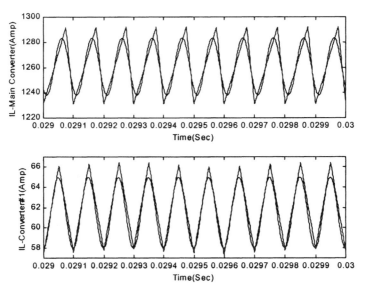

Figure 7.19 Current simulation of the steady state operation by the exact topological models (solid line) and generalized state space averaging method (dotted line).

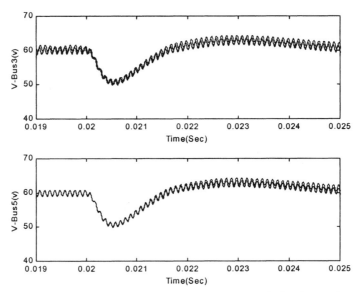

Figure 7.20 Simulation of the dynamic response to +30% load step change at bus #2 by the exact topological models (solid line) and generalized state space averaging method (dotted line).

Space Power Systems

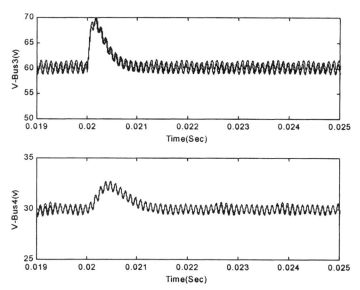

Figure 7.21 Simulation of the dynamic response to -20% load step change at bus #2 by the exact topological models (solid line) and generalized state space averaging method (dotted line).

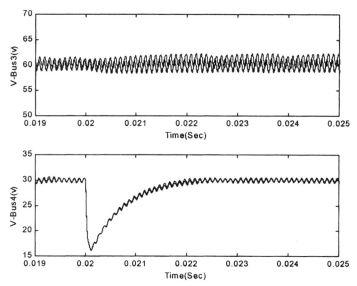

Figure 7.22 Simulation of the dynamic response to +50% load step change at bus #4 by the exact topological models (solid line) and generalized state space averaging method (dotted line).

Figures 7.18 and 7.19 show the steady state waveforms of voltages of bus #4 as well as bus #5 and the inductor current of the main Buck-Boost converter as well as converter #1, respectively. Figure 7.20 shows the voltage waveforms at bus #3 and bus #5 when there is a +30% load step change at bus #2. Furthermore, in Figure 7.21, effects of a –20% load step change at bus #2 on bus #3 and bus #4 are depicted. Therefore, this load step change at bus #2 affects other bus voltages in the system. Figure 7.22 also shows the voltage waveforms at Bus #3 and bus #4 when there is a +50% load step change at bus #4. These figures show the exact topological model together with the first order approximation results. As is depicted, the generalized state space averaging method works well in a multi-converter context. Therefore, it can be a convenient approach for studying the dynamics of multi-converter systems as well as designing controllers. In addition, these simulation results show that dynamics at the output of each converter affects other converters and bus voltages throughout the system. Some dynamics only affect a few converters and buses, others affect several converters and buses, and some dynamics may affect the system as a whole.

7.5 Real-Time State Estimation

In a multi-converter DC power electronic system, security of the system, which means ensuring continuity of the system operation in its normal condition, is of prime concern. The system is in the normal condition if there are no overloads, no overvoltages, and no undervoltages, and if loads are met. The control center of the system should maintain the system security. For security as well as control purposes, the control center must have complete information about the system, which is updated automatically in real time. Therefore, there is a need to monitor the system. However, measuring all the necessary signals may be impossible or may not be economically feasible. In addition, measurements are noisy. As a result, we need to obtain the complete set of necessary information from the incomplete noisy set of real time measurements for the control center. This is the duty of the state estimator.

Prof. Schweppe of MIT proposed and developed the idea of state estimation for conventional power systems in the late 1960s [28]. Since then, lots of research projects have been done and several hundreds of papers have been published on the state estimation of conventional power systems. However, to the best of our knowledge, very little work has been reported on the state estimation of multi-converter DC power electronic systems. It should be mentioned that, although, [29] and [30] have investigated different methods for estimating parameters and state variables in power electronic circuits, they are not for multi-converter DC power electronic systems with a system level point of view concerning control and security purposes.

State estimation in power electronic systems is much more complicated than in conventional power systems due to the switching effects. In fact, power electronic converters are non-linear time-dependent systems. Compared to conventional power systems, which only have constant frequency AC signals, in DC power electronic systems, not only DC values but also AC harmonics exist [31]. In other words, we need to estimate only amplitude and phase angle for each state variable in the conventional power systems; conversely, in a DC power electronic system, we need to estimate the DC value as well as amplitudes and phase angles of harmonics for each state variable. In this section, at first, we model the system using the averaging technique and generalized state space averaging method, in which we consider the weighted averages of the state variables as well as their harmonics [12]-[14], [24]-[27]. Then, we apply the weighted least squares (WLS) technique to the proposed models and estimate the state variables.

As we have studied in previous sections and chapters, more electric vehicles (MEV) and more electric hybrid vehicles (MEHV) [32]-[34], spacecraft and more electric aircraft (MEA) power systems [35], [36], the international space station (ISS) [22], [37]-[39], telecommunications, and advanced industrial DC power systems [31] are the best examples of multi-converter DC power electronic systems. In these systems, generating, distributing, and utilizing of electrical energy, in contrast with conventional power systems, are all in the form of power electronics. In fact, multiple DC/DC converters are used as source, load, and distributing converters to supply needed power at different voltage levels [31]. Figure 7.23 shows a schematic diagram of a representative multi-converter DC power electronic system.

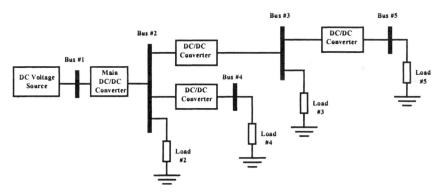

Figure 7.23 A representative multi-converter DC power electronic system.

The ISS electrical power system is a specific case of this kind of system. It is much larger in size and capacity than those of other multi-converter DC power electronic systems. In the ISS, power generated by the solar arrays is supplied through solar array switching units (SASU), which are power electronic converters, to the primary buses. The voltage at the primary buses is 160V DC. Some power is directed to or from batteries through the battery charge/discharge units (BCDU), which are again power electronic converters.

Power is provided to the secondary distribution systems through the DC/DC converter units (DDCU). The DDCUs convert 160V DC to lower voltage levels and provide the interface between the primary and secondary systems [31].

7.5.1 Role of State Estimation

Security is of primary concern in multi-converter DC power electronic systems that consist of several subsystems. Security of the system means ensuring continuity of the system operation in its normal condition in real time. The system is considered normal if it is in the nominal condition, i.e., there are no overloads, no overvoltages, or no undervoltages, and loads are met.

The control center of the system should maintain the system security in all situations. In fact, it must give control commands to the subsystems to ensure security in the presence of disturbance. For example, when a fault occurs in the system, the control center should identify it, isolate the faulted area, clear the fault, and restore the faulted area. To make the control actions properly, the control center must have complete information, which is updated automatically in real time, of the system. Therefore, there is a need to monitor the system for security and control purposes. The results of system monitoring for the control center are real time measurements, network topology information, and local converter controllers' information [39], [40]. However, measuring all the necessary signals may be impossible. In addition, obtaining all the possible measurements may not be economically feasible. As a result, we need to obtain the complete set of necessary information (all the state variables of the system) from the incomplete noisy set of real time measurement information for the control center.

Figure 7.24 shows the block diagram of the major subsystems of the control center. The objective of the state estimator is to estimate the states of the system, which are bus voltages and line currents, using the measurements of various line flows, bus voltages, and the input and output power of converters as well as the information about the status of the circuit breakers, switches, converters, and distribution lines. The state estimator solves a nonlinear optimization problem whose solution yields the state estimate for the entire system. In fact, state estimation is a mathematical procedure for computing the best estimate of the state variables based on the noisy data. Once the state is estimated, estimates for all other quantities of the interest such as the line flows can be computed.

In Figure 7.24, the network topology processor uses the raw network topology information and provides the online diagram of the system. The filter looks for obvious inconsistencies in the measurements. Other than the filter, there is a need for bad-data processing. The bad-data processor checks the measurements for possible bad data. If any of the measurements are flagged as bad data, they will be removed or corrected so that the state estimate will not be biased. Furthermore, the observability analyzer tests whether or not the available measurements are sufficient to estimate the entire system state. If the test fails, then observable islands and unobservable branches of the system will be identified. The output of the state estimator, as has been mentioned, is used for security analysis, contingency analysis, economic dispatching, load forecasting, and finally for triggering proper control actions.

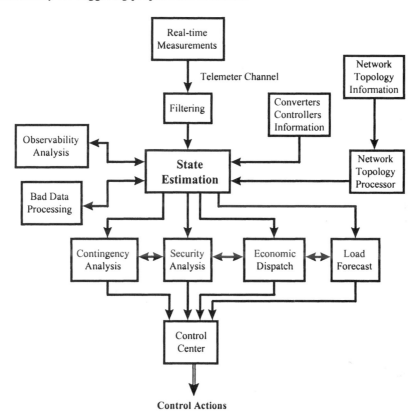

Figure 7.24 Block diagram of the major subsystems of the control center.

7.5.2 Electrical System Modeling

Figure 7.25 shows the interconnecting converters in a multi-converter system. Each converter receives its voltage source from the preceding converter, along with other converters in the stage, and supplies outputs to the following load converters, in addition to resistive load, all of which is represented by resistance R. We suppose that the converter of Figure 7.25 is a Buck converter, which is shown in Figure 7.26, operating in continuous conduction mode with switching period T and duty cycle d. Further, we suppose that the state-space variables are the Fourier coefficients of the circuit state variables, which are, in this case, i_L and v_o (i.e., inductors currents and capacitors voltages). Using the zero-order approximation of generalized state space averaging method in order to obtain i_L and v_o, we have two real state-space variables for the DC/DC converter of Figure 7.26, as follows:

$$\begin{aligned} <i_L>_0 = x_1 \\ <v_o>_0 = x_2 \end{aligned} \tag{7.23}$$

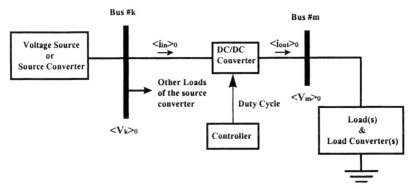

Figure 7.25 A typical DC/DC converter in a multi-converter DC power electronic system with the zero-order approximations of its inputs and outputs.

For the DC/DC converter of Figure 7.25, the signals that may be measured for monitoring and then state estimation purposes are input and output voltages, currents, and power. Therefore, we need to have the relations between these measurements and state variables as the input models for the state estimator. By applying the properties of the generalized state space averaging method, one comes to:

$$\begin{cases} <v_k>_0 = \dfrac{x_2 + L\dfrac{dx_1}{dt}}{d} \\ <v_m>_0 = x_2 \\ <i_{in}>_0 = d * x_1 \\ <i_{out}>_0 = x_1 - C\dfrac{dx_2}{dt} \\ <P_{in}>_0 = x_1(x_2 + L\dfrac{dx_1}{dt}) \\ <P_{out}>_0 = x_2(x_1 - C\dfrac{dx_2}{dt}) \end{cases} \tag{7.24}$$

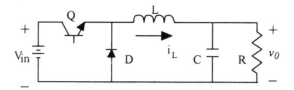

Figure 7.26 DC/DC Buck converter.

Using the first-order approximation to obtain i_L and v_o, we have four more real state space variables, compared to the conventional state space averaging technique, for the DC/DC Buck converter of Figure 7.26, as follows:

$$\begin{aligned}<i_L>_1 &= x_3 + jx_4 \\ <v_o>_1 &= x_5 + jx_6\end{aligned} \quad (7.25)$$

Figure 7.27 depicts a typical DC/DC converter in a multi-converter DC power electronic system with the first-order approximations of its inputs and outputs. Again, by applying the properties of the generalized state space averaging method in order to obtain relations between possible measurements and state variables as the input models for the state estimator, one comes to:

$$\begin{cases} <v_k>_0 = \dfrac{x_2 + L\dfrac{dx_1}{dt}}{d} \\ <v_m>_0 = x_2 \\ <i_{in}>_0 = d*x_1 + (\dfrac{1}{\pi}Sin(2*\pi*d))x_3 \\ \qquad -\dfrac{1}{\pi}(1 - Cos(2*\pi*d))x_4 \\ <i_{out}>_0 = x_1 - C\dfrac{dx_2}{dt} \\ <P_{out}>_0 = x_2(x_1 - C\dfrac{dx_2}{dt}) + 2\left[x_5(x_3 + C*\omega*x_6 - C\dfrac{dx_5}{dt}) \right. \\ \qquad \left. + x_6(x_4 - C*\omega*x_5 - C\dfrac{dx_6}{dt})\right] \end{cases} \quad (7.26a)$$

$$\begin{cases} <P_{in}>_0 = <v_k>_0 \left[d*x_1 + (\frac{1}{\pi} Sin(2*\pi*d))x_3 \right. \\ \qquad\qquad \left. -\frac{1}{\pi}(1-Cos(2*\pi*d))x_4 \right] \\ \qquad + 2\left[\operatorname{Re}\{<v_k>_1\}\left(d*x_3 + (\frac{1}{2\pi} Sin(2*\pi*d)x_1\right) \right. \\ \qquad\quad \left. + \operatorname{Im}\{<v_{k>_1}\}\left(d*x_4 - \frac{1}{2\pi}(1-Cos(2*\pi*d))x_1\right) \right] \end{cases} \quad (7.26b)$$

$$\begin{cases} <v_k>_1 = \frac{1}{d}\left[-\frac{1}{2\pi} Sin(2*\pi*d) <v_k>_0 + x_5 \right. \\ \qquad \left. + L\frac{dx_3}{dt} - L*\omega*x_4 \right] + j\frac{1}{d}\left[\frac{1}{2\pi}(1-Cos(2*\pi*d)) \right. \\ \qquad \left. *<v_k>_0 + x_6 + L\frac{dx_4}{dt} + L*\omega*x_3 \right] \\ <v_m>_1 = x_5 + jx_6 \\ <i_{in}>_1 = \left[d*x_3 + (\frac{1}{2\pi} Sin(2*\pi*d))x_1 \right] \\ \qquad + j\left[d*x_4 - \frac{1}{2\pi}(1-Cos(2*\pi*d))x_1 \right] \\ <i_{out}>_1 = \left[x_3 + C*\omega*x_6 - C\frac{dx_5}{dt} \right] \\ \qquad + j\left[x_4 - C*\omega*x_5 - C\frac{dx_6}{dt} \right] \\ <P_{in}>_1 = \left[<v_k>_0 (d*x_3 + (\frac{1}{2\pi} Sin(2*\pi*d))x_1) \right. \\ \qquad + d*x_1 * \operatorname{Re}\{<v_k>_1\}\right] + j\left[<v_k>_0 (d*x_4 - \frac{1}{2\pi} \right. \\ \qquad \left. *(1-Cos(2*\pi*d))x_1) + d*x_1 * \operatorname{Im}\{<v_k>_1\} \right] \end{cases} \quad (7.27a)$$

Space Power Systems

$$\begin{cases} <P_{out}>_1 = \left[x_2(x_3 + C*\omega*x_6 - C\frac{dx_5}{dt}) + x_5(x_1 - C\frac{dx_2}{dt}) \right] \\ \qquad + j\left[x_2(x_4 - C*\omega*x_5 - C\frac{dx_6}{dt}) + x_6(x_1 - C\frac{dx_2}{dt}) \right] \end{cases} \qquad (7.27b)$$

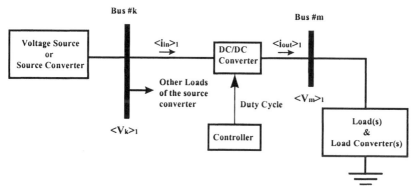

Figure 7.27 A typical DC/DC converter in a multi-converter DC power electronic system with the first order approximations of its inputs and outputs.

7.5.3 Weighted Least Squares State Estimator

The weighted least squares (WLS) technique is commonly used as a basis for state estimation. In this section, we apply it as a state estimator for multi-converter DC power electronic systems. Consider the set of measurements given by the vector Z:

$$Z = h(x) + e \qquad (7.28)$$

where:

Z: vector of m measurements.

x: vector of n system state variables.

$h(.)$: nonlinear vector function relating measurements Z to the state vector x.

e: vector of m measurement errors.

In addition, consider the following assumptions regarding the statistical properties of the measurement errors:

$\forall i = 1,2,...,m: \qquad E[e_i] = 0$

$\forall i, j = 1,2,...,m: \qquad E[e_i * e_j] = 0$ (independent measurement errors)

Hence, $E[e.e^T] = R = diag\{\sigma_1^2, \sigma_2^2, ..., \sigma_m^2\}$, where σ_i is the standard deviation of measurement i, and it is calculated to reflect the expected accuracy of the corresponding meter used. The WLS estimator will minimize the following objective function:

$$J(x) = [z - h(x)]^T .R^{-1}.[z - h(x)] \qquad (7.29)$$

To minimize the objective function, we must satisfy the first order optimality conditions:

$$\frac{dJ(x)}{dx} = H^T(x).R^{-1}.[z - h(x)] = 0 \qquad (7.30)$$

where:

$$H(x) = \frac{dh(x)}{dx} \qquad (7.31)$$

In the $m \times n$ matrix $H(x)$, each row represents a measurement and each column represents a state variable. The above nonlinear equation can be solved using the Newton iterative method as in the following:

$$x_{k+1} = x_k + \left[H^T(x_k).R^{-1}.H(x_k)\right]^{-1} .H^T(x_k).R^{-1}[z - h(x_k)] \qquad (7.32)$$

The final solution is obtained when $|\Delta x_{k+1}| = |x_{k+1} - x_k|$ becomes less than the convergence tolerance ε. ε is defined based on the required accuracy.

7.5.4 Test Examples

The proposed state estimation method has been tested on the representative system of Figure 7.23. Both zero order and first order approximations of the generalized state space averaging method are tested. Figures 7.28 and 7.29 show the associated measurement configurations for zero order and first order approximations, respectively. These measurements are simulated by using the exact analysis of the system. Noise is added randomly to each measurement. Noise is in the range of $\pm 0.3\%$.

As was mentioned, the relation between the measurement vector Z and the state vector x is as below:

$$Z = h\left(x, \left[\frac{dx}{dt}\right]\right) + e \qquad (7.33)$$

where $h(.)$ is a non-linear vector function. Since we are studying the system at the steady state, i.e., static state estimation, we have:

$$\left[\frac{dx}{dt}\right] = 0 \tag{7.34}$$

Therefore,

$$Z = h(x) + e \tag{7.35}$$

Relation (7.35) is, in fact, the same as relation (7.28). Therefore, a WLS based state estimator can be used to estimate the state variables from the measurements. The results of estimating state variables in vector x from measurements given in vector Z using the WLS technique for zero order and first order methods are shown in Tables 7.1 and 7.2, respectively.

In summary, in this section, the concept of state estimation in multi-converter DC power electronic systems has been presented. The modular modeling approach and the generalized state space averaging technique have been used to build large-signal models of the system. Algorithms based on zero order and first order approximations have been developed. These algorithms use an incomplete noisy set of real time information of the system to obtain an estimation of the complete set of real time system state variables based on the WLS method. Results of applying the proposed method to a representative system have also been presented. The results of the state estimator can be used for security analysis, contingency analysis, economic dispatching, load forecasting, and finally for making proper control actions by the control center in the system. Detailed investigation is required in order to determine other state estimation techniques that may improve the accuracy in the steady state and transient modes. The observability analyzer and the bad-data processor require further study.

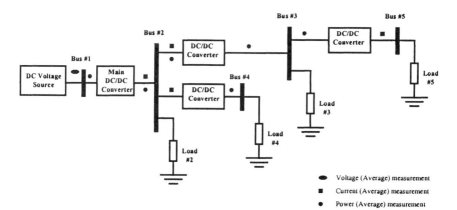

Figure 7.28 Representative system with the measurement configuration for the zero order estimation.

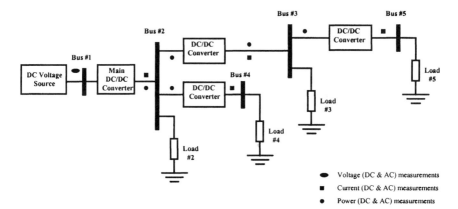

Figure 7.29 Representative system with the measurement configuration for the first order estimation.

Table 7.1 Results of the WLS state estimator with zero order models.

Bus	Estimated Voltage (Average)	Error (%)
#1	159.96V	0.025
#2	119.95V	0.038
#3	60.03V	0.046
#4	59.97V	0.042
#5	30.02V	0.049

Table 7.2 Results of the WLS state estimator with first order models.

Bus	Estimated Voltage (Average)	Error (%)	Estimated Voltage (magnitude of first harmonic)	Error (%)	Estimated Voltage (phase of the first harmonic)	Error (%)
#1	159.96V	0.028	0.00V	0.000	0.00°	0.000
#2	119.96V	0.036	1.73V	0.125	23.18°	0.314
#3	60.03V	0.051	1.05V	0.179	39.64°	0.397
#4	59.97V	0.048	0.89V	0.162	41.75°	0.372
#5	29.98V	0.055	0.52V	0.203	34.89°	0.418

7.6 Stability Assessment

Stability is one of the most fundamental issues in multi-converter power electronic space systems. This section concentrates on the considerations involved in stability analysis of these systems. The question of the basis of analysis and simulation of these systems, prior to implementation, has not been fully answered. Past works in simulating these systems have utilized averaged, linearized, and reduced models of the components of the systems [41]-[48]. This was required mainly due to overall system complexity and nonlinearity, and also due to the inability of handling simulations of exact models of such interconnected systems.

Therefore, the common approach is to develop small-signal linearized equivalent circuits and then simulate the system to determine system stability. System stability is thus defined as the absence of instabilities on the main bus and on the smaller buses. However, a question of accuracy and validity of small-signal frequency domain modeling and simulation for these systems exists. Furthermore, since ripples and rapid dynamics cannot be followed by the averaging methods, the state space averaged models could not be used without additional considerations. This leads to the conclusion that the effect of using various modeling techniques on an overall system stability analysis is quite profound and can lead to erroneous conclusions. As a result, for large-signal stability analysis, time domain simulations using reliable large-signal models are inevitable.

In this section, definitions of large-signal and small-signal stability, which are applicable to these systems, are addressed. In addition, conventional methods of system modeling, simulation, and stability analysis are discussed. Furthermore, this section deals with the features which help to improve stability. Recommendations on methods of stability analysis for these systems are also given. Finally, we summarize the results obtained in this study.

7.6.1 System Considerations

Exact simulation of the total system at once would entail modeling every converter and control circuit as well as the load equivalent circuits. This would be a massive program with time constants ranging from nanoseconds to seconds. This clearly would require very long computational time and much computer memory. Furthermore, it would probably require that a custom software package be developed for each massive system of this kind. More realistically, smaller subsystems would be modeled. Reference [41], which covers design aspects of the ISS, indicates that the system behind the source DC/DC converters is robust. The dividing points of the smaller systems would then best be located to include a source and all loads following. This division would allow the inclusion of the effects of interactions between source and loads and between the individual load modules on the smaller buses. Here, the term "source" is defined as all parallel sources supplying power to the same main bus. "Loads" hereafter referenced will refer to individual load modules, such as water pumps, fans, and motor drives. A real-life example of a smaller subsystem would be part of the overall system on the ISS. This might include a DC/DC source converter supplying 120V DC power to 3 smaller buses through electronically controlled switch boxes and 120 feet of cable. One of the three smaller buses provides power to 8 parallel loads, another provides 12, and a third provides 10 [42]. Typically, the design requirements on the DC/DC source converter include a specific maximum output impedance magnitude and an acceptable phase window, while the loads include a minimum impedance magnitude requirement and phase window. Reference [49]-[51] show the source and load magnitude impedance requirements. The significance of these impedance requirements will be explained in the next sections.

7.6.2 Definition of Stability

There are many different definitions of stability for linear, non-linear, discrete, and continuous systems. Stability analysis of linear systems is commonly performed by using one of the following classical methods: Routh-Hurwitz, root locus, Bode plot, and Nyquist criterion [60], [61]. Another well-known approach for stability analysis of linear or non-linear systems is Lyapunov's second method. Difficulty arises, however, in developing Lyapunov functions for many non-linear systems. Thus, in determining the stability of a large system, more state-specific definitions are desirable. Study of stability in non-linear systems requires consideration of initial conditions, external inputs, and their effects on non-linear components of the system. To the best of our knowledge, there is no general analytical approach to solving non-linear systems. Previous works done on the study of different multi-converter power electronic systems were based on linear system stability analysis techniques [43]-[46]. In [46], the R, L, C, and DC component values were varied within small-signal linearized models of switching converters using a linearized PWM switch model, and the resulting system phase margin was used to determine the stability of the system. In [43], design considerations for the studied system are presented, and as derived from [62], a small-signal system stability analysis based on source and load impedance relations is utilized. Reference [43] completes the stability analysis with large-signal stability analysis through simulations and laboratory testing of prototypes and actual hardware.

The definition of stability in this section considers the asymptotic stability as in the case of Lyapunov. However, we will not search for a suitable Lyapunov function for each individual converter, or for a larger subsystem containing the source and load converters, as previously described, but instead consider the conditions of voltages and currents on the inputs and outputs of the converters. One form of instability we consider is the divergence of one or more of these input or output voltages or currents (signals) from a desired value to some value which is outside the zone of tolerance, and the failure of the signal to return to within the zone of tolerance within a predefined period of time, i.e., unboundedness. Unboundedness in power electronic circuitry cannot always be predicted, as in the case of a switching failure in a boost converter which can practically cause a short across the supplying bus [42]. This type of non-linear instability cannot be predicted with conventional linear system phase and gain margin methods. Without proper protection and control circuitry, these types of instabilities usually will result in system failure. Chattering is another non-linearity which is not detectable by conventional control system stability measures and must be found through examining actual system hardware and circuitry. Chattering essentially is the absence of a hysteresis mechanism [42]. Chattering can also cause component failure over time due to the heating effects. Chaos is an instability in power electronic systems which presents itself as a pseudo-random functioning of one or more elements in the system. One form of chaos, as illustrated in Figure 7.30, is the deterministic harmonic noise generated by the switching in the system. This coupled with external noise on the control signal in the feedback loop of a DC/DC converter PWM circuit, causes the PWM duty cycle to vary randomly.

As previously stated, the types of systems considered in this chapter consist of DC/DC source converters supplying main DC buses, and multiple load modules containing different power electronic converters that produce different voltage levels as required for each individual load. A possible method of determining the stability and characteristics of this system may be to simulate the previously defined subsystem. The most important questions then become what kind of models should be developed for the individual source DC/DC converters and load modules, and whether frequency or time-domain methods should be used. Some general types of simulations which should be considered are the large-signal, small-signal, and DC simulations. The most prevalent method of modeling is the use of state space averaging and small-signal linearization.

It is well known that most common DC/DC converters (i.e., Buck, Boost, Buck-Boost) have two modes of operation when operating in a continuous conduction mode, and three when operation is discontinuous. If the switching frequency is high enough, when compared to the control dynamics of the converter, the state space equations describing these modes of operation can be averaged. Commonly, the converter gain crossover frequency is designed to be at least an order of magnitude below the switching frequency, and thus the state space averaged equivalent circuit model is valid for purposes of performing small-signal simulations of DC/DC converters. This state space averaged equivalent circuit is now a non-linear, continuous-time equivalent circuit which represents the original non-linear pulsed (discrete) system. This model is relevant in simulations for large and small signals; however, usually only the converter and input filter are modeled, and not the dynamics and limitations of the control circuitry. Large-signal deviations may occur due to bus faults, mechanical load malfunctions, extreme signal variations on the bus due to load changes, electronic component failure, etc. Without the consideration of large-signal deviations, the norm is to linearize the non-linear state space averaged system. The result of this linearization is a system model which is valid for only small-signal analysis, meaning the model is good for only small perturbations of the system's signals, i.e., bus voltage, current, electronic components, and load variations. How small is small is determined by the other system parameters in the model equations.

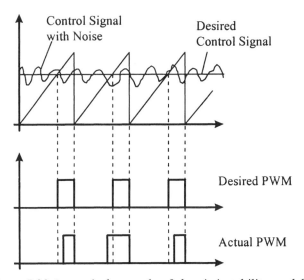

Figure 7.30 A practical example of chaotic instability modeling.

The linearization removes terms which contain the product of two perturbations. Obviously then, when one or more of these terms become significant in the model of the system, due to large-signal variations, the model becomes invalid. Small-signal models are the models of choice during design of control circuitry and filtering in DC/DC converters. However, there are limitations as to their use in system stability analysis, as will be discussed in the next section.

7.6.3 Simulation Considerations for Stability Analysis

The use of the linearized state space model, which is common in stability analysis of these systems, relies on the reliability of the protection, the limiting circuitry, and the assumed stability margins of each individual source and load in the overall system to prevent large-signal instabilities within the individual loads. The common approach is to develop small-signal linearized equivalent circuits and simulate the system to determine system stability. The question arises as to the need for this type of simulation at all. From the classical work in [62], which has been considered as a foundation for simulation, this type of simulation is indeed not even necessary in these systems. The paper by Middlebrook [62] shows that the system source output impedance and the aggregate load input impedance are all that are required for the determination of

Space Power Systems

stability. Using two-port theory, he showed that if a factor of minor loop gain satisfied the Nyquist stability criterion, the system would be stable. Figure 7.31 shows the general definition of the impedance of this small-signal stability criterion. The small-signal stability criterion is that $T(s)=Z_S/Z_L$ satisfies the Nyquist stability criterion. Moreover, Ref. [43] states that with phase margins of $T(s)$ less than 30 degrees, system performance degradation becomes more noticeable.

In considering a method of determining the small-signal stability of the larger power subsystem considered, there are no more accurate impedance characteristics of a system than those taken by actual measurements on the system. Characteristics determined from a simulation model always provide some error, in comparison to the actual system. This would then lead to the conclusion that the individual source and load fabricator's test data (i.e. output and input impedance plots) are all the data necessary to determine small-signal stability of these systems, and with greater accuracy than with inexact models. Stability can be determined using the data obtained directly from individual fabricators testing of the actual converters to be used in the system. The method is fundamentally simple.

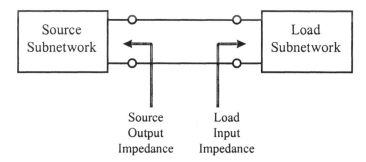

Figure 7.31 Source output and load input impedance defined.

Using the input impedance magnitude and phase plots of the actual loads of the previously defined subsystems, the total parallel input impedance of the loads and cabling, as seen by the main bus, can be calculated on a point-by-point basis, with for example, 200 points taken between 10 and 10 kHz. Once the total parallel input impedance of the loads and the total parallel output impedance of the sources, for an individual subsystem, are known, the method described by Middlebrook [62] can be utilized to determine the small-signal stability of the system. This simply means plotting the resulting $T(s)$ and determining the phase and gain margins. System source, load, and cable impedances for this small-signal stability investigation are shown in Figure 7.32.

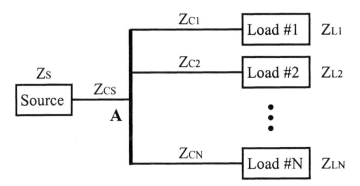

Figure 7.32 Impedance defined for numerical small-signal analysis.

Now, if point A of Figure 7.32 is taken as the point of common coupling, the small-signal analysis of this subsystem's bus is dependent on $T(s)$ satisfying the Nyquist stability criterion, where $T(s)$ is

$$T(s) = \frac{Z_{ST}}{Z_{LT}} = \frac{Z_s + Z_{cs}}{(Z_{C1} + Z_{L1}) \| (Z_{C2} + Z_{L2}) \| \cdots \| (Z_{CN} + Z_{LN})} \qquad (7.36)$$

If the system is defined as a large-signal model, then clearly linearized state space models are invalid. Time domain simulation of non-linear system modules, which includes the protection circuitry and system control dynamics and limitations, must be used for accuracy of overall system performance simulation. Failure to accurately include the non-linear protection circuitry and actual feedback control dynamics and limits (dynamic range) of each individual source and load converter could render the large-signal study results incorrect.

7.6.4 Large-Signal Stability Analysis

The impedance requirements govern small-signal stability of a multi-converter power electronic system; but, they do not guarantee large-signal stability for the entire system. Large-signal stability refers to the ability of the system to move from one steady state operating point following a disturbance to another steady state operating point.

Large-signal stability involves major disturbances such as negative interactions between converters, loss of generation, line-switching operations, faults (short circuits, open circuits), and sudden load changes. Following a disturbance, bus voltages undergo transient deviations from nominal values. The objective of a large-signal stability study is to determine whether or not the converters will return to nominal values with new steady state power conditions.

It is possible for controls to affect dynamic stability of the system. The action of local power converter controllers, and controls from a main power system dispatch center, can interact to stabilize or destabilize the multi-converter

Space Power Systems

power electronic system several minutes after a disturbance has occurred. Dynamic behavior of a multi-converter system can be improved and its stability margins can be increased by considering the following issues:

- Using suitable fast-response protection devices.
- Avoiding heavily loaded power electronic converters.
- Overrating the source converters to ensure an appropriate reserve in delivering power.
- Managing the loads properly according to the operating conditions of source and distribution converters to avoid overloading.
- Controlling the loads via the control commands of the management center in such a way that the system always operates around its nominal power.
- Avoiding weakening the network by the simultaneous outage of a large number of lines and power converters.
- Avoiding overvoltage and undervoltage.

In practice, financial considerations determine the extent to which any of these features can be implemented. In addition, there must always be a compromise between operating a system close to its stability limit and operating a system with an excessive reserve of generation and transmission. The risk of losing stability can be reduced by using additional elements inserted into the system to help smooth the system's dynamic response.

7.6.5 Conclusion

In summary, for small-signal stability studies, a small-signal analysis by modeling with averaged, linearized, and reduced circuit equivalencies of an actual system is performed. Furthermore, a numerical method, as is described in this section, would save a great amount of time and, thus, be a much more economical solution. This would, however, require the knowledge of the source converter output impedances, the load converter input impedances, and the cable characteristic impedances. These are easily attainable in a laboratory environment if the systems are available, and they are usually available from the manufacturers on demand.

On the other hand, for large-signal stability studies, time domain simulations using reliable large-signal models are inevitable. In addition, stability analysis of these systems after designing and building will be best performed via time domain simulations. The limits of these simulations depend on the actual control and protection circuit dynamics, which may include but are not limited to undervoltage lockout, overvoltage and overcurrent protections, cycle-by-cycle limiting, and nonlinearities due to magnetic saturation, leakage, semiconductor operation, temperature variations, aging, and catastrophic failures. Large-signal stability simulations for large systems, such as space power systems, require vast amounts of time in modeling and even more in simulation.

A conservative approach to designing multi-converter power electronic systems may be to over-design the ratings of the source and load converters by conservatively defining their margins of stability and enhancing the protection mechanisms to react to any unforeseen instability during the operation. Alternatively, a full-scale multi-converter power system, such as the ISS, should be built and tested, in mock assembly. This would provide nearly 100% confidence in the system, and could still cost less than future redesigns, repairs, and potentially hazardous conditions to the users. However, even an actual mock system assembly is not necessarily optimizable for weight, volume, and performance, since no reliable and accurate system simulation and optimization models and methodologies are available.

Nevertheless, more optimal space power system designs are possible in which the use of materials and components ratings would be minimized, with better overall system performance and reliability. This would require system dynamics and stability simulation in the time domain. The methodology for such a system modeling is not widely known. However, in Chapters 12 and 13, major large-signal dynamics and instabilities associated with the multi-converter environments are studied in detail.

7.7 References

[1] NASA Facts, "Powering the future – NASA Glenn contributions to the International Space Station (ISS) electrical power system," see http://www.grc.nasa.gov/.

[2] A. Emadi and M. Ehsani, "Multi-converter power electronic systems: Definition and applications," in *Proc. 32nd IEEE Power Electronics Specialists Conf.*, Vancouver, BC, Canada, June 2001, vol. 2, pp. 1230-1236.

[3] International Space Station Evolution Data Book, "Baseline design," vol. 1, Revision A, Oct. 2000.

[4] E. Gholdston, J. Hartung, and J. Friefeld, "Current status, architecture, and future technologies for the international space station electric power system," *IEEE Aerospace and Electronics Systems Magazine*, vol. 11, no. 2, pp. 25-30, Feb. 1996.

[5] E. B. Gietl, E. W. Gholdston, B. A. Manners, and R. A. Delventhal, "The electric power system of the international space station – A platform for power technology development," in *Proc. IEEE Aerospace Conf.*, Big Sky, Montana, March 2000, vol. 4, pp. 47-54.

[6] L. Truong, F. Wolff, and N. Dravid, "Simulation of the interaction between flywheel energy storage and battery energy storage on the international space station," in *Proc. 35th Intersociety Energy Conversion Engineering Conf.*, Las Vegas, Nevada, July 2000, vol. 2, pp. 848-854.

[7] T. W. Kerslake and L. P. Gefert, "Solar power system analyses for electric propulsion missions," *IEEE Aerospace and Electronics systems Magazine*, vol. 16, no. 2, pp. 3-9, Jan. 2001.

[8] "Energy and power," *IEEE Aerospace and Electronics systems Magazine*, vol. 15, no. 10, pp. 19-26, Oct. 2000.

[9] A. Emadi and S. S. Williamson, "Status review of power electronic converters for fuel cell applications," *Journal of Power Electronics*, vol. 1, no. 2, pp. 133-144, Jan. 2002.

[10] D. J. Keys, "Earth observing system (EOS) TERRA spacecraft 120 volt power subsystem," in *Proc. 35th Intersociety Energy Conversion Engineering Conf.*, Las Vegas, Nevada, July 2000, vol. 1, pp. 197-206.

[11] M. H. Moody and C. A. Maskell, "Electrical power systems for space based radar satellites," in *Proc. 24th Intersociety Energy Conversion Engineering Conf.*, Arlington, Virginia, July 1989, vol. 1, pp. 571-577.

[12] R. D. Middlebrook and S. Cuk, "A general unified approach to modeling switching converter power stages," in *Proc. IEEE Power Electronics Specialist Conf.*, June 1976, pp. 18-34.

[13] P. T. Krein, J. Bentsman, R. M. Bass and B. Lesieutre, "On the use of averaging for the analysis of power electronic systems," *IEEE Trans. on Power Electronics*, vol. 5, no. 2, pp. 182-190, 1990.

[14] J. Sun and H. Grotstollen, "Averaged modeling of switching power converters: reformulation and theoretical basis," in *Proc. IEEE Power Electronics Specialist Conf.*, June 1981, pp. 1165-1172.

[15] M. O. Bilgic and M. Ehsani, "Analysis of single flying capacitor converter by the state space averaging technique," in *Proc. IEEE International Symposium on Circuits and Systems*, Helsinki, Finland, 1988, pp. 1151-1154.

[16] M. O. Bilgic and M. Ehsani, "Analysis of inductor-converter bridge by means of state space averaging technique," in *Proc. IEEE Power Electronics Specialist Conference*, June 1988, pp. 116-121.

[17] B. R. Needham, P. H. Eckerling and K. Siri, "Simulation of large distributed DC power systems using averaged modeling techniques and the Saber simulator," in *Proc. IEEE Applied Power Electronic Conf.*, Orlando, FL, Feb. 1994, pp. 801-807.

[18] K. S. Tam and L. Yang, "Functional models for space power electronic circuits," *IEEE Trans. on Aerospace and Electronic Systems*, vol. 31, no. 1, pp. 288-296, Jan. 1995.

[19] K. S. Tam, L. Yang, and N. Dravid, "Modeling the protection system components of the space station electric power system," *IEEE Trans. on Aerospace and Electronic Systems*, vol. 30, no. 3, pp. 800-808, July 1994.

[20] K. J. Karimi, A. Booker, and A. Mong, "Modeling, simulation, and verification of large DC power electronics systems," in *Proc. IEEE Power Electronics Specialist Conference*, Baveno, Italy, June 1996, pp. 1731-1737.

[21] B. Cho, "Modeling and analysis of spacecraft power systems," *Ph.D. Dissertation*, Virginia Polytechnic Institute and State University, Blacksburg, VA, 1985.

[22] J. R. Lee, H. H. Cho, S. J. Kim, and F. C. Lee, "Modeling and simulation of spacecraft power systems," *IEEE Trans. on Aerospace and Electronics Systems*, vol. 24, no. 3, pp. 295-303, May 1988.

[23] T. L. Skvareniana, S. Pekarek, O. Wasynczuk, P. C. Krause, R. J. Thibodeaux, and J. Weimer, "Simulation of a more electric aircraft power system using an automated state model approach," in *Proc. 31st Intersociety Energy Conversion Engineering Conf.*, Washington, DC, Aug. 1996, pp. 133-136.

[24] S. R. Sanders, J. M. Noworoski, X. Z. Liu, and G. C. Verghese, "Generalized averaging method for power conversion circuits," *IEEE Trans. on Power Electronics*, vol. 6, no. 2, April 1991.

[25] G. Verghese and U. Mukherdji, "Extended averaging and control procedure," in *Proc. IEEE Power Electronics Specialist Conf.*, June 1981, pp. 329-336.

[26] J. Mahdavi, A. Emadi, M. D. Bellar and M. Ehsani, "Analysis of power electronic converters using the generalized state space averaging approach," *IEEE Trans. on Circuits and Systems I: Fundamental Theory and Applications*, vol. 44, no. 8, pp. 767-770, Aug. 1997.

[27] J. Xu and J. Yu, "An extension of time averaging equivalent circuit analysis for DC-DC converters," in *Proc. IEEE International Symposium on Circuits and Systems*, 1989, pp. 2060-2063.

[28] M. B. Do Coutto Filho, A. M. Leite da Silva, and D. M. Falcao, "Bibliography on power system state estimation," *IEEE Trans. on power systems*, vol. 5, no. 3, pp. 950-961, Aug. 1990.

[29] L. A. Kamas and S. R. Sanders, "Parameter and state estimation in power electronic circuits," *IEEE Trans. on Circuits and Systems I: Fundamental Theory and Applications*, vol. 40, no. 12, pp. 920-928, Dec. 1993.

[30] M. P. Kudisch, G. C. Verghese, and J. H. Lang, "Off-line parameter and state estimation for power electronic circuits," in *Proc. 1988 IEEE Power Electronics Specialist Conf.*, April 1988, pp. 509-516.

[31] A. Emadi, "Modeling, analysis, and stability assessment of multi-converter power electronic systems," *Ph.D. Dissertation*, Texas A&M University, College Station, TX, Aug. 2000.

[32] J. G. Kassakian, "Automotive electrical systems – the power electronics market of the future," in *Proc. 15th IEEE Applied Power Electronics Conf.*, New Orleans, LA, Feb. 2000, pp. 3-9.

[33] J. M. Miller, A. Emadi, A. V. Rajarathnam, and M. Ehsani, "Current status and future trends in more electric car power systems," in *Proc. IEEE Vehicular Technology Conference*, Houston, TX, May 1999.

[34] C. C. Chan and K. T. Chau, "An overview of power electronics in electric vehicles," *IEEE Trans. on Industrial Electronics*, vol. 44, no. 1, pp. 3-13, Feb. 1997.

[35] R. E. Quigley, "More electric aircraft," in *Proc. IEEE Applied Power Electronics Conf.*, San Diego, CA, March 1993, pp. 609-911.

[36] A. Emadi and M. Ehsani, "Aircraft power systems: technology, state of the art, and future trends," *IEEE Aerospace and Electronic Systems Magazine*, vol. 15, no. 1, pp. 28-32, Jan. 2000.

[37] M. F. Rose, "Space power technology," in *Proc. International Power Modulator Symposium*, Boca Raton, FL, June 1996, pp. 9-14.

[38] F. C. Berry, R. F. Gasser, and H. M. Chen, "ADEPTS: Space-based power system management using parallel architecture," *IEEE Trans. on Aerospace and Electronics Systems,* vol. 30, no. 1, pp. 275-280, Jan. 1994.

[39] G. L. Kusic, W. H. Allen, E. W. Gholdston, and R. F. Beach, "Security for space power systems," *IEEE Trans. on Power Systems*, vol. 5, no. 4, pp. 1068-1075, Nov. 1990.

[40] N. Balu et al, "On-line power system security analysis," *Proceedings of the IEEE*, vol. 80, no. 2, pp. 262-280, Feb. 1992.

[41] R. E. de Gaston, "Space station freedom electrical power system stability assessment," Office of Safety and Mission Quality, NASA HQ, Houston, TX, June 1992.

[42] A. Emadi, J. P. Johnson, and M. Ehsani, "Stability analysis of large DC solid-state power systems for space," *IEEE Aerospace and Electronic Systems Magazine*, vol. 15, no. 2, pp. 25-30, Feb. 2000.

[43] E. W. Gholdstone, K. Karimi, F. C. Lee, J. Rajagopalan, Y. Panov, and B. Manners, "Stability of large DC power systems using switching converters, with application to the International Space Station," in *Proc. 31st Intersociety Energy Conversion Engineering Conf.*, Washington, DC, Aug. 1996, pp. 166-171.

[44] I. Lazbin and B. R. Needham, "Analysis of the stability margins of the space station freedom electrical power system," in *Proc. IEEE Power Electronics Specialist Conference,* Seattle, WA, June 1993, pp. 839-845.

[45] Y. V. Panov and F. C. Lee, "Modeling and stability analysis of a DC power system with solid state power controllers," in *Proc. IEEE Applied Power Electronics Conf.*, San Jose, CA, March 1996, pp. 685-691.

[46] K. J. Karimi, A. J. Booker, A. C. Mong, and B. Manners, "Verification of space station secondary power system stability using design of experiment," in *Proc. 32nd Intersociety Energy Conversion Engineering Conf.*, Honolulu, HI, Aug. 1997, pp. 526-531.

[47] G. S. Thandi, R. Zhang, K. Xing, F. C. Lee, and D. Boroyevich, "Modeling, control and stability analysis of a PEBB based DC DPS," *IEEE Trans. on Power Delivery*, vol. 14, no. 2, pp. 497-505, April 1999.

[48] T. C. Wang and J. B. Raley, "Electrical power system stability assurance for the international space station," in *Proc. 32nd Intersociety Energy Conversion Engineering Conf.*, Honolulu, HI, Aug. 1997, pp. 246-252.

[49] C. M. Wildrick, "Stability of distributed power supply systems," *Masters Thesis*, Virginia Polytechnic Institute and State University, Blacksburg, VA, Jan. 1993.

[50] C. M. Wildrick, F. C. Lee, B. H. Cho, and B. Choi, "A method of defining the load impedance specification for a stable distributed power system," *IEEE Trans. on Power Electronics*, vol. 10, no. 3, pp. 280-285, May 1995.

[51] S. E. Schulz, "System interactions and design considerations for distributed power systems," *Masters Thesis*, Virginia Polytechnic Institute and State University, Blacksburg, VA, Jan. 1991.

[52] B. H. Cho, J. R. Lee, and F. C. Y. Lee, "Large-signal stability analysis of spacecraft power processing systems," *IEEE Trans. on Power Electronics*, vol. 5, no. 1, pp. 110-116, Jan. 1990.

[53] B. Choi, "Dynamics and control of switchmode power conversions in distributed power systems," *Ph.D. Dissertation*, Virginia Polytechnic Institute and State University, Blacksburg, VA, May 1992.

[54] V. Grigore, J. Hatonen, J. Kyyra, and T. Suntio, "Dynamics of a Buck converter with constant power load," in *Proc. IEEE 29th Power Electronics Specialist Conf.*, Fukuoka, Japan, May 1998, pp. 72-78.

[55] S. F. Glover and S. D. Sudhoff, "An experimentally validated nonlinear stabilizing control for power electronics based power systems," *Society of Automotive Engineers (SAE) Journal*, Paper No. 981255, 1998.

[56] S. D. Sudhoff, K. A. Corzine, S. F. Glover, H. J. Hegner, and H. N. Robey, "DC link stabilized field oriented control of electric propulsion systems," *IEEE Trans. on Energy Conversion*, vol. 13, no. 1, pp. 27-33, March 1998.

[57] A. S. Kislovski, "Optimizing the reliability of DC power plants with backup batteries and constant power loads," in *Proc. IEEE Applied Power Electronics Conf.*, Dallas, TX, March 1995, pp. 957-964.

[58] A. S. Kislovski and E. Olsson, "Constant-power rectifiers for constant-power telecom loads," in *Proc. 16th International Telecommunications Energy Conf.*, Nov. 1994, pp. 630-634.

[59] I. Gadoura, V. Grigore, J. Hatonen, and J. Kyyra, "Stabilizing a telecom power supply feeding a constant power load," in *Proc. 10th International Telecommunications Energy Conf.*, Nov. 1998, pp. 243-248.

[60] H. A. Khalil, *Nonlinear Systems*, Prentice-Hall, Upper Saddle River, NJ, 1996.

[61] J. E. Slotine and W. Li, *Applied Nonlinear Control*, Prentice-Hall, Upper Saddle River, NJ, 1991.

[62] R. D. Middlebrook, "Input filter considerations in design and application of switching regulators," in *Proc. IEEE Industry Application Conf.*, Oct. 1976, pp. 366-382.

8
Sea and Undersea Vehicles

Based on the concept of more electric vehicles (MEV), conventional mechanical, hydraulic, and pneumatic power transfer systems are replaced by electrical systems in different sea and undersea vehicles. Considering different levels of power requirements of various electrical loads and for the achievement of compact, light, safe, and efficient power supplies, implementation of power electronics based power systems is the most feasible option in development of these vehicles.

8.1 Power System Configurations in Sea and Undersea Vehicles

8.1.1 Segregated Power System (SPS)

Navies utilize surface combatants and aircraft carriers as sea vehicles and submarines as undersea vehicles. Traditionally, segregated power system (SPS) configurations have been used in these types of vehicles where, as shown in Figure 8.1, separate prime movers are used to supply power to the propulsion system through geared drives, which consume almost 80-90% of the total power [1]. The prime movers are also used to drive generators. Generators supply power to the electrical distribution systems that contain transformers, switchboards, and circuit breakers to supply power to various electrical loads such as ship service and combatant loads. However, this type of power system configuration has proved inefficient, as the high-speed propulsion is not always the prime requirement of these vehicles. Therefore, tactical diversion of this surplus mechanical power to electricity was needed for the procurement of all distinct advantages of the MEV approach.

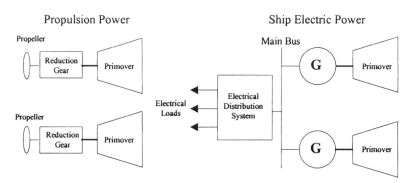

Figure 8.1 Segregated power system.

8.1.2 Integrated Power System (IPS)

In recent sea and undersea vehicles, implementation of integrated power system (IPS) configurations has brought revolution, where, as shown in the Figure 8.2, a common set of generators is used to supply power to ship service and combatant loads as well as electric propulsion systems [1]. This configuration has improved fuel efficiency of these vehicles for different ranges of speed while ensuring quieter operation. IPS has also made the modular equipment approach more realistic. As shown in Figure 8.2, compact and efficient power generation systems can be obtained by a primary generation source like a permanent magnet alternator driven by gas or diesel turbines generating voltage at the level of 440V or 4160V integrated with secondary generation sources like fuel cell and energy storage devices like flywheels and batteries to meet various load demands at normal sea load, maneuvering, and emergency conditions. IPS implements ring-type electrical distribution system, which allows any generator in the system to provide power to any load, giving top priority of supply continuity for important loads in critical condition.

As shown in Figure 8.2, the power electronic interface contains modular high or medium power AC or DC converters with inbuilt control assemblies that handle high generated electrical power. These converters supply power to main AC or DC bus in sea and undersea vehicles via isolation transformers and generator switch boards (SBs). Then, the power conversion and distribution system contains multiple interconnected power electronic converters and inverters working in an individual module. They operate in parallel, supplying required power to various intra-zonal electrical loads in various forms of AC and DC voltages. These modules have the ability to actively control the coupling in various parts of the system.

In IPS, different configurations are possible based on the propulsion system incorporated in the network. It can be powered through advance electric drives connected to a high power AC bus or main AC or DC bus in sea and undersea vehicles [2]. In IPS, as both propulsion and ship service electrical

Sea and Undersea Vehicles

systems are powered from a common electrical bus, electrical generation and distribution systems must be designed to handle the propulsion transients associated with ship maneuvering, while maintaining acceptable power quality. It can be achieved by implementing advanced control techniques and using intelligent power management systems with the introduction of a zonal nature in electrical distribution systems [2].

IPS accommodates hierarchical control system with distributed functionality for individual modules. As shown in Figure 8.3, power management center coordinates IPS operation with electric generation and propulsion modules by sending the commands to lower-level modules through higher-level modules according to the feedback variables. These commands are accompanied with power flow management according to the mission scenarios and functions associated. Thus, it makes autonomous fault detection, isolation, and reconfiguration of the power system at different operating profiles more reliable.

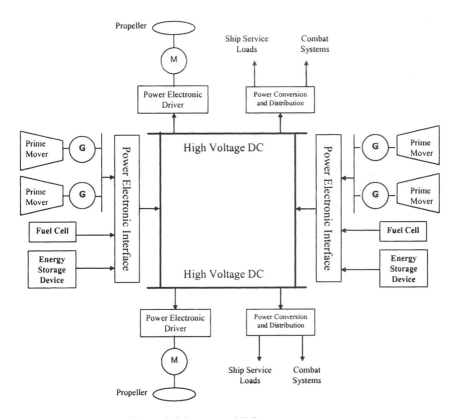

Figure 8.2 Integrated DC power system.

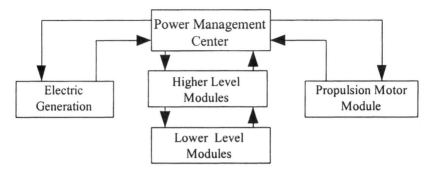

Figure 8.3 Control system chain of commands in sea and undersea vehicles.

8.2 Power Electronics Building Blocks (PEBB)

In power electronic circuits, there are very few universal and flexible solutions that can be applied for the medium and higher-power ranges of application because this type of system is designed only for particular operations. This approach serves as a deterrent for large-scale implementation of power electronic based circuits in various fields due to the long time of development and higher cost [3]. Therefore, navies have started research in the development of power electronics building blocks (PEBB), with the goal to get a flexible and scalable system architecture and family of modules, which significantly reduce acquisition and life cycle costs. This technology is more enhanced as it is capable of meeting the critical needs of increasing power density and inclusion of intelligence within power devices that can provide electric power control, conditioning, and very high speed switching operations of power switching elements [3]. In addition, it makes a constructive step towards fault tolerant power distribution systems in sea and undersea vehicles.

PEBB is a single package, multi-function controller that can replace complex power electronic circuits with a single device, which simplifies development and design of electrical power systems [4]. As shown in Figure 8.4, PEBB comprises power-switching elements along with snubbers and resonant components, which are configured via gate drives. It is controlled by the software program in a control module. Required feedback of variables for control can be given from sensors through series or parallel ports. These ports are accompanied by analog, digital, and communication buses [4]. A control module utilizes distributed or integrated control architecture for monitoring modulation index as well as inner loop parameters such as voltage, current, and frequency. Along with PEBB, a separate filter is provided to eliminate harmonic currents and to get stable control operation. Therefore, the same PEBB can be used for different applications such as steering control, automatic bus transfer (ABT), and fire-fighting pumps as shown in Figure 8.3 by making certain changes externally in the circuit.

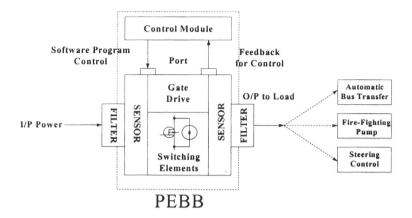

Figure 8.4 Power electronics building block.

It is necessary to reduce thermal stress on the switching elements, allowing them to work at high current density and efficiency while increasing utilization of devices [3]. Therefore, in the development phase of PEBB, special attention is given to packaging of devices related to the issues such as electromagnetic interferences (EMI), thermal, interconnections, communication interface, switching schemes, reliability, and controller architecture [4]. Therefore, PEBB can bring standardization in the communication and interface between different power electronic based circuits.

8.2.1 PEBB Applications in the System

PEBB is integrated in the systems in terms of application level PEBB (APEBB) in a multi-layered form as shown in Figure 8.4. It indicates the application of APEBB as an electrical drive for propulsion system. In this case, PEBB 1 (DC/DC converter) supplies power to the electric motor through intrazonal PEBB 2 (DC/AC inverter) from the main DC bus. Field-programmable application level controller is implemented for getting the desired load control strategy from the given feedback variables and communicating with the zonal-level or central system controller.

Power electronic converters create harmonics at particular frequency ranges. Therefore, properly designed filters with active control and standard specification of dynamic impedance are required to facilitate parallel operation of PEBBs for supplying high power electrical loads.

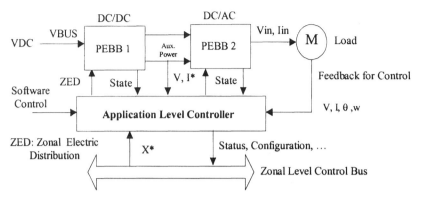

Figure 8.5 Implementation of PEBB in the system.

8.3 Controller Architecture for Power Electronic Circuits

In the electrical distribution system of sea and undersea vehicles, multiple interconnected power electronic based converters operate simultaneously. They must be controlled efficiently for getting stable operation of the system at various configurations and operating conditions. Analog methods of control for converters are cheaper and require simple algorithms. Yet, these methods appear to be inadequate for newly developed complex power electronic converters, as they offer less flexibility in changing algorithms and, thus, they are incapable of realizing different switching schemes on common platform with fine-tuning [5]. Therefore, microprocessor, microcontroller, and digital signal processor (DSP) control implementation for power electronic converters have positive edge over analog control, as they possess all the required qualities mentioned above [5].

In general, the controller architecture of a power electronic converter has the following four parts: (1) gate driver, (2) modulator with inner loop, (3) load controller, and (4) system level controller. Each of these parts operates in a different time domain. A gate driver controls the switches with time constants of about 1us; modulator and inner loops take care of the internal process within about 10us; a load controller operates at time constants of 10ms coordinated with the system level controller which performs system level monitoring.

8.3.1 Centralized Digital Controller (CDC)

Recently, centralized digital controller (CDC) has been used in high and medium power applications in the digital control method of converters [6], [7]. In this method of control, a single controller performs all control and monitoring tasks regardless of topology, size, and power rating of the converters. Consider the example of a single-phase rectifier supplying power to a universal motor, as shown in Figure 8.6. It has 10 point-to-point links with a centralized controller and single DSP processor and operates four gate drivers through a digital I/O interface according to voltage and current feedbacks at the output of the rectifier.

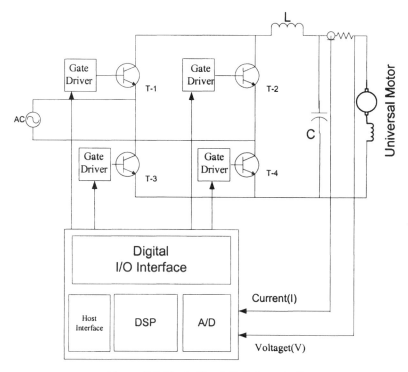

Figure 8.6 Centralized digital controller.

For advanced converters, the interface between power stage and controller appears very complex as the number of point-to-point links increases significantly and fault in the controller brings the complete system to standstill [6]. In sea and undersea vehicles, as most of the converters and inverters operate in parallel, design of communication and control architecture becomes a tedious task and it makes control software architecture difficult to build, debug, and maintain [9]. Thus, a centralized controller architecture fails to bring standardization and modularization in power electronic converters. It can be possible by implementing distributed control methods using PEBBs that accommodate various power levels uniformly and consistently [8].

8.3.2 Distributed Digital Controller (DDC)

Different power levels in the power electronic based system accompanies various levels of hardware, control, and data distribution. Therefore, efforts must be taken in the development of the controller to get assimilated with the PEBB system to achieve a high level of integration, a simple software and hardware reconfiguration, and flexible and multifunctional system. A distributed digital controller (DDC) has the ability to accommodate all the features mentioned above while working within a PEBB system. It can also modularize and standardize the overall system [8]. This controller is divided into four basic parts: (1) gate driver, (2) hardware manager, (3) application manager, and (4) system level controller, as shown in Figure 8.7.

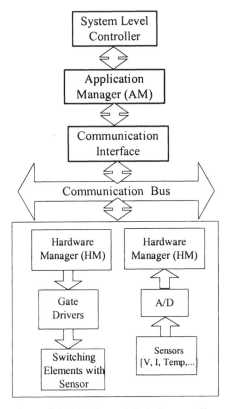

Figure 8.7 Distributed digital controller.

The application manager (AM) performs higher level and supervisory controls that are independent of hardware related tasks. The AM consists of the DSP processor, programmable logic device (PLD), and serial I/O ports to communicate with the hardware manager (HM) through the interface. By providing different libraries of software programs for HM and AM, DDC can be reconfigured for various applications. Therefore, it significantly reduces the time and cost of development of the controller.

Sea and Undersea Vehicles

HM working as an integral part of PEBB performs hardware oriented tasks such as developing gate pulses for switching elements through gate drivers. Furthermore, it provides over-current protection and performs voltage, current, and temperature measurements through various sensors. In DDC, the system level controls responses to the commands from the power management center (PMC). It also monitors the system level execution by coordinating various converters operating in a module. In state of the art technology, a daisy-chained serial optic link is used as a communication interface [8] between AM and HM. Because the distributed controller operates on faster time scales, it is susceptible to noise and electromagnetic interference as communication interface comes very close to the vicinity of power converters. Therefore, considerable research work is done to establish communication control protocols based on local area network (LAN) architecture [7].

8.4 Power Management Center (PMC)

The power management center (PMC) works like a heart for sea and undersea vehicles by acting as the main controller in the system. As shown in Figure 8.8, a PMC receives input data from arm analysis and radar systems to operate weaponry systems by regulating propulsion power. It monitors automatic bus transfer (ABT) switches and fire fighting pumps according to the information from casualty analysis and fire control systems, respectively. The PMC performs steering control from the feedback of navigation system. In each of these operations, it co-ordinates the generated power and electric distribution system operation.

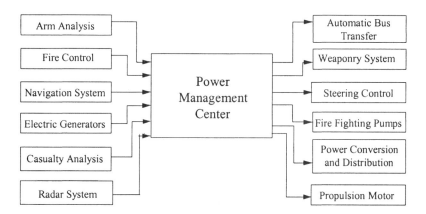

Figure 8.8 Power management center.

The PMC coordinates all resources, considering factors such as available generation power, allocated power, present load, and surplus power available for unpredicted loads. This system performs according to the operating scenario subjected to the constraints presented by the system fault, casualty, and other various configurations. The power management can be possible by various methods such as auctioneering power switching elements, active control, or voltage drop.

8.5 Electrical Distribution System in Sea and Undersea Vehicles

In the current sea and undersea vehicles, AC radial or ring-type distribution systems are implemented where, as shown in Figure 8.9, generators at the center supply power for various electrical loads via remote power cables through generator switchboards and load center [4]. However, this electrical distribution system proves uneconomical due to the high cost of power cables and switchboard feeder circuits [2] Furthermore, it is susceptible in terms of survivability of these vehicles during battle situations. It is also unable to accommodate new, sophisticated weapons to improve war-fighting capabilities of these vehicles.

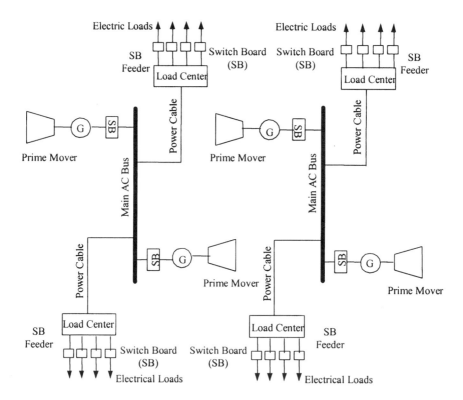

Figure 8.9 Conventional AC electric distribution system in sea and undersea vehicles.

Sea and Undersea Vehicles

Since the advancement in power electronics (in terms of increase in power density, reliability of power switching elements with reduced cost, and the invention of advance control technique), navies have started looking for new options to change orthodox architecture of electrical distribution systems. Many have replaced conventional switchboards and distribution panels with individual power electronic modules installed on a power bus in the vicinity of respective loads, which is known as zonal electrical distribution system [2].

8.5.1 AC Zonal Electrical Distribution System

As mentioned before, the main bus in sea and undersea vehicles can be AC or DC. Let us first analyze one part of AC zonal distribution, as shown in Figure 8.10, that appears as a radial system.

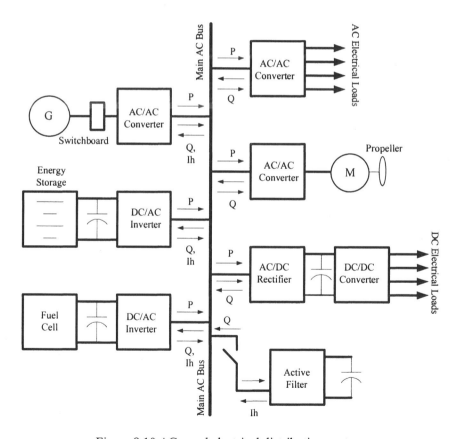

Figure 8.10 AC zonal electrical distribution system.

As shown in Figure 8.10, AC electrical loads receive power from the AC/AC converters at various voltages and frequencies especially at higher levels for weaponry loads. AC/AC converters are also used as electric drives for single and three-phase AC motors in sea and undersea vehicles. Different DC electrical loads in the system receive supply from single or three-phase controlled AC/DC rectifiers. Power quality can be maintained in AC zonal distribution system with active filters by harmonic elimination.

However, this type of distribution system is susceptible in terms of superior performance at various operating profiles. It faces problems of voltage dips at various points in the system due to the transients resulting from any casualty, switching or lighting stroke, or sudden increase in power demand [11]. Moreover, variation of voltage and frequency propagates its effect on the entire system, affecting performance of individual equipment and instruments as well as power electronic converters. Thus, reliability of the overall system is reduced. Therefore, AC zonal electrical distribution systems are unable to meet all distinct features of power electronic based systems.

8.5.2 DC Zonal Electrical Distribution System

Navies have been doing extensive research for implementing DC zonal electrical distribution systems in sea and undersea vehicles. As shown in Figure 8.11, in this type of distribution system, AC power is rectified from a high power AC bus to DC at the voltage levels of 1200-2000V via three-phase boost rectifiers to supply power to main DC bus. Then, the voltage at the main bus is stepped down to 850-950 V through DC/DC buck converters. They operate in parallel in power conversion modules (PCM) and act as buffers between main DC bus and various zones separated by watertight bulkheads [2] to supply single and three-phase inverters and other DC loads at suitable voltages. These inverters operate in parallel in power distribution modules (PDM). A solid-state frequency changer is used to supply power at 400 Hz frequencies to gyros, radar, sonar, and weapon systems. These power conversion and distribution modules can actively control the couplings between various parts of the system and, thus, manage to isolate and prevent propagation of fault. DC zonal electrical distribution systems are related with serious stability issues because of the presence of constant power propulsion loads in the system [13], [14]. Stability issues are presented in the last two chapters of this book.

As most of the electrical loads are of a nonlinear nature, superior control techniques such as distributed digital controller (DDC) should be provided for facilitating parallel operation of these converters and inverters to maintain stability in each zone. In each zone, automatic bus transfer (ABT) switches are implemented which ensure supply continuity to vital loads despite any casualty damage within zone [10]. ABT switches employ power switching elements, digitally controlled sensing, and switching operation. Micro-controllers used in ABT perform fault detection, identification and alarm, critical event data logging, and remote monitoring and control through serial communication ports.

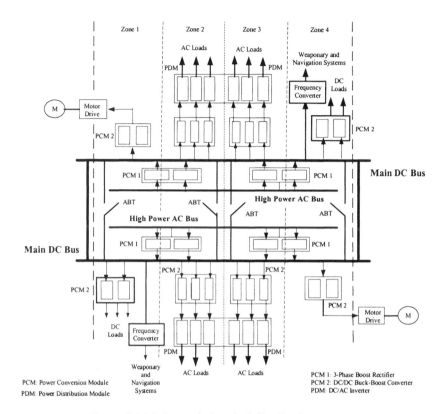

Figure 8.11 DC zonal electrical distribution system.

The implementation of DC zonal distribution systems has reduced the requirement of large AC switchgear and magnetic components, which drastically curtails the size and weight of the overall system. It is estimated that, with IPS configuration and a DC zonal system, the U.S. Navy has saved almost $14.113 million per ship in terms of labor, material and fuel cost [2]. Reactive power losses are quite lower compared to the AC systems. Algorithms required for detection and restorations of faults also become simple with the DC load flow analysis. Furthermore, in this type of distribution system, transients can be restricted to particular zones only with superior control techniques used for individual modules [12]. As the required power conversion for electrical loads is made at separate zones, harmonic distortion is lower than in AC distribution systems and the effects of voltage and frequency variations are limited.

8.6 Advanced Electrical Loads in Sea and Undersea Vehicles

The implementation of the MEV approach in sea and undersea vehicles has introduced many new, sophisticated electrical loads, especially in the weaponry systems. These loads are categorized as vital and non-vital loads, as shown in Figure 8.12, according to the safety of the power system, battle damage control, and supply continuity requirement for improving combat system performance [2]. They are subcategorized as low, medium, and high power loads for the reliable operation of power conditioning and power management systems providing proper coordination between them.

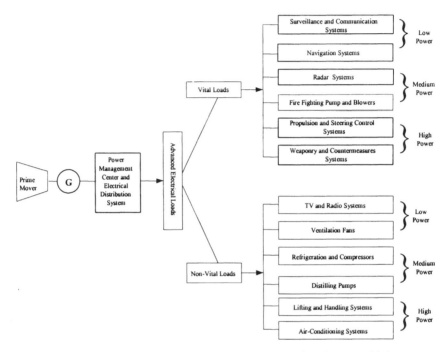

Figure 8.12 Advanced electrical loads in sea and undersea vehicles.

The power requirement of these loads varies at different voltage levels between 440V, 115V, and 4160V at a frequency 60 Hz or 400 Hz. Electrical loads requiring 400 Hz frequency are associated with command, surveillance, weapon system, and aircraft and aviation equipment. DC loads consumes power at 640 to 900V. The aircraft carriers generally contain electrical loads that consume power at 4160V [10].

The electrical distribution system takes care of various voltage levels and frequency modes of operation by accommodating different types of power converters, inverters, and static solid-state frequency converters. PMC works for ensuring supply continuity in the system and voltage regulation in the critical conditions of large variations in power demand, component failures, internal errors, and hostile system disruptions.

Considering their average weight of 70,000 to 80,000 tons, one of the main vital high power loads is the electric propulsion and steering control system that drives sea and undersea vehicles. Electric guns and direct energy weapons in weaponry and countermeasure systems consume an average power of tens of megawatts through pulse forming networks. Super conducting inductive storage and high voltage capacitors supported with magneto-dynamic storage (MDS) are the likely candidates for supplying power for future weapons [15]. Therefore, it is possible to make the best use of kinetic energy stored in vehicle and propulsion systems [16], [17].

One of the major dangers for sea and undersea vehicles is fire, which can be caused by short circuits and battle damages. Therefore, fire-fighting pumps are installed on board, with centrifugal compressors and variable speed motor drives that consume almost 6 MW of power. Lifting and handling systems contain electrical loads like hoists, cranes, lifts, and electromagnetic launcher, consuming power in the range of 5-7 MW.

8.7 Advanced Electric Drives in Sea and Undersea Vehicles

In the case of all electric sea and undersea vehicles, integrated electric drive propulsion systems underscore the importance of IPS configuration compared to the conventional mechanical drive systems. They give significant benefits in terms of improved war fighting capabilities by increasing available power for future weapon systems and reducing the vulnerability of these vehicles to critical conditions by rapid restoration of power. Integrated electric propulsion systems are economical and reliable in terms of control and operation. They have optimal maneuvering and positioning properties and their implementation reduces vibration and noise levels drastically.

For many years, DC motors maintained a monopoly on electric propulsion systems because of their large starting and braking torque and good reversion ability. DC motors cannot work at high voltage due to the restriction on commuter voltage. New trends require high power and speed as well as wide speed range for propulsion. Therefore, DC motors have limited application in advance propulsion systems [18], [20].

Advances in solid-state devices made AC motors such as multi-phase induction or field wound synchronous machines viable candidates for propulsion systems. With power electronic converters, various torque-speed characteristics can be achieved for AC motors. Multilevel power electronic converters are used for AC motors to meet high power propulsion demands of 20-30MW [2]. However, variable frequency AC drives suffer from harmonic induced vibration and low speed torque pulsation.

Extensive research is done in the development of switched reluctance machines (SRM) and permanent magnet (PM) motors such as PM brushless DC (PM-BLDC) motors and PM synchronous motors (PMSM) to be used in electric propulsion in sea and undersea vehicles. These motors offer many advantage in terms of efficiency, torque-to-weight ratio, maintenance, dynamic performance, reliability, and durability compared to the conventional motors [19]-[21]. DC motors with slotless armatures and multiplex windings can be used in propulsion systems as well [18]. These motors can be of multistage forms and single or separate power electronic converters can control each stage. Consider the example of PM synchronous motor as shown in Figure 8.13; the drive is supplied through DC/AC inverter via the main DC bus in sea and undersea vehicles.

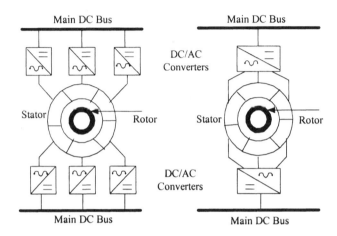

Figure 8.13 Propulsion drive of PM synchronous motor.

In the above case, the reliability of supplying power from separate converters to each stator sector is greater and it is possible with proper coordination among them. Even if a failure occurs in one of the windings of the stator sector, the motor can operate at low power during emergency and, thus, the chance of survival of sea and undersea vehicles during battle is increased.

The U.S. Navy is also looking at the option of implementing high temperature superconducting (HTS) motors in propulsion systems. It is possible to get a high power density with HTS motors. It can reduce machinery spaces substantially without creating the electromagnetic interference problem found in PM motors. Because propulsion load is high constant power load, its integration with the electric bus brings stability issues in power systems of sea and undersea vehicles. DC motor drives have traditional been used in various applications such deck receptacles, compartment ventilation fans, and pumps. An electromagnetic launcher employs a linear synchronous motor with a cycloconverter in sea and undersea vehicles [22]. This book also presents advanced electric motor drives for vehicular applications including sea and undersea vehicles.

The explanation given in this chapter gives a complete idea about the type of advanced electric distribution and electric loads that exist in modern sea and undersea vehicles. It can be noticed that electric motors are one of the most important electrical loads in the complete system. They are installed in fire fighting pumps, drilling system, lifting and handling systems, electromagnetic launcher, emergency door openers, excavators, cranes and winches, air conditioning systems, blowers, compressors, heaters, and prominently in electric propulsion system of these vehicles. Therefore, various types of conventional and advanced electric motors are the main candidates for variable speed power electronics based drives in sea and undersea vehicles.

As mentioned above, electric propulsion is the major electric load in sea and undersea vehicles and, therefore, optimization of power in this application can save huge amounts of power with implementation of advanced power electronics based drives. Also, in that case, the selection of an electric motor plays an important role. In general, vehicles like cruise liners or ferries have a couple of main propellers that contain a synchronous motor and several smaller thrusters that use asynchronous motors. Basically, the requirements of the motor used in the propulsion system of these vehicles are a wide range of variable low speed (150-200 rpm) motors with power from 20-40 MW to drive the single propeller. In the case of undersea vehicles, voltage requirement is also variable [18].

8.7.1 Slotless Armature DC Machines

The propulsion power requirements in the new vehicles have increased, since the trend these days is to use a single propeller, instead of a double propeller, in propulsion systems [20]. Implementation of DC motors in such systems requires a complex mechanical drive with a large, heavy, expensive propeller shaft and a gearbox. Another serious problem that hinders the application of DC motors in propulsion systems is a commutation problem in high power applications because it is mandatory to keep the voltage within limit between adjacent commutator bars. Also, it is needed to restrict reactance emf induced in the winding [20]. Therefore, it is difficult to implement DC motors in the electric propulsion applications of sea and undersea vehicles.

Slotless armature along with multiplexed winding in the DC motor can be the solution to meet high propulsion power demands because the multiplex winding limits the voltage between commutator bars even at higher induced emf in coil. Also, because of slotless nature of the armature, leakage inductance of the armature coil is reduced and current carrying capacity is increased without adding any spark at the brushes [18]. As in the case of the multiplexed winding on the armature, several commutator bar separators share the voltage induced in the single coil; therefore, the bar-to-bar voltage can be kept within limits. The other issue one needs to ensure is that, while designing multiplexed winding for the armature winding for motor, current is shared equally in parallel circuits under each pole. The implementation of the equalizer connection and interpoles helps to reduce the circulating current and the effect of the armature reaction [18].

For high power applications, usage of a DC motor with a slotless armature and multiplexed winding makes it possible to produce high field mmf if the field system is strong enough for the given large effective airgap. But, in that case, field system would have large power losses.

8.7.2 High Temperature Superconducting (HTS) Motor

Many navies are also looking for option of implementing high temperature superconducting (HTS) motors in propulsion systems [23]. It is possible to get a high power density with an HTS motor because field winding shows zero resistance at temperatures above 25K. It can reduce machinery spaces substantially without creating any electromagnetic interference problem and with least noise production.

Figure 8.14 shows the basic structure of typical HTS machines for slotless armature, where superconducting field winding is installed on the rotor and copper winding is installed on the stator. The field winding is cooled through refrigerators in the cryostat. These motors operate efficiently compared to the conventional motors under full load and partial load conditions by reducing resistance losses in the field winding. In addition, HTS motors exhibit low synchronous reactance, which leads to small voltage regulation between no load and full load conditions.

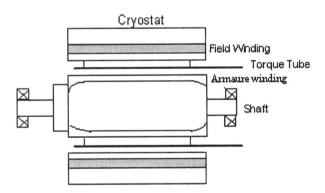

Figure 8.14 General layout of the HTS machine.

8.7.3 Permanent Magnet (PM) Machines

As mentioned above, extensive research is going on in the development of permanent magnet (PM) motors to be used in electric propulsions to replace currently used three-phase induction/synchronous motor in sea and undersea vehicles. PM motors offer many advantages in the form of efficiency, torque-to-weight ratio, maintenance, dynamic performance, reliability, and durability compared to the conventional motors. They can be available in types of PM brushless DC (BLDC) motor or PM synchronous motor (PMSM). Some of these

Sea and Undersea Vehicles

types of motors are designed especially for podded electric propulsion drives and other applications.

It is discussed in the previous section that in high power applications such as propulsion, we need to implement compact and very efficient electric motors, which have high torque-to-weight ratio. Axial flux PM motor with its improved design shows its suitability for the propulsion systems [19]. These motors are generally used in direct driving ship propulsions and in the compressor drive, which are current trends in sea and undersea vehicles.

Figure 8.15 shows the basic structure of this kind of axial flux PM synchronous motor. In this case, a slotless toroidal wound strip iron core is sandwiched between two permanent magnet rotor discs. The stator iron core contains a three-phase concentrated winding. As shown in Figure 8.15, this permanent magnet rotor disc produces flux into the airgap. The current flowing through stator winding interacts with flux produced by the permanent magnet rotor disk and tangential force produced with the contribution of all forces acting on the working surface of the toroidal core.

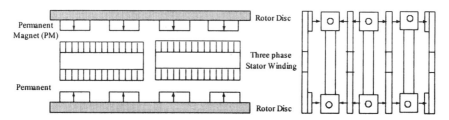

Figure 8.15 Basic structure of the axial flux permanent magnet motor.

The electromagnetic torque is a function of the outer diameter of the machine. This kind of motor is available in a multistage type, which can be accommodated in less space, and, hence, is a great advantage for sea and undersea vehicles. For n stator windings, there exist n+1 PM rotor discs. The given n stator windings can be connected in series or in parallel and can be supplied by one or separate converters [19]. Therefore, torque-speed characteristics of this kind of motor are adjusted through a normal PWM inverter or multilevel inverter, discussed in the next sections.

Transverse flux PM machines have a different kind of structure, but they operate on the same principal that electromagnetic torque is produced because of the transverse interaction of stator and rotor flux. The circular type of the stator phase winding of these motors produces homoplaner mmf distribution in the airgap. It is governed by the stator teeth pattern for producing a high order spatial harmonic of flux [24].

This flux interacts with the pattern of magnets on the rotor to produce torque. It has higher torque densities, as compared to the radial flux machine.

The reason for the high specific torque is the decoupling of magnetic flux paths and armature coils, which allows a very high current loading and, consequently, a high force density by simply enlarging the number of pole pairs.

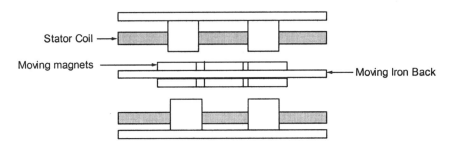

Figure 8.16 Basic structure of the transverse flux machine.

Furthermore, radial flux PM machines are based on the traditional and conventional operation of radial flux flow principle for rotating electrical machines. They can be designed efficiently for lower or higher number of pole numbers.

Linear synchronous motors (LSM) contain same kind of stator coils as in the rotary synchronous motor, but instead of circular structure, they are laid out in a flat pattern. Also, in the case of the rotor, instead of rotating the rotor, a carriage on rails suitably holds magnet is provided to propel along the stator coils. For controlling LSM, we need a variable frequency drive, such as cyclo-converter and sensors to monitor the lateral displacement of the magnet, to change coil polarity with switching in time with the magnets progress, and to accelerate. These controls are the most complex section of an LSM system; but, once established, the system is reasonably efficient and powerful.

The LSM discussed above has been implemented in the electromagnetic launcher. However, it can face the drawback of electromagnetic interference (EMI) caused due to the permanent magnet motor associated with electronic equipment [25]. Also, the size of the motor increases when the thrust-to-weight ratio requirement is high. Therefore, HTS wires are applied as primary coil in order to increase thrust applying for large current through it to develop superconducting linear BLDC motor [25].

8.7.4 Adjustable Speed Drives

In general, the field of power electronics has gone through rapid evolution in the last decade especially after the arrival of new type of devices such as GTO, IGCT, and IGBT. Therefore, many high power applications have appeared feasible [26], [27]. Besides, a lot of improvement has taken place in software and hardware control, signal processing methods, and control and

Sea and Undersea Vehicles 315

estimation techniques [26]. Therefore, in the last few decades, adjustable speed drive (ASD) technology has emerged as an important concept for reduction in energy consumption in various processes in different power ranges, which constitute heavy and large size electromechanical assemblies. They were replaced by compact and efficient power electronic equipment. It uniquely integrates electronics and controls within a motor to perform specific motor control functions. Many types of adjustable speed drives exist in sea and undersea vehicles for the continuous speed and torque control of AC and DC motors. They are classified according to the voltage and power ratings of the required motors. It is important to consider the torque-speed characteristic of the load to be imposed on the drive.

8.7.5 Variable Frequency Drives

In sea and undersea vehicles, AC motors especially dominate systems compared to DC motors. Therefore, the main focus will be on the AC drives, specifically on the variable frequency drives [28]. In this type of drive, changing the input frequency supplied to the motor can control speed of the drive. Also, it regulates output voltage according to the output frequency essential for the characteristic of AC motors for the production of adequate torque. The variable frequency drive receives input supply from a main DC bus in sea and undersea vehicles. Therefore, it is necessary to analyze various types of inverters, which appears in the power conversion and distribution system of these vehicles. They can be classified as follows.

8.7.5.1 Variable Voltage Inverters (VVI)/Voltage Source Inverters (VSI)

Voltage source inverters (VSI) are generally used in low and medium power applications in sea and undersea vehicles where they can operate efficiently in the range of 15-200% of motor operating speed [29]. As shown in Figure 8.17, they receive DC power from a variable voltage source like a DC chopper and then adjust voltage and frequency according to the motor requirements. VVI inverters, also known as six-step inverters, control voltage in a separate section from the frequency generation output. They use six solid-state switching devices in combination with six diodes. These solid-state switches are controlled to produce a six-step voltage waveform for each phase. Conducting time for each of the six switches is changed, resulting in a change in the frequency of the output voltage.

VVI is the simplest adjustable frequency drive and the most economical; however, it has the poorest output waveform as shown in Figure 8.18. In addition, the power factor decreases with speed and load. Besides, it requires a bigger filter for the inverter compared to other inverters. Therefore, it is operated within a particular speed range to get the high power factor used in applications such as distilling pumps and ventilation fans in sea and undersea vehicles.

8.7.5.2 Current Source Inverters (CSI)

Current source inverters (CSI) operate as shown in Figure 8.19, much the same as the six-step variable voltage inverter except that solid-state switching devices construct a six-step current wave for each phase instead of a voltage wave as shown in Figure 8.20. In this type, current is controlled through a large inductor at the input of the inverter while voltage is varied to meet the requirements of the motor. Generally, hysteresis control is used in these applications.

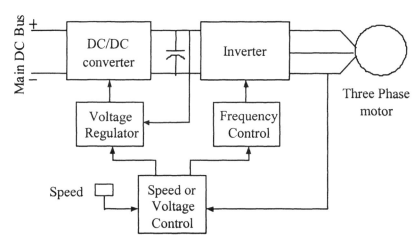

Figure 8.17 Application of voltage source inverter in sea and undersea vehicles.

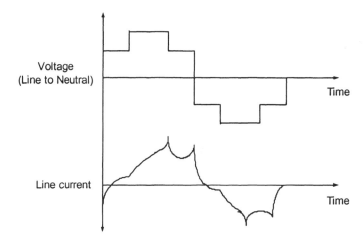

Figure 8.18 Output of voltage source inverter.

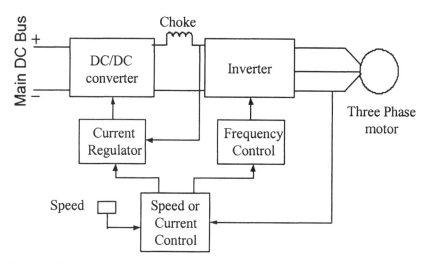

Figure 8.19 Application of current source inverter in sea and undersea vehicles.

CSIs have many advantages over VSIs. They have short circuit protection at the output stage. They have simple control circuits and work at high efficiency. Also, they have the ability to control current and, therefore, can efficiently be implemented in variable torque applications. They have disadvantages as well. They are bulky because of the large filter inductor; they require complex regulator; and they cannot provide open circuit protection.

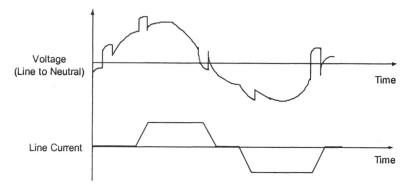

Figure 8.20 Output of current source inverter.

CSIs can be used in the medium to high power applications such as pumps, fans, and compressors in the electric loads of sea and undersea vehicles, where the torque requirement is variable [30]. In such types of loads, torque is proportional to the cube of the speed (T α n^3). Therefore, CSIs are suitable in the applications for fire fighting pumps and compressors present in sea and undersea vehicles. Also, as they have regenerative capability, they can feed power back to the AC power system to recharge the batteries when they are used in the applications of overhauling loads such as cranes and winches.

8.7.5.3 Load Commutated Inverters (LCI)

Load commutated inverters (LCI) are current source inverters. They are used to control blowers and large fans in the air conditioning system of sea and undersea vehicles, which implement a synchronous motor in the power range of 400-600 hp applications [31]. As shown in Figure 8.21, LCIs utilize natural power factor controlling ability of synchronous motors through the field circuit by increasing or decreasing the excitation current to commutate power-switching elements such as thyristors. They offer advantages in terms of cost/performance compared to the VSIs driving induction motors when torque pulsation is not of primary concern.

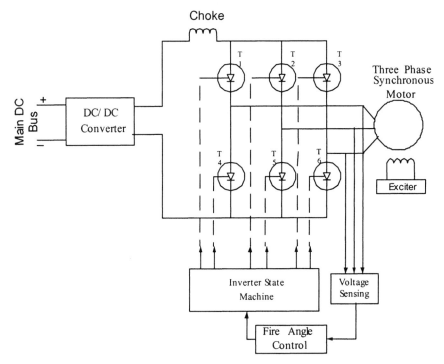

Figure 8.21 Application of load commutated inverter in sea and undersea vehicles.

This inverter, while operating in the standard six-step mode, supplies motor with three-phase line currents, which contain permissible switching harmonic contents. It synchronizes the rotor position of the motor with frequency and phase of the line currents. In the complete control, LCIs use sensors for the terminal voltage sensing to trigger inverter signals for synchronizing rotor positions [32]. Therefore, they compensate the phase advance produced due to the armature reaction. In the effective operation of LCIs, we only need information of zero crossing and frequency of the line current, which enhances simplicity and reliability.

In all of these inverters (CSI, VSI, and LCI), pulse width modulation (PWM) control is used [33]. This type of control exists in different types such as sinusoidal PWM, sinusoidal PWM with third harmonic injection, and space vector modulation depending upon the type of the application.

8.7.5.4 Pulse Width Modulation (PWM) Inverters

PWM inverters are one of the most commonly used inverters in sea and undersea vehicles for various applications in the power range of 5000 kW. As shown in Figure 8.22, a PWM inverter utilizes solid-state switching devices to produce a series of constant voltage pulses of various widths to produce an AC output. The timing and number of pulses are varied to produce the varying frequency. PWM inverters are generally used in motors in refrigerators, air conditioner, navigation, and radar systems of sea and undersea vehicles.

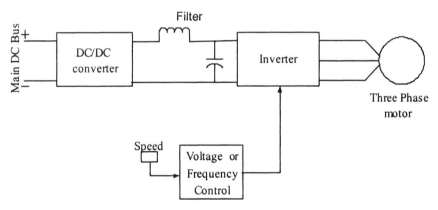

Figure 8.22 Application of PWM inverter in sea and undersea vehicles.

In PWM inverters, the speed reference command is usually directed by a digital signal processor (DSP) or microcontroller, which simultaneously optimizes the carrier frequency and inverter output frequency to maintain a proper V/Hz ratio and higher frequency throughout the normal speed. Therefore, it is possible to operate at a high power factor over the entire speed range at the rated load. However, in the case of PWM inverters, power distortion is more than voltage source inverters, as shown in Figure 8.23, which produces an adverse effect on motor insulation systems while working at higher frequency.

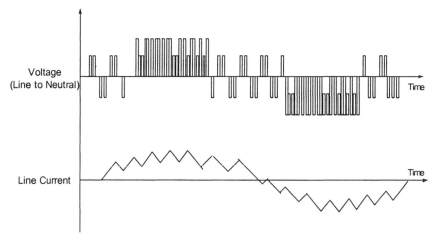

Figure 8.23 Output of PWM inverter.

8.7.6 Cyclo-Converters

These types of converters are used both in slow speed and high power applications such as propulsion systems and electromagnetic launchers driving induction motors and synchronous motors in sea and undersea vehicles. They are usually phase controlled and use switching devices such as thyristors in phase commutation. They are divided into two types such as AC/AC matrix converters and high frequency AC/AC converters. Unlike other converters, cyclo-converters do not contain any storage device such as an inductor or capacitor. These converters are designed in various types such as single-phase to single-phase cyclo-converters, three-phase to single-phase cylco-converters, and three-phase to three-phase cyclo-converters according to the application requirement.

In advanced sea and undersea vehicles, an electromagnetic launcher replaces steam catapults because it offers many advantages in terms of controllability, reliability, weight, volume, and efficiency and also it requires less maintenance [22]. Therefore, it requires electric drive, which can adjust efficient operation at the time of launch by controlling the time duration of the power supply applied to the coil of a linear synchronous motor and disk alternator. Therefore, the electromagnetic launcher can operate effectively at all speeds with variable voltage and frequency supply. Generally, in this type of application, a three-phase to single-phase cyclo-converter is used, which is discussed below.

Sea and Undersea Vehicles

8.7.6.1 Principles of Operation

Three-phase to single-phase cyclo-converters are available in two types: half-wave and bridge cyclo-converters. As shown in Figure 8.24, both positive and negative converters generate voltage of either polarity. However, positive converters can supply positive current only and negative converters can supply negative current only.

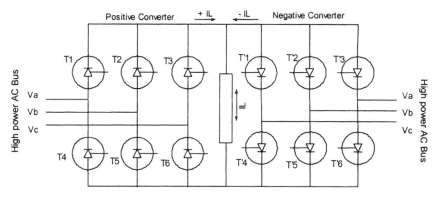

Figure 8.24 Three phase to single phase cyclo-converter.

In the application of the electromagnetic launcher, the output of one bridge is connected in series or parallel to attain at the required power level which eliminates the need of current sharing reactors. When it is used in the application of electromagnetic launcher, it gets supply from the high power AC bus present in the electrical distribution system of the vehicle. The output of this kind of cyclo-converter varies in the range of 0-644 Hz frequencies and 0-1520 V voltage.

8.7.6.2 Matrix Converters

This is one of the types of three-phase AC/AC forced commuted cyclo-converters and it shows potential to be implemented in the electric propulsion system of sea and undersea vehicles [34]. They use nine bi-directional switches, arranged in three groups, that connect the input phase directly to the output phase as shown in Figure 8.25.

It is possible to have 512 different combinations of switches, out of which 27 combinations are permitted. It does not require any kind of passive elements for intermediate energy storage. Therefore, matrix converters show high power density and high reliability. They directly perform voltage and frequency AC/AC conversion and, as all the switches are bi-directional, reactive energy can be circulated between the output phases and the displacement factor can be maintained at unity. Similar to the current source inverters, matrix converters can regenerate energy back into the high power AC bus present in the electrical distribution system of the vehicle from the load side.

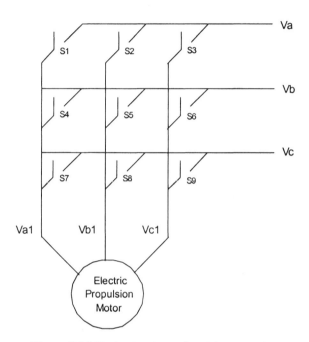

Figure 8.25 Basic structure of matrix converter.

Three configurations of bi-directional switches are possible, as shown in Figure 8.26, for use in the matrix converter.

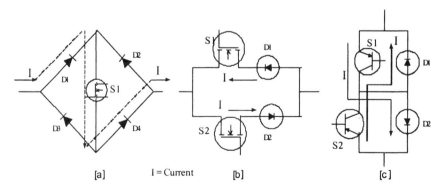

Figure 8.26 Bi-directional switch configurations.

Sea and Undersea Vehicles 323

The first type of configuration is simple, which is shown in Figure 8.26(a). But, in this case, semiconductor losses are high as current conducts through the path provided by three switches, which results into high power loss [34].

In the second type of configuration as shown in Figure 8.26(b), two anti-parallel switching devices are connected in series with blocking diodes. Therefore, it is difficult to control, as it is required to monitor current polarity and, at the same time, switching devices need to operate at different voltage potential [34].

In the third type of configuration of bi-directional switches, as shown in Figure 8.26(c), control becomes simpler as the two switches are connected with one diode in parallel. This is most reliable combination available [34].

Because the number of switches in a matrix converter is very high, they need very complicated control strategy. It is because current is commuted from one switching form to another and there is no freewheeling path for the load current. On the contrary, in conventional inverters, freewheeling diodes are associated with switches to provide a path for inductive load currents. The current commutation problem in bi-directional switches has an impact on protecting power circuit against faulty conditions. Because in a faulty condition such as overcurrent, if the device is turned off to protect the system, there is a high possibility of overvoltage and destruction of the power circuit, since the nature of the load is inductive. Therefore, while deciding the modulation strategy of matrix converters, the first precaution we need to take is to avoid connecting two different input lines to the same output line. The second precaution we need to take is to avoid opening the output line circuits at the same time [35].

8.7.6.3 Control Strategy of Matrix Inverters

In order to take into consideration the precautions discussed above, it is necessary to analyze special modulation strategies such as the following.

In overlap current commutation, the incoming switch is fired before the outgoing switch is turned off. This makes a temporary short circuit of input lines, which can cause overcurrent in the circuit. It can be limited by introducing extra chokes at the inputs. This method is infeasible because chokes are bulky and expensive [36].

In dead time current commutation, the outgoing switch is turned off when the incoming switch is not yet connected. This can cause overvoltage at the output side and, therefore, the clamp circuit is provided at the output to provide continuity of current at the load side. Both of the strategies discussed above require reactive elements and suffer from high power losses [36].

Semi-soft commutation technique, where it is possible to control the direction of current using bi-directional switch cells as shown in Figure 8.27, is the most reliable method. This method contains commutation and logic circuits receiving information from the current detection unit and signals from the previous gate driver circuit through the input port. Accordingly, it sends signals to the gate driver unit of the next bi-directional switch cell from transmitter unit through the output port [36].

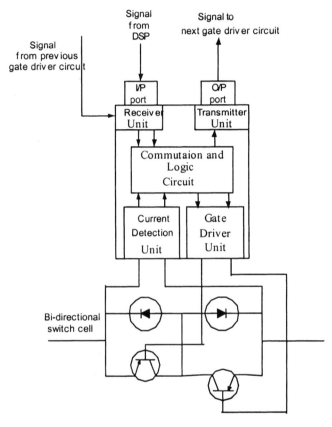

Figure 8.27 Semi-soft commutation technique.

Considering the advancements in power electronics building blocks (PEBB) discussed before, it is possible to integrate power modules with bi-directional switch cells and programmable logic devices (PLD) in order to develop matrix converter power electronics building blocks (MC-PEBB) [35]. Also, research is going on to develop novel topologies for matrix converters to reduce the number of switching elements and to improve the control strategies. It is found that indirect matrix converters (IMC) or sparse matrix converters (SMC) are the promising alternatives to the current matrix converters, since they use the space vector modulation technique to provide natural zero current commutation, reducing complexity in the control strategy [37].

Sea and Undersea Vehicles

8.7.7 Multilevel Inverters

In high power and high voltage applications, multilevel inverters are receiving lots of attention. We look at their suitability for propulsion systems in sea and undersea vehicles with the following discussion. In adjustable speed drives, discussed in previous sections, inverters convert input DC voltage to a variable AC voltage. However, the industry has experienced insulation failures in the stator windings of the motors, because of a high rate of change of voltage at the motor input (dv/dt) [33]. These cause large circulating currents and corona layers between winding layers. This also leads to bearing failure of the motors. Also, at high frequency, switching losses are significant in high power applications such as propulsion systems.

Multilevel inverters have emerged as one of the important adjustable speed drives in sea and undersea vehicles, since they have the ability to synthesize the output waveforms with better harmonic spectrum and reduced total harmonic distortion (THD) and for higher voltage levels using switching elements of low power ratings [38]. Because in the case of multilevel inverters, as the number of levels increases, output waveforms have more steps in the staircase pattern, reducing the harmonic distortion. Therefore, it is one of the promising candidates for electric propulsion systems. Multilevel inverters reduce the stress on each of the switching elements and eliminate electromagnetic interface problems considerably. The stress reduction on the switching elements is proportional to the number of levels [38].

Multilevel inverters have evolved with various configurations as explained below.

8.7.7.1 Multilevel Diode Clamped Inverter

This is also called a neutral point clamped (NPC) pulse width modulated inverter. As shown Figure 8.28, these act as a multiplexer where output voltage level varies between +/-V and zero with respect to the neutral point by different switching schemes of the given switches for different phases. Back to back connected clamp diodes maintain the desired output voltage level at the output in the staircase pattern and provide bi-directional current path for inductive loads [39], [40]. Clamped diodes can be replaced by controlled switches such thyristors, GTOs, and IGBTs to improve the spectral performance of the output voltage. By adding more levels with clamped diodes, it is possible to increase the voltage and power levels and it is the best interface with the onboard generation source for supplying a high power electric propulsion load in sea and undersea vehicles [39], [40].

Furthermore, with the addition of resonant circuits in the original topology, it is possible to reduce switching losses significantly. The multilevel diode clamped inverter has one disadvantage: that required blocking voltage of the clamping diodes increases with the number of levels.

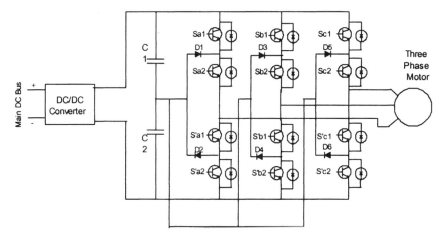

Figure 8.28 Three level diode clamped converter.

8.7.7.2 Flying Capacitor Multilevel Inverter

In this case, clamped diodes are replaced by dedicated capacitors. In this topology, we can maintain a balanced voltage across the capacitors. However, at the same time, this requires a complicated control strategy to control the floating capacitor voltage [38].

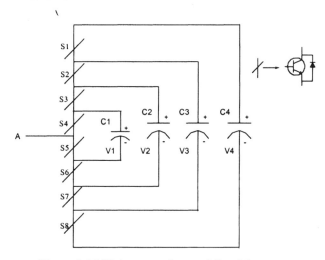

Figure 8.29 Flying capacitor multilevel inverter.

It is possible to use a combination of both topologies (diode clamped and flying capacitor multilevel inverters) mentioned above, as shown in Figure 8.30 [38].

Sea and Undersea Vehicles

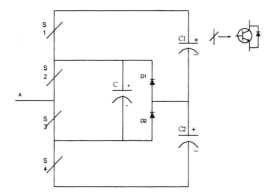

Figure 8.30 Hybrid diode clamped and flying capacitor multilevel inverter.

8.7.7.3 Cascaded Single-Phase H-Bridge Inverter

In this type of multilevel inverter, several single-phase full bridge inverters are cascaded in series together with input supply as independent voltage source or dedicated isolated bus capacitors. Consider one of the phases in a three-phase multilevel inverter; three such single-phase inverters are connected together. In this type, we eliminate the problem of the capacitor voltage balancing [40].

Therefore, in this case, we can get 2n +1 voltage levels such as +/-3V, +/-2V, +/-V, and 0V, where n is the number of cascaded single full bridge inverters for one leg of the multilevel inverter if the independent source supplies voltage V. In this topology, we can increase the number of voltage levels without adding much complexity to the structure [38], [40]. The only problem with this kind of structure is that it requires multiple dedicated DC buses. Implementation of this kind of multilevel inverter in sea and undersea vehicles is therefore difficult. Fuel cells or batteries can be used as independent voltage sources for every stage of the multilevel inverter. However, it is difficult to cascade inverters back to back in series with independent voltage sources at the input for higher power applications. However, it can be applied in multi-propulsion systems with double propellers of these vehicles where individual propulsion requirement is substantially lower compared to the main single propulsion.

8.7.7.4 Hybrid H-Bridge Multilevel Inverter

In the modified H-bridge multilevel inverters, we have lots of flexibility in increasing the number of levels. In this type, we use switching devices with high voltage blocking capability, such as GTOs operating at the fundamental frequency of the inverter output at higher voltage levels and switching devices such as IGBTs at lower voltage levels operating at the frequency of the inverter output [41], as shown in Figure 8.31.

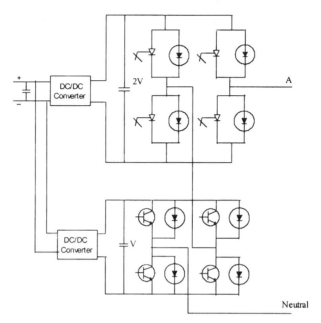

Figure 8.31 Hybrid H-bridge inverter.

8.7.7.5 Three-Phase Multilevel Inverter

It is possible to supply electric propulsion motor through a three-phase multilevel inverter from the main DC bus in the electric distribution system of sea and undersea vehicles. Two full bridge cells are connected in series per phase of the electric propulsion motor. As explained in the previous section, a hybrid H-bridge is implemented to get high efficiency. These full bridge cells are connected in series and positive and negative output voltages are added together by implementing proper modulation techniques to produce high voltage output for each phase of the three-phase motor [42].

8.7.7.6 Modulation Techniques

In pulse width modulation (PWM) method, two triangular functions are compared with a sinusoidal signal for each half bridge to generate positive and negative steps, ensuring that none of the half bridges (positive or negative) produces output of same polarity. However, this method suffers from high commutation and switching power losses [42]. On the other hand, the step modulation technique is based on the load sharing methods. Here, it avoids unequal power distribution in each half bridge.

However, this method also suffers from unequal distribution of switching losses among the half-bridge, in each leg of the inverter. The switching losses can be distributed equally by implementing the step modulation technique in a rotating pattern in continuous sequence for each half-bridge (positive and negative) from the higher level to the lower level in each leg of the multilevel inverter. It can be observed that positive half-bridge switching is delayed and the other positive half-bridge switching is advanced simultaneously after the first cycle and it is repeated for other cycles continuously in a sequence [42]. When the requirement is accurate torque control, the vector modulation technique is used for the multilevel inverter.

8.7.8 DC Drives

Even though AC motors are used to a greater extent in sea and undersea vehicles, there are some applications where DC motors still remain the strong contender to get replaced by other motors. Therefore, in this final section, we will just take the overview of DC drives. These drives are widely used in applications requiring adjustable speed drive, good speed regulation and frequent starting, braking, and reversing.

As shown in Figure 8.32, one application of DC drives in sea and undersea vehicles is a DC series motor in the crane system. This offers advantages in terms of lower cost, reliability, and simple control. Similar kinds of drives exist in excavators in these vehicles by using separately excited or shunt motors.

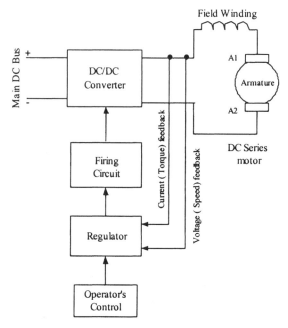

Figure 8.32 Application of DC drives in sea and undersea vehicles.

8.8 References

[1] J. A. Momoh, S. S. Kaddah, and W. Salawu, "Security assessment of DC zonal naval-ship power system," in *Proc. Large Engineering Systems Conference on Power Engineering*, 2001.

[2] J. G. Ciezki and R. W. Ashton, "Selection and stability issues associated with a navy shipboard DC zonal electric distribution system," *IEEE Trans. on Power Delivery*, vol. 15, no. 2, April 2000.

[3] J. D. Van Wyk and F. C. Lee, "Power electronics technology at the dawn of the new millennium - status and future," in *Proc. IEEE Power Electronics Specialist Conference*, 1999.

[4] T. Ericsen and A. Tucker, "Power electronics building blocks and potential power modulator applications," in *Proc. the Twenty-Third International Power Modulator Symposium*, 1998.

[5] J. G. Ciezki and R. W. Ashton, "The application of a customized DSP board for the control of power electronic converters in a DC zonal electric distribution system," in *Proc. Thirty-Second Asilomar Conference on Signals, Systems & Computers*, vol. 2, pp. 1017-1021, 1998.

[6] J. A. du Toit, A. D. le Roux, and J. H. R. Enslin, "An integrated controller module for distributed control of power electronics," in *Proc. IEEE 13th Applied Power Electronics Conference and Exposition*, 1998.

[7] I. Celanovic, N. Celanovic, I. Milosavljevic, D. Boroyevich, and R. Cooley, "A new control architecture for future distributed power electronics systems," in *Proc. IEEE 31st Power Electronics Specialists Conference*, pp. 113-118, 2000.

[8] I. Celanovic, I. Milosavljevic, D. Boroyevich, R. Cooley, and J. Guo, "A new distributed digital controller for the next generation of power electronics building blocks," in *Proc. IEEE 15th Applied Power Electronics Conference and Exposition*, pp. 889-894, 2000.

[9] J. Guo, I. Celanovic, and D. Borojevic, "Distributed software architecture of PEBB-based plug and play power electronics systems," in *Proc. IEEE 16th Applied Power Electronics Conference and Exposition*, pp. 772-777, 2001.

[10] K. L. Butler, N. D. R. Sarma, C. Whitcomb, H. Do Carmo, and H. Zhang, "Shipboard systems deploy automated protection," *IEEE Computer Applications in Power*, vol. 11, no. 2, April 1998, pp. 31-36.

[11] I. Jonasson and L. Seder, "Power quality on ships - a questionnaire evaluation concerning island power system," in *Proc. Power Engineering Society Summer Meeting*, pp. 216-221, 2001.

[12] R. W. Ashton, J. G. Ciezki, and C. Mak, "The formulation and implementation of an analog/digital control system for a 100-kW dc-to-dc buck chopper," *IEEE Trans. on Circuits and Systems II: Analog and Digital Signal Processing*, vol. 46, no. 7, July 1999, pp. 971-974.

[13] J. G. Ciezki and R. W. Ashton, "The design of stabilizing controls for shipboard DC-to-DC buck choppers using feedback linearization

techniques," in *Proc. IEEE 29th Power Electronics Specialists Conference*, pp. 335-341, 1998.

[14] M. Belkhayat, R. Cooley, and A. Witulski, "Large signal stability criteria for distributed systems with constant power loads," in *Proc. IEEE 26th Power Electronics Specialists Conference*, pp. 1333-1338, 1995.

[15] J. Biebach, P. Ehrhart, A. Muller, G. Reiner, and W. Weck, "Compact modular power supplies for superconducting inductive storage and for capacitor charging," *IEEE Trans. on Magnetics*, vol. 37, no. 1, Jan. 2001, pp. 353-357.

[16] D. A. Clayton, S. D. Sudhoff, and G. F. Grater, "Electric ship drive and power system," in *Proc. 25th International Power Modulator Symposium*, pp. 85-88, 2000.

[17] G. F. Grater and T. J. Doyle, "Propulsion powered electric guns-a comparison of power system architectures," *IEEE Trans. on Magnetics*, vol. 29, no. 1, Jan. 1993, pp. 963-968.

[18] E. Spooner, N. Haines, and R. Bucknall, "Slotless-armature DC drives for surface warship propulsion," *IEE Proc. Electric Power Applications*, vol. 143, no. 6, Nov. 1996, pp. 443-448.

[19] F. Caricchi, F. Crescimbini, and O. Honorati, "Modular, axial-flux, permanent-magnet motor for ship propulsion drives," in *Proc. IEEE International Electric Machines and Drives Conference*, pp. WB2/6.1-WB2/6.3, 1997.

[20] R. Xiuming, Y. Bingchuan and H. Hai, "High-power multi-phase permanent magnet (PM) propulsion motor," in *Proc. the 5th International Conference on Electrical Machines and Systems*, pp. 835-837, 2001.

[21] M. Rosu, A. Arkkio, T. Jokinen, J. Mantere, and J. Westerlund, "Permanent magnet synchronous motor for ship propulsion drive," in *Proc. IEEE International Electric Machines and Drives Conference*, pp. 776-778, 1999.

[22] M. R. Doyle, D. J. Samuel, T. Conway, and R. R. Klimowski, "Electromagnetic aircraft launch system-EMALS," *IEEE Trans. on Magnetics*, vol. 31, no. 1, Jan. 1995, pp. 528-533.

[23] B. B. Gamble, S. Kalsi, G. Snitchler, D. Madura, and R. Howard, "The status of HTS motors," in *Proc. IEEE 2002 Power Engineering Society Summer Meeting*, 2002, pp. 270-274.

[24] A. J. Mitcham, "Transverse flux motors for electric propulsion of ships," *IEE Colloquium on New Topologies for Permanent Magnet Machines*, Digest no. 1997/090, 1997, pp. 3/1-3/6.

[25] W.-S. Kim, S.-Y. Jung, H.-Y. Choi, H.-K. Jung, J.-H. Kim, and S.-Y. Hahn, "Development of a super conducting linear synchronous motor," *IEEE Trans. on Applied Superconductivity*, vol. 12, no. 1, pp. 842-845, March 2002.

[26] B. K. Bose, "Power electronics and motor drives - recent technology advances," in *Proc. IEEE 2002 International Symposium on Industrial Electronics,* 2002, pp. 22-25.

[27] B. K. Bose, "Power electronics and drives - technology advances and trends," in *Proc. IEEE 1999 International Symposium on Industrial Electronics,* 1999.

[28] M. J. V. Wimshurst, "Variable frequency drives - application to ships propulsion systems," in *Proc. IEEE 2002 Power Engineering Society Summer Meeting,* 2002, pp. 265-269.

[29] A. Weber and S. Eicher, "10kV power semiconductors - breakthrough for 6.9 kV medium voltage drives," in *Proc. 14th International Symposium on Power Semiconductor Devices and ICs,* 2000, pp. 45-48.

[30] A. Khaizaran, M. S. Rajamani, and P. R. Palmer, "The high power IGBT current source inverter," in *Proc. IEEE 36th Industry Applications Society Annual meeting,* 2001, pp. 879-885.

[31] S. D. Sudhoff, E. Lzivi, and T. Dcollins, "Start up performance of load-commutated inverter fed synchronous machine drives," *IEEE Trans. on Energy Conversion,* vol. 10, no. 2, pp. 268-274, June 1995.

[32] H. Le-Huy, A. Jakubowicz, and R. Perret, "A self-controlled synchronous motor drive using terminal voltage system," *IEEE Trans. on Industry Applications.,* vol. IA-18, pp. 46-53, Jan./Feb. 1982.

[33] D. N. Zmood and D. G. Holmes, "Improved voltage regulation for current-source inverters," *IEEE Trans. on Industry Applications,* vol. 37, no. 4, pp. 1028-1036, July/Aug. 2001.

[34] S. Siinter and J. C. Clare, "Development of matrix converter induction motor drive," in *Proc. 7th Mediterranean Electrotechnical Conference,* 1994, pp. 833-836.

[35] C. Klumpner, F. Blaabjerg, and P. Nielsen, "Speeding-up the maturation process of the matrix converter technology," in *Proc. IEEE 32nd Power Electronics Specialists Conference,* 2001, pp.1083-1088.

[36] L. Empringham, P. W. Wheeler, and J. C. Clare, "Intelligent commutation of matrix converter bi-directional switch cells using novel gate drive techniques," in *Proc. IEEE 29th Power electronics Specialists Conference,* 1998, pp. 707-713.

[37] J. W. Kolar, M. Baumann, F. Schafmeister, and H. Ertl, "Novel three-phase AC-DC-AC sparse matrix converter," in *Proc. IEEE 17th Applied Power Electronics Conference and Exposition,* 2002, pp. 777-791.

[38] S. B. Seok, G. Sinha, M. D. Manjrekar, and T. A. Lipo, "Multilevel power conversion - An overview of topologies and modulation strategies," in *Proc. 6th International Conference on Optimization of Electrical and Electronic Equipments,* 1998, pp. AD/11-AD/24.

[39] L. M. Tolbert, F. Z. Peng, and T. G. Habetler, "Multilevel inverters for electric vehicle applications," *Power Electronics in Transportation,* 1998, pp. 79-84.

[40] L. M. Tolbert and F. Z Peng, "Multilevel converters for large electric drives," in *Proc. IEEE 13th Applied Power Electronics Conference and Exposition,* 1998, pp. 530-536.

[41] M. D. Manjrekar and T. A. Lipo, "A hybrid multilevel inverter topology for drive applications," in *Proc. IEEE 13th Applied Power Electronics Conference and Exposition,* 1998, pp. 523-529.

[42] N. P. Schibli, N. Tung, and A. C. Rufer, "Three-phase multilevel converter for high-power induction motors," *IEEE Trans. on Power Electronics,* vol. 13, no. 5, pp. 978-986, Sep. 1998.

9
Fuel Cell Based Vehicles

Fuel cells have emerged as one of the most promising technologies for meeting the new energy demands. They are environmentally clean, quiet in operation, and highly efficient for generating electricity. This shining new technology provides the impetus towards a huge market for power electronics and its related applications.

This chapter aims at discussing the basic operation of fuel cells for vehicular applications. Furthermore, this chapter covers various types of fuel cell systems and their structures along with their typical power electronic converter topologies. Primarily, power electronics plays a vital role in improving flexibility of utilizing fuel cells in different vehicles.

9.1 Structures, Operations, and Properties of Fuel Cells

A fuel cell is typically similar in operation to a conventional battery, although they have some distinct physical differences. Primarily, a fuel cell is an electro-chemical device wherein the chemical energy of a fuel is converted directly into electric power [1]-[5]. The main difference between a conventional battery and a fuel cell is that, unlike a battery, a fuel cell is supplied with reactants externally. As a result, whereas a battery is discharged, a fuel cell never faces such a problem as long as the supply of fuel is provided. As is depicted in Figure 9.1, electrodes and electrolyte are the main parts of a fuel cell. The most popular type of fuel cell is the hydrogen-oxygen fuel cell.

As is shown in Figure 9.1, hydrogen is used as the fuel to be fed to the anode. The cathode, on the other hand, is fed with oxygen, which may be acquired from the air. The hydrogen atom is split up into protons and electrons, which follow different paths, but ultimately meet at the cathode. For the splitting up process, we need to use a suitable catalyst. The protons take up the path through the electrolyte, whereas the electrons follow a different external path of their own. This, in turn, facilitates a flow of current, which can be used to supply an external electric load. The electrode reactions are given as follows:

Anode: $2H2 \rightarrow 4H^+ + 4e^-$
Cathode: $O2 + 4e^- \rightarrow 2O^-$
Overall Reaction: $2H2 + O2 \rightarrow 2H2O$

From these simple and basic expressions describing the operation of a typical fuel cell, we can see that there is absolutely no combustion and, hence, no production of emissions. This makes the fuel cell environmentally suitable.

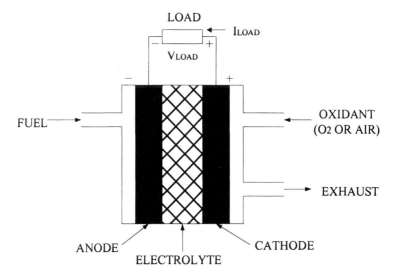

Figure 9.1 Typical schematic diagram of a fuel cell.

A typical i-v curve of fuel cells is shown in Figure 9.2. The output voltage decreases as the current increases. Moreover, the efficiency of a fuel cell is defined as the ratio of electrical energy generated to the input hydrogen energy. Generally, cell efficiency increases with higher operating temperature and pressure [1].

Fuel cells have many favorable characteristics for energy conversion. As explained earlier, they are environmentally acceptable due to a reduced value of carbon dioxide (CO_2) emission for a given power output. Moreover, the usage of fuel cells reduces transmission losses, resulting in higher efficiency. Typical values of efficiency range between 40%-85%. Another advantage of fuel cells is their modularity. They are inherently modular, which means that they can be configured to operate with a wide range of outputs, from 0.025-50 MW for natural gas fuel cells to 100 MW or more for coal gas fuel cells. Another unique

Fuel Cell Based Vehicles

advantage of fuel cells is that hydrogen, which is the basic fuel used, is easily acquirable from natural gas, coal gas, methanol, and other similar fuels containing hydrocarbons. Lastly, the waste heat/exhaust can be utilized for co-generation and for heating and cooling purposes. This exhaust is useful in residential, commercial, and industrial co-generation applications. Basically, for co-generation, the fuel cell exhaust is used to feed a mini or a micro turbine-generator unit. These turbines are generally gas turbines. Since the waste thermal energy is recovered and converted into additional electrical energy, the overall system efficiency is improved. The gas turbine fulfills this role suitably. The typical sizes of such systems range from 1 to 15MW.

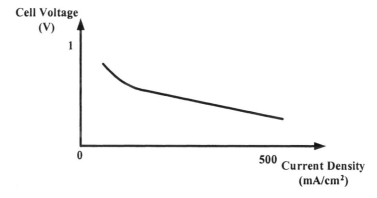

Figure 9.2 Typical *i-v* curve of a fuel cell.

Fuel cells are generally characterized by the type of electrolyte that they use. Main fuel cell systems under development for practical applications are phosphoric acid (PA), proton exchange membrane (PEM), molten carbonate (MC), solid oxide (SO), direct methanol (DM), and alkaline fuel cells [5]-[9]. In this section, we explain different types of fuel cells, their individual structures, and operations. The main applications are also addressed.

9.1.1 Phosphoric Acid Fuel Cell (PAFC)

These fuel cells are known as the first generation fuel cells. They are also the closest to commercialization. The power generating efficiency of these types of fuel cells are typically in the range of 35–45%. They operate at a temperature of about 200°C. The schematic diagram of a typical PAFC is depicted in Figure 9.3. As seen from the reactions, water molecules have to be removed from the cathode to retain the cell water balance. An ion is transported through the electrolyte from the anode to the cathode. The electrodes are connected through an external load to complete the electrical circuit [2]. The chemical reactions taking place at the anode and cathode can be summarized as follows:

Anode: $2H_2 \rightarrow 4H^+ + 4e^-$
Cathode: $O_2 + 4H^+ + 4e^- \rightarrow 2H_2O$
Overall Reaction: $2H_2 + O_2 \rightarrow 2H_2O + POWER$

Since these fuel cells operate at temperatures around 200°C, their exhaust can be considered for the purpose of co-generation [1]. Nearly 85% of the produced steam is used for co-generation. The waste heat generated may also be used for space heating or heating of water.

9.1.2 Proton Exchange Membrane (PEM) Fuel Cell

The electrolyte of a PEM fuel cell consists of a layer of solid polymer, which allows protons to be transmitted from one side to the other. It basically requires hydrogen and oxygen as its input, though the oxidant may also be ambient air.

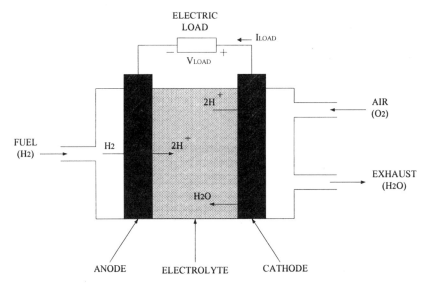

Figure 9.3 Schematic diagram of a typical phosphoric acid fuel cell.

These gases must be humidified. A very simple depiction of this type of fuel cell is shown in Figure 9.4. The anode conducts the electrons that are freed from the hydrogen molecules so that they can be used in an external circuit. The cathode, on the other hand, has channels etched into it, which distribute the oxygen to the surface of the catalyst. It also conducts the electrons back from the external circuit to the catalyst, where they can recombine with the hydrogen ions and oxygen to form water. The electrolyte is the proton exchange membrane. This specially treated material only conducts charged ions. The membrane

Fuel Cell Based Vehicles

blocks electrons. In fact, the catalyst is a special material that facilitates the reaction of oxygen and hydrogen. It is usually made of platinum powder coated very thinly onto carbon paper or cloth. The catalyst is rough and porous so that the maximum surface area of platinum can be exposed to the hydrogen or oxygen. The platinum-coated side of the catalyst is placed facing the membrane.

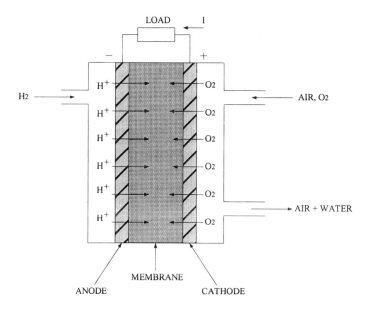

Figure 9.4 Schematic diagram of a typical PEM fuel cell.

The anode reaction, which is basically an oxidation reaction, is as follows:

Anode: $2H_2 \rightarrow 4H^+ + 4e^-$

On the other hand, the cathode reaction is a reduction reaction wherein the end product is water. This can be expressed as follows:

Cathode: $O_2 + 4H^+ + 4e^- \rightarrow 2H_2O$

Hence, the overall reaction in the cell can be written as:

Overall Reaction: $2H_2 + O_2 \rightarrow 2H_2O$

These fuel cells can be used for a variety of different applications, especially in automotive systems. Typically, these fuel cells can operate with efficiencies of about 55% at a temperature of about 90°C [1], [6].

9.1.3 Molten Carbonate Fuel Cell (MCFC)

This fuel cell, as its name suggests, uses a molten alkali carbonate mixture as the electrolyte. They operate at a temperature of about 650°C, which means that useful heat is produced. The basic structure of a MCFC is shown in Figure 9.5.

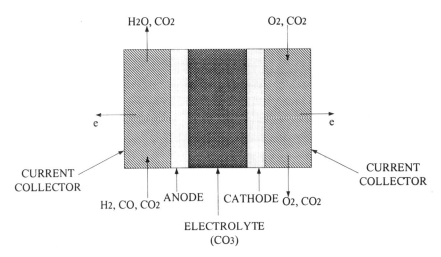

Figure 9.5 Schematic diagram of a typical molten carbonate fuel cell.

A single cell is a laminate of two porous electrodes made of Ni and NiO, an electrolyte, and separators. A fuel cell stack is made by alternating these single cells with cooling plates [1]. The electrolyte is a porous plate of $LiAlO_2$ impregnated with carbonates. In this case the cathode must be supplied with carbon dioxide, which reacts with the oxygen and electrons to form carbonate ions, which carry the ionic current through the electrolyte. At the anode, these ions are consumed in the oxidation of hydrogen, which also forms water vapor and carbon dioxide to be transferred back to the cathode. The chemical reactions can be written as:

Anode: $H_2 + CO_3^{=} \rightarrow H_2O + CO_2 + 2e^-$
Cathode: $\frac{1}{2}O_2 + CO_2 + 2e^- \rightarrow CO_3^{=}$
Overall Reaction: $H_2 + \frac{1}{2}O_2 + CO_2$ (Cathode) $\rightarrow H_2O + CO_2$ (Anode)

Carbon monoxide (CO) is not directly used by the electrochemical oxidation, but produces additional H_2 when combined with water. Carbon dioxide and oxygen from air together with electrons from the cathode are converted to carbonate ions, as can be seen in the expression above. These carbonate ions react with the fuel (hydrogen and carbon monoxide) at the anode, releasing electrons and producing carbon dioxide and water [2], [3].

Fuel Cell Based Vehicles

This type of fuel cell finds applications as a stationary source of power, such as stationary power generators. As mentioned earlier, these fuel cells operate at temperatures as high as 650°C. In fact, it is at these temperatures that they have been found to provide a good conducting nature. Typical values of efficiencies for this type of fuel cell are about 45% [1].

9.1.4 Solid Oxide Fuel Cell (SOFC)

The SOFC has emerged as an alternative high temperature technology contender. The most striking quality is that the electrolyte is in solid state and is not a liquid electrolyte, which would mean having material corrosion and electrolyte management problems. The schematic diagram of an SOFC showing how oxidation of the fuel generates electric current to the load is shown in Figure 9.6.

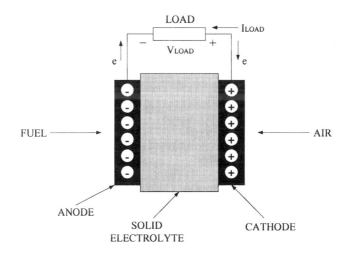

Figure 9.6 Schematic diagram of a typical SOFC.

As illustrated in the figure, the cell readily conducts oxygen ions from the air electrode (cathode), where they are formed through the zirconia-based electrolyte to a fuel electrode (anode). Here they react with fuel gas CO or H_2, or any other mixture, and deliver electrons to an external circuit to produce electricity [7]. A number of different fuels can be used, from pure hydrogen to methane to carbon monoxide. The nature of the emissions from the fuel cell varies correspondingly with the fuel mixture. The electrochemical reactions occurring in a SOFC are:

Anode: $H2 + O^{=} \rightarrow H2O + 2e^{-}$
Cathode: $\frac{1}{2}O2 + 2e^{-} \rightarrow O^{=}$
Overall Cell Reaction: $H2 + \frac{1}{2}O2 \rightarrow H2O$

Carbon monoxide (CO) and hydrocarbons such as methane (CH_4) can be used as fuels in an SOFC. It is also feasible that the water gas shift involving CO and the steam occurs at the high temperature environment of an SOFC to produce H_2 that is easily oxidized at the anode. The reaction of CO is considered as a water gas shift rather than an oxidation.

The SOFC is a very attractive solution for electric utility and industrial applications [4]. The major advantage of this cell lies in its operating temperature and efficiency. It has shown efficiencies ranging from 55-60% and temperatures of roughly 1000°C.

9.1.5 Direct Methanol Fuel Cell (DMFC)

As we have explained, the primary fuel used in fuel cells is hydrogen, which can be easily obtained from reformation of hydrocarbon fuels such as methane and methanol. However, such a method of electricity generation immediately suggests a low efficiency of fuel utilization as well as increased cost of operation. This limitation can be overcome by the use of the DMFC wherein methanol is directly used as the primary source of fuel. Therefore, the operation of the DMFC is based on the oxidation of an aqueous solution of methanol in a PEM fuel cell without the use of a fuel processor [8]. A typical DMFC structure is shown in Figure 9.7.

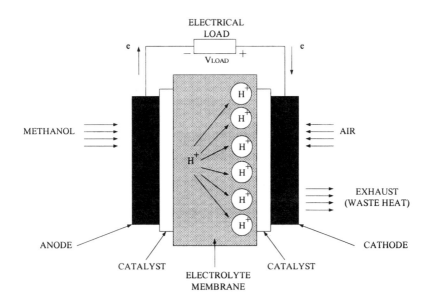

Figure 9.7 Schematic diagram of a typical DMFC.

Fuel Cell Based Vehicles 343

As is clear from the figure, the methanol fuel, which may be stored in a reservoir, is directly fed to the anode. As is the usual practice, O_2 from air is fed to the cathode. The catalyst used at the cathode is typically a platinum-ruthenium (Pt-Ru) combination, while only platinum (Pt) is used at the anode [9]. The controlled oxidation, taking place at the anode, can be expressed as:

$$CH_3OH + H_2O \rightarrow CO_2 + 6H^+ + 6e^-$$

The cathode reaction can be depicted as follows:

$$1.5O_2 + 6H^+ + 6e^- \rightarrow 3H_2O$$

As is clear from the above expressions, there is a controlled oxidation taking place at the anode, whereas a reduction of oxygen to water is observed at the cathode. The entire conversion process takes place at lower temperatures (200°C), as compared to other hydrocarbon fuels [4].

The DMFC is being considered to play an important role in the development of integrated power systems. In addition, there has been an effort to create miniature fuel cells to replace rechargeable batteries. Moreover, it is expected that the DMFC will be more efficient than the polymer electrolyte fuel cells (PEFC) such as PEM fuel cell for automobiles that use methanol as a fuel. There are already working DMFC prototypes used by the military for powering different electronic equipments. Recently, there has been much concern about the poisonous effects of methanol. Hence, manufacturers are considering the usage of ethanol instead of methanol.

Having discussed the DMFC, it is worthwhile to mention briefly about the indirect methanol fuel cells (IMFC). Primarily, the IMFC is similar in structure to a PEM fuel cell. Each cell stack uses an assembly of membrane and electrodes with thin layers of catalysts merged with the membrane [10]. Furthermore, a Pt-Ru combination is used as a catalyst at the anode, while Pt is used at the cathode.

There are basically two primary interactions within the IMFC system. They are the interaction between the fuel processor and stack and the interaction between air supply and stack. Finally, there exists the water and thermal management (WTM) system, which is closely related to the two major interactions. The WTM has two major functions, firstly controlling the thermal condition of the components and, at the same time, maintaining water levels in the system per the operating requirements [10]. As of now, the IMFC is being considered for transportation applications.

9.1.6 Alkaline Fuel Cell (AFC)

The AFC has recently attracted the attention of most fuel cell manufacturers. These fuel cells are also known as ambient temperature fuel cells wherein the electrolyte is circulated in order to transfer heat and water and, at the same time, aid in removal of the carbonate [3]. Schematic representation of a typical AFC, showing all the essential parts, is depicted in Figure 9.8.

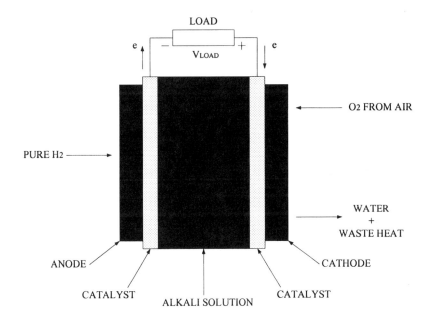

Figure 9.8 Schematic diagram of a typical AFC.

Referring to Figure 9.8, we see that the electrolyte used in an AFC is, as the name suggests, an alkaline solution, generally being potassium hydroxide solution (KOH). To prevent carbonation of the electrolyte, scrubbers are used to scrub both the air and the fuel gas inlets. Also, in this case anode and cathode electrodes are catalyzed with small quantities of platinum (Pt) and silver (Ag), respectively. A summary of the chemical reactions taking place at the cathode and anode of the AFC is as follows:

Anode: $H_2 + 2OH^- \rightleftharpoons 2H_2O + 2e^-$

Cathode: $\frac{1}{2}O_2 + H_2O + 2e^- \rightleftharpoons 2OH^-$

Despite the problems associated with scrubbing CO_2 from the fuel gas and air inlets, the AFC seems to be a very attractive solution for large-scale power generation systems. The AFC is being used extensively in the U.S. for various space programs since the early 1960s. The cost of the AFC is comparatively much higher than any other fuel cell at the moment. This high cost arises due to the fact that the reactant processing cost to handle the carbonation problem is quite high.

9.1.7 Zinc Air Fuel Cell (ZAFC)

The ZAFC belongs to the class of regenerative type of fuel cells. It uses zinc pellets and atmospheric oxygen to produce electricity. The ZAFC differs from the conventional zinc air battery with respect to the refueling aspect. Whereas in a conventional battery the unit has to be dismantled completely for the purpose of refueling, the ZAFC is refueled without having to dismantle it. A typical structure of a ZAFC is shown in Figure 9.9. As is clear from the figure, the ZAFC uses a zinc oxide anode and an air cathode.

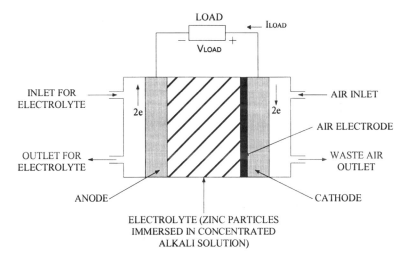

Figure 9.9 Schematic diagram of a typical ZAFC.

The electrolyte consists of zinc particles through which an alkaline solution, such as KOH, is passed. The zinc particles undergo electrolytic dissolution and, hence, the volume of the particles reduces [11]. At the same time, additional pellets of zinc are stored in a hopper, which is positioned just above this electro-active area. As the zinc particles dissolve, the zinc bed volume decreases and a fresh set of zinc pellets are introduced via this hopper. This process continues until the hopper is empty, at which point the cell is completely discharged and must be refueled. Due to the circulating KOH solution, the zincate is removed constantly after each reaction ends and, as a result, clogging due to the mixtures of zinc and reaction products is avoided.

The chemical reactions taking place at anode and cathode electrodes can be summarized as follows:

Anode: $Zn + 4OH^- \rightarrow Zn(OH)_4^{--} + 2e^-$

This zincate is converted to zinc oxide (ZnO) in the electrolyte management unit. It is removed from the electrolyte via the reaction:

$$Zn(OH)_4^- \rightarrow ZnO + H_2O + 2OH^-$$

At the air cathode the reaction is as follows:

Cathode: $\frac{1}{2}O_2 + H_2O + 2e^- \rightarrow 2OH^-$

In the above reaction, O_2 is supplied as air. Thus, the overall reaction is zinc plus oxygen reacting to form zinc oxide.

$$Zn + \frac{1}{2}O_2 \rightarrow ZnO$$

As far as the applications of these types of fuel cells are concerned, they are being considered for DC battery backup applications, uninterruptible power supply (UPS) systems, and transportation applications.

9.1.8 Characteristics of Different Types of Fuel Cells

Each of the above mentioned types of fuel cells are different in some respect from each other. We shall discuss some of the major differences between the various types of fuel cells with regard to their operating temperatures, specific applications, range of power output, and fuel efficiency. This is depicted in a tabular format in Table 9.1.

From the above fuel cells, the SOFC is most likely to be used for large and small-scale plants of greater than 1kw. The DMFC most probably will replace the conventional batteries for a variety of portable applications. The PAFC is being produced at the moment for medium sized power plants, while the AFC has been produced in limited quantities for decades now. Moreover, the SOFC is widely considered to be superior compared to the PAFC and, hence, would substitute for it in time.

9.1.9 Fuel Cell Based Power Processing Systems

Having discussed the structures of different types of fuel cells and having seen their respective properties and applications, it is essential to know how a fuel cell produces the required power for a particular application. A typical fuel cell based power processing system showing the major plant processes is depicted in Figure 9.10. There are three major steps involved in the generation of power from a fuel cell. The first and foremost step is to achieve purity of the available hydrogen gas. This is done with the help of a fuel processor. As seen previously, a carbonaceous fuel is fed to the fuel processor, which, in turn, produces a hydrogen rich gas at its output. This hydrogen rich gas is then fed to the anode electrode of the fuel cell.

Table 9.1 Characteristics of different types of fuel cells.

Type	Operating Temp. (°C)	Efficiency (%)	Output (W)	Applications
PAFC	160-220	Up to 40%, when used for distributed generation, efficiencies greater than 40% are possible.	Ranging between 100kW-5000kW, suitable for use in small power generating applications.	Utility, co-generation and vehicular applications.
PEMFC	80-90	Has lower costs and shows efficiencies up to 45%, making it suitable for supplying peak and emergency loads.	Output ranges from 1kW-1000kW, which is comparatively lower than that for PAFC.	Transportation and space and underwater applications.
MCFC	≈650	Has reasonably higher efficiencies ranging between 50-75%, making it suitable for larger applications.	Output ranges from 1000kW-100MW, slightly lesser than SOFC.	Stationary power generators.
SOFC	800-1000	Nearly 50%, system efficiency can be increased to 60%, if used with micro-turbine generator systems.	Has widest range of output, between 100kW-100MW, suitable for co-generation.	Industrial co-generation and utility applications.
DMFC	50-85	Has lower efficiency, ranging between 45-50%, due to higher stack operating temperature.	Lowest output, ranging from 100W-5kW, since it has highest contamination problem.	Portable electronics, emergency power generators and transportation.
AFC	60-90	Shows efficiencies ranging from 40-50%, thus, making them applicable for small power supply applications.	Has power output, ranging between 10kW-100kW, even though it is smaller in size.	Transport applications, standby UPS and for defense and space applications.

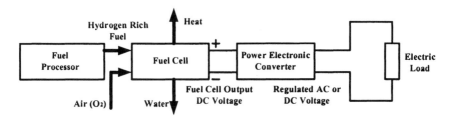

Figure 9.10 Block diagram of a typical fuel cell based power processing system.

The second step involves the fuel cell operation itself. The fuel cell is fed with the hydrogen rich gas at its anode and a supply of air to the cathode. The hydrogen atoms at the anode get split up into positive protons and negative electrons. These electrons follow an external path on their way to the cathode, thus supplying power to an external load in the process. For facilitating the splitting up of the hydrogen atoms, suitable catalysts are used. As seen previously from the chemical reactions, there is no combustion and, hence, no emissions are produced.

The third and final step is the power-conditioning step, which includes power electronic converters. Power electronic converters add more flexibility to the operation of the system. There may be two stages of converters. Firstly, it consists of a DC/DC converter, which converts the low voltage DC output from the fuel cell to a level at which the next stage, i.e., a DC electric load or an inverter can safely operate. The inverter is basically used to invert the DC output from the DC/DC converter to a suitable AC voltage, if the electric load is of AC type. Ideally, the power conditioner must have minimal losses, thus leading to a higher efficiency. Power conditioning efficiencies are typically higher than 90%, and go up to 99%.

9.2 Important Properties of Fuel Cells for Vehicles

Various advantages of fuel cells, including reliability, simplicity, quietness of operation, and most importantly low pollution, have made the fuel cells attractive in different low and medium power applications. One such major application is the automotive industry, where all the above-mentioned factors play a vital role. Generally, hydrogen is considered to be the primary fuel to be used for automotive fuel cell based applications. However, due to the danger of carrying hydrogen aboard the vehicle, the automotive industry is considering alternate fuel sources. These alternate fuels are generally natural gases, which can be reformed to get nearly pure hydrogen.

In conventional cars, electric loads are currently about 1.2kW. Hence, small sized batteries are sufficient to supply these loads. But, in the newer cars, known as more electric vehicles (MEVs), electrical systems are replacing mechanical and hydraulic systems. There are also many newly introduced electrical loads, such as electrically assisted power steering, X-by wire,

Fuel Cell Based Vehicles

integrated starter/generator, active suspension, and AC power point. Therefore, electrical power in an advanced car can be more than 10kW. As a result, higher voltage batteries, such as the proposed 36V storage system, are required to meet the higher power demands.

On the other hand, hybrid electric vehicles (HEVs) have been developed wherein an electric machine is coupled to the internal combustion engine (ICE) to propel the vehicle. An electric machine is fed from a DC source and its power rating might be up to 80kW. The DC source could be a battery or a fuel cell of suitable capacity. Typical DC link voltage is either about 300V or 140V. Fuel cells could also be considered an advanced version of the battery powered vehicles in automobiles with all-electric drive train.

Fuel cells can be refueled quickly and, at the same time, provide longer operating range. Research results show that the fuel economy of a direct methanol fuel cell vehicle (DMFCV), which is considered the best option for a fuel cell powered car, is approximately 2 times greater than that of a conventional ICE car. In addition, compared to ICE efficiencies of about 10–30%, fuel cells have shown efficiencies ranging between 30–40% for automobile applications. Even higher efficiencies are attainable when fuel cells make use of direct hydrogen as a fuel source. In this case, efficiencies as high as approximately 50% are attainable. Thus, compared to an ICE powered car, the fuel cell car can do about twice the amount of work and, hence, can cover almost double the distance.

This chapter aims at discussing the fuel cell applications from the point of view of various vehicular applications, such as small vehicles like passenger cars and heavy-duty vehicles like buses and trucks. Furthermore, a comparison between the most popular zero emission vehicles (ZEVs) including battery electric vehicles (BEVs), hybrid electric vehicles (HEVs), direct hydrogen vehicles (DHVs), and direct methanol fuel cell vehicles (DMFCVs) are presented. Current status of fuel cell vehicles and a brief overview regarding their future are also discussed.

In this section, the various properties of fuel cells for automotive applications will be investigated. A brief overview of various fuels being considered for fuel cell vehicles (FCVs) and fuel cell hybrid vehicles (FCHVs) will be presented. Furthermore, these fuel options will be compared with each other on the basis of cost and fuel economy.

9.2.1 Various Alternate Fuels for Fuel Cell Vehicles

As is well known, hydrogen is the primary fuel source for FCVs. It is a certain fact that direct hydrogen fuel cell vehicles (DHFCVs) have zero tailpipe emissions and no emissions from evaporation either. The major problem lies in its transportation and storage. Current methods of on-board storage include cryogenic liquid storage and compressed gas storage [12]. These practices impose additional problems in terms of energy requirements, safety, and space requirements, which in turn increase the fuel costs further. Thus, for the moment, DHFCVs are not considered cost-competitive options, as compared to conventional IC engine vehicles (ICEVs). Most automotive companies are rooting for their own brand of DHFCVs, but these remain innovative concepts, upon which research and development can be carried out. Also, not long back, the alkaline fuel cell (AFC) was used for space applications. However, when considered for vehicular applications, it was found that air supply for this fuel cell is not easily available. In addition, the phosphoric acid (PA) and solid oxide (SO) fuel cells have been considered reasonable options.

Due to the above-mentioned difficulties, methanol fuel cells are considered as promising replacements for hydrogen for FCV/FCHV applications. Methanol, being a liquid fuel, can make use of the already well-established distribution system used for gasoline, with minor modifications [12]. A slight drawback of methanol is that it requires an on-board reformer to get converted into a hydrogen-rich fuel, which then makes the indirect methanol fuel cell vehicle (IMFCV) a near zero-emission vehicle. This process again involves huge costs and reduction in efficiency, not to mention a much more complex vehicular system design. A simple solution has been found to this drawback in recent times in the form of direct methanol fuel cell vehicles (DMFCVs). In the DMFCV, the methanol is oxidized inside the fuel cell stack itself, thus eliminating the need for an on-board fuel reformer [12]. Other critical advantages in favor of this type of fuel cell are high current densities and operating conditions, matching those required for vehicular applications [12], [13].

The most important issue in all of the above discussions is the fuel cost and the subsequent system cost. The costs of the most likely fuel contenders for FCV/FCHV, namely, reformulated gasoline, methanol, and hydrogen, are comparatively depicted in Figure 9.11.

In addition, these fuels also depict quite varying fuel economies when considered for fuel cell vehicular technology [12]. This fact is clearly illustrated in Figure 9.12.

It can be learned from Figure 9.12 that the IMFCV has a fuel economy of 50 mpg of gasoline-equivalent (geq) and is predicted to reach a value of up to 70 mpg, geq. Furthermore, the DMFCV and DHFCV have initial fuel economies of 65 mpg, geq, with DHFCV predicted to reach values as high as 85 mpg, geq, which is higher compared to any of the other ZEV technologies. Also, due to its late introduction, the gasoline FCV is assumed to depict a mileage of about 60 mpg [12]. For comparative purposes, it is notable that hybrid ICE/EV possesses a mileage of 45 mpg, geq, which could ultimately reach values of about 65 mpg, geq.

Apart from these major issues, much lower start-up times are demanded from these systems, and system costs must also be kept to an absolute minimum [13]. Most of the materials available for manufacturing the membranes and catalysts today are extremely costly. However, all costs have to be reduced and major performance barriers will have to be overcome in order for fuel cells to replace the IC engines in modern vehicles.

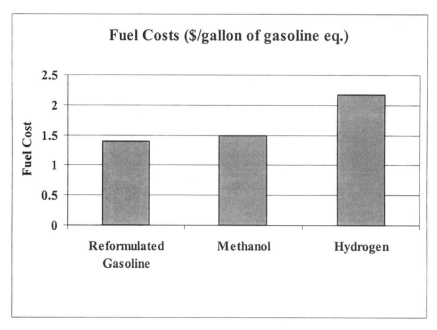

Figure 9.11 Fuel cost comparison for FCVs.

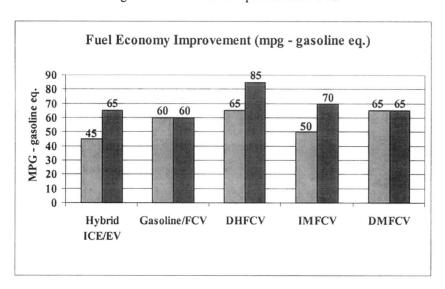

Figure 9.12 Fuel economies of popular zero emission vehicles.

Although most advantages of using DMFCVs have been highlighted above, there exists a minor drawback in their operation, which involves the crossover of unreacted methanol from the anode to the cathode [14]. Leading fuel cell manufacturers are still seeking solutions to reduce losses incurred due to this methanol crossover. This crossover effect also has a direct impact on the voltage efficiency of the DMFC.

9.2.2 Electrical Characteristics of Fuel Cells

Typical *i-v* curve of a DMFC is depicted in Figure 9.13. Characteristics of the DMFC considering the crossover methanol are presented in [14]. These curves are useful in predicting the operation of the DMFC under various different values of cell current. They also emphasize the impact of increasing cell current on the value of crossover current.

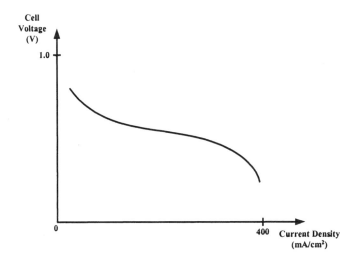

Figure 9.13 Typical *i-v* curve of a DMFC.

Crossover current decreases for higher values of cell current. Thus, fuel cell developers consider operating the system on higher values of cell power densities in order to achieve increased values of cell current. This, in turn, reduces the crossover current losses due to the crossover current. One of the popular methods adopted by manufacturers in order to reduce the crossover currents is to choose different fuel concentrations and flow rates to the anode [14]. This allows the manufacturer to come up with optimized values of fuel concentration and flow rate. The result of this optimization is that higher conversion efficiency is attainable over a wide range of power density [14]. A typical plot, showing cell efficiency against power density as a result of this optimization, is depicted in Figure 9.14.

Fuel Cell Based Vehicles

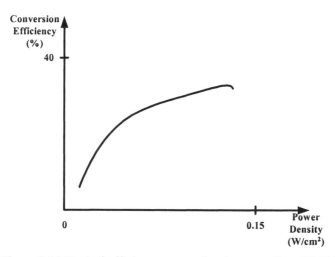

Figure 9.14 Typical efficiency-power density curve for a DMFC.

As is indicated in Figure 9.14, the efficiency curve seems almost flat over a particular range of power density. The typical values of maximum cell efficiency range between 30–40% depending upon optimum values of fuel concentration and flow rate.

9.3 Light-Duty Vehicles

As mentioned earlier, the main reasons for the development of fuel cell powered cars are the greater fuel efficiency and lower emissions. However, there exists one major drawback, that hydrogen cannot be easily stored aboard the vehicle. Hence, the best option being considered by manufacturers is the DMFC. In this section, the basic factors, which influence the use of fuel cells for cars, will be discussed. Moreover, a brief description of a typical arrangement of a fuel cell powered car is presented and the layout of a fuel cell power system for HEVs will be explained.

9.3.1 Fuel Cell Based Drive Trains

Conventional cars have electrical load demands of approximately 1kW. A 14V DC power system with 12V batteries is sufficient to satisfy these loads. With the gradual increase in the amount of electrical loads in conventional commercial automobiles up to 10kW, 42V DC power systems are proposed to handle the high power loads. In hybrid electric cars, higher voltages, such as 300V with electrical systems capable of delivering tens of kilowatts, are required to drive traction and non-traction electrical loads. Fuel cells can replace the high voltage batteries. A typical schematic diagram of the power system of an automobile with a fuel cell based drive train is shown in Figure 9.15.

The fuel cell is fed with fuel and air at the anode and cathode, respectively. A low voltage DC is produced, which is made usable by an electric machine by passing it through a power electronic converter. The electrical machine output is a mechanical output used to drive the wheels of the vehicle.

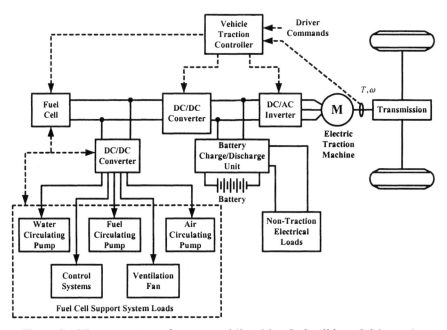

Figure 9.15 Power system of an automobile with a fuel cell based drive train.

The arrangement shown in Figure 9.15 includes a battery pack, making it a hybridized arrangement, and hence, it facilitates regenerative braking [14]. It is worthwhile to present here a brief analysis, focusing on battery electric vehicles (BEVs) and fuel cell vehicles (FCVs) that are hybridized with battery packs to reduce the size of the fuel cell system [15]. Table 9.2 summarizes the comparison between BEV, DHFC/EV, and DMFC/EV characteristics with respect to battery energy capacities and achievable driving ranges. The comparative values of Table 9.2 are illustrated pictorially in Figures 9.16 and 9.17 below. From Table 9.2, it is clear that hybrid FCV/EV requires much lesser capacity batteries and achieves much higher driving ranges compared to pure electric vehicles (BEVs).

Additional R&D results have predicted that a hybrid DHFC/EV is less costly to manufacture initially than a pure FCV when fuel cell system costs are high [15]. However, the costs of pure FCVs become comparable to those of hybrid FCVs when fuel cell systems are mass-produced.

Fuel Cell Based Vehicles

Table 9.2 Comparison of characteristics of BEV, DHFC/EV, and DMFC/EV.

Characteristic	BEV	DHFC/EV	DMFC/EV
Battery Capacity	30 kWh	5.2 kWh	5.2 kWh
Driving Range	125 miles	300 miles	300 miles

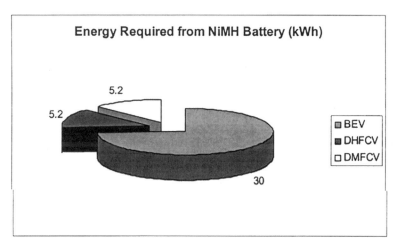

Figure 9.16 Comparison of battery capacities for BEV, DHFC/EV, and DMFC/EV.

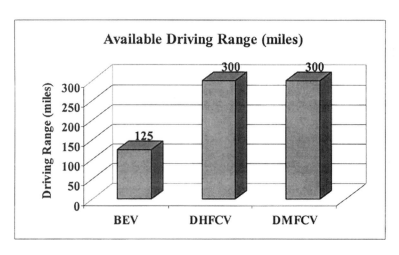

Figure 9.17 Comparison of achievable driving ranges for BEV, DHFC/EV, and DMFC/EV.

Another FCV configuration, comprising an on-board fuel processor system, is as shown in Figure 9.18. A feedback from the fuel processor is given to the system controller, which, in turn, controls the fuel cell based system. This controller also receives a feedback from the vehicle and accordingly gives control signals to the fuel cell stack. The power conditioning unit (PCU) provides the appropriate power to the vehicle loads. The control scheme of the PCU generally involves a PWM control method.

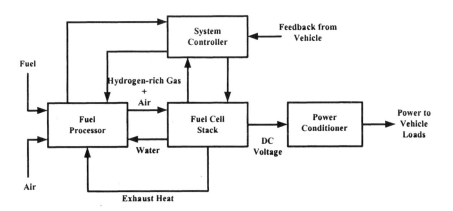

Figure 9.18 Block diagram of a fuel cell system with fuel processor.

Recently, neural networks have also found application in fuel processor control. They are basically hybrid control systems, consisting of a cerebellar-model artificial neural network in parallel with a conventional proportional-integral-derivative (PID) controller [16]. This is basically a dynamic control scheme that controls the amount of carbon monoxide (CO) that is produced, and hence helps minimize the poisonous effects it has on the electrodes in case of DMFC vehicles.

9.4 Heavy-Duty Vehicles

As far as automotive applications for fuel cells are concerned, this is considered by far the most popular application. Most transit authorities in the U.S.A. and Europe have proposed usage of fuel cell powered buses. Again, as in cars, most buses use either direct hydrogen fuel cells and direct methanol fuel cells. In addition to these, phosphoric acid fuel cells have also been successfully demonstrated by transit bus agencies. In fact, buses using all of the above-mentioned types of fuel cells have already been developed and tested [17]. The world's first fuel cell powered city transit bus was developed and demonstrated in Vancouver, Canada. The fuel cells used in this demonstration were proton exchange membrane (PEM) versions of direct methanol fuel cells, developed by Ballard Power Systems Inc., Vancouver, Canada. Fuel cell powered transit

Fuel Cell Based Vehicles

buses, when used in urban areas, offer advantages of low environmental emissions and quiet operation. Furthermore, as a major step towards reducing emissions, most transit authorities have also opted to use compact natural gas (CNG) fueled buses.

9.4.1 Transition from Diesel to Fuel Cell Engines

The emission standards are of prime importance when heavy-duty vehicles, such as buses and trucks, are considered for urban transit applications. These standards are set in terms of average emissions from the engine on a prescribed engine test cycle. Standards are set for NO_x, hydrocarbons, carbon monoxide (CO), and particulate materials [17]. Over the years, bus manufacturers have significantly reduced the NO_x and particulate material emissions. In order to do so, conventional buses using diesel and CNG utilize an oxidation catalyst for controlling emissions. On the other hand, the diesel hybrid electric buses make use of additional regenerative particulate controls apart from the oxidation catalysts [17].

The diesel hybrid electric buses use batteries for electrical energy storage. These batteries are essentially lead-acid batteries, which provide the option of regenerative braking and, hence, energy recovery in the process [17]. Research results have shown that this results in significantly reduced NO_x and hydrocarbon emissions. Moreover, their fuel economy is comparatively higher, mainly due to the fact that there exists the energy recovery due to regenerative braking. Typical values of fuel economy for diesel hybrid electrical buses have been found to be 50–60% higher than those of conventional diesel and CNG fueled buses [17]. When fuel cells are used instead of diesel engines, the fuel economy value is found to be even greater. This is obvious due to the direct oxidation of the hydrogen fuel. Typically, the fuel economy of a DHFC bus is about 30% higher than that of buses using a DMFC with a reformer. Thus, as with smaller vehicles, heavy-duty vehicle manufacturers, such as bus and truck developers, also encourage the use of the direct hydrogen fuel cell (DHFC). However, keeping in mind the drawbacks of storage and handling of hydrogen fuel in case of DHFCVs, at least for the moment, bus manufacturers, similar to light-duty vehicle manufacturers, also favor the usage of direct methanol fuel cells (DMFCs) and, as mentioned previously, to a great extent with a high degree of success, phosphoric acid fuel cells (PAFCs) as well. Furthermore, as discussed, strict standards are set for emissions from vehicles and, hence, the fuel cell technology provides an attractive solution.

9.4.2 Fuel Cell Transit Bus Technology

As stated earlier, the three major fuel cell bus technologies make use of PAFC, DMFC, and DHFC. Among these, the PAFC and DMFC make use of hybridized power system topologies, wherein a lead-acid or nickel/cadmium (NiCd) battery pack is connected in parallel with the fuel cell in order to supply the peak power demands and to take advantage of regenerative braking. A typical arrangement of a hybrid fuel cell/battery power system, applicable to both the PAFC/battery hybrid bus and the DMFC/battery hybrid bus, is as shown in Figure 9.19 below.

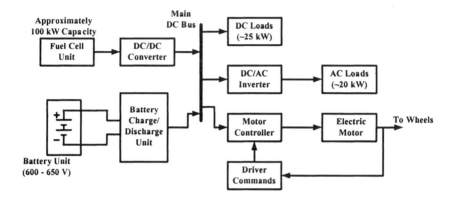

Figure 9.19 Basic block diagram representation of a fuel cell/battery hybrid power system.

As is clear from Figure 9.19, the DC/DC converter raises the level of the voltage from the fuel cell stack up to the level of the main DC bus voltage. The initial peak power during transients, such as start-up and acceleration, is supplied by the battery pack. The various DC and AC loads are fed from the main DC bus through appropriate DC/DC converters and DC/AC inverters. Furthermore, the electric motor (DC motor, brushless DC motor [BLDC], switched reluctance motor [SRM], or AC induction motor) is controlled from a motor-controller system, which in turn drives the wheels of the bus. Whenever the driver changes the speeding and braking commands, a suitable signal is fed back to the controller, which accordingly controls the speed and torque delivered by the motor.

As discussed earlier, either lead-acid or nickel/cadmium (NiCd) battery packs are used in these hybrid fuel cell/battery topologies. These battery packs provide the additional power required during acceleration and store energy recovered during regenerative braking. The heavy-duty lead-acid batteries provide the necessary power and charge acceptance to meet the performance requirements, but their overall lifetime has been found to be low. Based on test results, these batteries require replacement within 1 year of their installation [18]. On the other hand, NiCd batteries offer good performance and possess much lower weight than lead-acid batteries. Although their initial costs are higher, they are neutralized by their longer life. Upon testing these batteries for fuel cell/battery hybrid bus technology, it has been found that a replacement has not been necessary for approximately 3 years of operation, thus proving their

Fuel Cell Based Vehicles

superiority over lead-acid batteries. Some additional performance parameters of fuel cell hybrid electric buses to be recorded during testing include speed, time/distance of travel, power delivered by fuel cell/battery, battery state of charge (SOC), and voltage/current/temperature of major vehicular parts [19].

Georgetown University (Washington) introduced the first commercially viable hybrid PAFC/battery transit bus in 1998. They later upgraded this bus to a PEMFC/battery hybrid bus in 2001 [20]. Some useful technical data with regard to this 40-foot hybrid PEMFC/battery transit bus is summarized in Table 9.3 below.

Table 9.3 Technical specifications of Georgetown 40-foot PEMFC/battery bus.

Characteristic	Specification
Total vehicular weight	Nearly 40,000 lbs.
Propulsion system	PEMFC/NiCd hybrid
Fuel	Methanol
Motor drive	AC induction motor
Motor rating	250 HP
Acceleration (0-30 mph)	14.5 seconds
Maximum speed	66 mph
Driving range	350 miles
Noise level	10 dB below ICE

Another configuration for fuel cell buses includes non-hybridized topologies. In this case, direct hydrogen fuel cells (DHFCs) are used to provide the entire power to the propulsion system. XCELLSiS Fuel Cell Engines, jointly owned by Daimler-Chrysler, Ford Motor Company, and Ballard Power Systems, are involved in the manufacture of such fuel cell engines for transportation applications [21]. A brief summary of the technical data of such a DHFC bus is as shown in Table 9.4.

Table 9.4 Technical specifications of XCELLSiS 40-foot DHFC bus.

Characteristic	Specification
Fuel cell technology	Direct hydrogen
Fuel cell engine weight	Nearly 5000 lbs.
Net shaft power	205 kW
Net efficiency	Nearly 40%
Driving range	225 miles
Fuel cell operating temperature	70 – 80 °C
System voltage	600 – 900 VDC
Power conditioning	Liquid cooled IGBT inverter
Motor drive	Liquid cooled BLDC motor

Currently, extensive field tests are being carried out on the DHFC bus by XCELLSiS engineers to verify its commercial viability [21]. This transit bus is just a concept vehicle as of now and is not ready for revenue service.

At this point, it is appropriate to study how the fuel economies of these fuel cell transit buses compare with those of conventional diesel engine and compact natural gas (CNG) buses. Various transit authorities have conducted comprehensive test procedures in order to calculate the fuel economy (mpg – diesel equivalent) for each of the described types of fuel cell buses. For a clear view in this regard, a comparative illustration is as shown in Figure 9.20 below. As is clear from Figure 9.20, the fuel economy of the DHFC bus is significantly higher than in a hybrid diesel-electric or hybrid fuel cell/battery bus. In general, various transit agencies believe that DHFC buses can achieve about 25 – 30 % higher fuel economy compared to other engine-battery or fuel cell battery hybrid technologies.

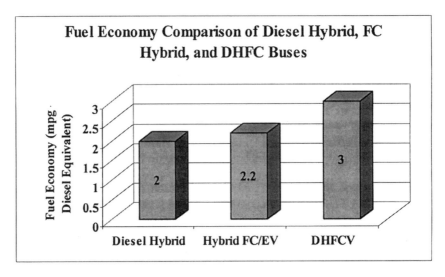

Figure 9.20 Fuel economy comparison of various alternate fuel buses.

But, here again, the simple fact remains that the automobile industry still awaits results of R&D being conducted by fuel cell companies in an effort to find an easier way of storing and handling hydrogen in its gaseous or liquid form.

9.5 Current Status and Future Trends in Fuel Cell Vehicles

In this chapter, the major motivations towards opting for the fuel cell powered vehicles have been discussed. In addition, various types of fuel cells

Fuel Cell Based Vehicles 361

and their properties for automotive applications have been presented. Moreover, a comparison between different types of zero emission vehicles has also been given. This section will cover issues regarding the current status of the fuel cell technology for vehicular applications and will look at the future prospects of such vehicular systems.

Pure zero emission vehicles hold distinct air quality advantages over technologies that use conventional fuels such as gasoline. From the time that electric vehicle technology was introduced, the battery electric vehicles have been considered the best option for meeting the zero emission requirements. More recently, however, the fuel cell technology has gained the attention of major vehicle manufacturers. This is due to the fact that, historically, the inability of batteries to store sufficient amount of energy at a reasonable cost has limited the market for battery electric vehicles. Moreover, as mentioned earlier, fuel cell vehicles are much quieter in operation and have no harmful gas emissions. In addition to these advantages, they also have much higher efficiencies.

While there are several different fuel cell technologies available for use in vehicular systems, the best option according to scientists and automobile developers is the PEM version of the DMFC. The PEMFC stack technology has improved over the past few years and has been found to achieve performances close to that of conventional cars. Significant advances have been made in the development of smaller sized fuel processors for methanol and gasoline [22]. Furthermore, fuel cell stacks of various different structures are now available, which depict high efficiencies over a wide range of power densities.

Although hydrogen gas is considered to have the highest conversion efficiency when used directly in a DHFC, vehicles using these fuel cells cannot be easily commercialized. This is because it is not easy to store hydrogen gas in a compact space with the required amount of safety. Moreover, as highlighted before, hydrogen gas is extremely expensive and is not readily available [22]. The main advantage of the DHFC is that its usage eliminates the need for an expensive reformer. When compared with the DHFC, the DMFC has a comparatively simpler design and operation, higher reliability, less maintenance, and lower capital and operating costs [23].

Hence, due to the lower energy densities and resulting high mass of battery electric vehicles, it is fairly clear that vehicle manufacturers are turning their attention to hybrid electric vehicles and fuel cell hybrid vehicles [24]. Also, the major focus in the U.S.A., Europe, and Japan is on introducing fuel cell hybrid vehicles for testing purposes. Research results have predicted a fair share of the automobile market for fuel cell vehicles and fuel cell hybrid vehicles. A pictorial idea, as to what percentage of the market these vehicles will share by the year 2010, is shown in Figure 9.21.

As depicted in Figure 9.21, almost 17% of the share is expected to belong to the other fuels for automobiles. These fuels are used in methanol fuel cell vehicles, diesel hybrid electric vehicles, compact natural gas vehicles, and battery electric vehicles (BEVs).

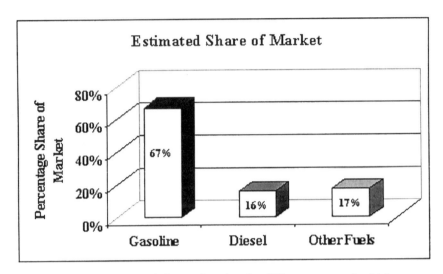

Figure 9.21 Estimated share of market for different types of vehicles.

It is expected that hybrid electric vehicles will share about 10% of the market out of the 17% shown. This means that upon further development towards reduction in overall costs of materials, components, and the fuel cell technology itself, the hybrid electric vehicles using fuel cells can definitely be introduced to the market. Thus, it is likely that the 21st century will witness the development and usage of fuel cell vehicular systems, which will make transportation much more environmentally friendly, unlike the current transportation systems [24].

9.6 Aerospace Applications

For long space missions, it has been identified that energy storage and conversion based on hydrogen technology is of critical importance [25]. Therefore, fuel cell technology has gained considerable momentum for aerospace applications. One of the first fuel cell technologies considered worthwhile for such applications was the AFC. The AFC has high efficiency and low overall mass, thus making it suitable for long space missions. In addition, on comparing the AFC with batteries, the AFC has higher power density and much more energy storage capacity [25].

The AFC, with an immobile electrolyte, is considered one of the best options for powering space shuttles. This technology has been developed for space missions in the U.S. by NASA [25]. In the immobile electrolyte AFC, the

Fuel Cell Based Vehicles

electrolyte is kept in a fixed position by capillary forces. Furthermore, for the water intake and removal, certain membrane techniques, which are explained in [25], are used.

In recent years, the PEMFC has gained favor over the AFC for space shuttle applications. For example, NASA is considering upgrading its fuel cell program by replacing the existing AFC units with PEMFC units. This is being considered to provide power for the space shuttle orbiters [26]. According to the results, it has been proved that use of the PEMFC helps reducing life cycle costs considerably. Moreover, the projected life of a PEMFC is comparatively longer (approximately 10,000 hours). PEMFC units have also shown to have higher power densities, which enables them to produce approximately 1.5 times the power produced by an AFC unit.

Apart from these advantages, the PEMFC has higher system stability and safety, thus leading to significantly lower lifecycle cost of the power plants [26]. In addition, the usage of PEMFC can improve flexibility of space missions. Therefore, missions that cannot be supported by the AFC can be undertaken by the PEMFC [26].

During selection of a particular design of PEMFC, important issues need to be considered. Some of them are related to safety, maintenance and replacement costs, and overall performance. Furthermore, the PEMFC plant must work in an integrated manner with all other shuttle programs and must be compatible with them. Hereafter, a brief comparison between the above mentioned regenerative fuel cell technologies for aerospace applications are presented.

As was mentioned earlier, the PEMFC has reached a higher level of performance compared to the AFC, which was used in earlier space missions. Moreover, it has comparatively longer life, higher safety and reliability, lower capital cost, and lower life cycle cost. One of the main reasons for this is that recently there has been a major technology innovation in terms of membrane structure for the PEMFC. In addition, the usage of the PEMFC eliminates the use of the corrosive alkali solution contained in the AFC. Typical performance and efficiency curves for the immobile AFC and the PEMFC are as shown in Figures 9.22 and 9.23.

As is shown in Figures 9.22 and 9.23, the performance of the AFC is slightly better, compared to the PEMFC, at lower current densities.

However, at higher values of current density, the performance of the PEMFC is better, which is emphasized by the fact that it has a lower value of cell voltage. In fact, at higher current densities, the PEMFC surpasses the AFC in terms of total efficiency. From the above observations, it is clear that the PEMFC holds a distinct advantage over the AFC for space shuttle power plant applications.

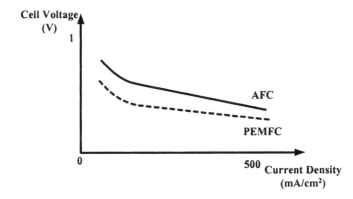

Figure 9.22 Typical *i-v* curves of AFC and PEMFC.

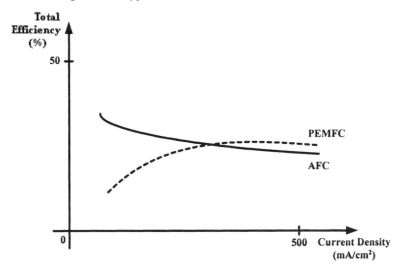

Figure 9.23 Typical electrolyzer system efficiency curves of AFC and PEMFC.

9.7 Other Applications of Fuel Cells

9.7.1 Utility Applications

In this section, utility applications of fuel cells are described. These new applications provide substantial improvements in performance and flexibility for the utility grids. Issues dealing with distributed generation and cogeneration systems are discussed. Emphasis is given to the role of fuel cells in these utility applications. In addition, fuel cell based power generation systems for residential and commercial buildings are explained. The main challenge for these systems is to operate in accordance with the utility grid, which is investigated in details. Power electronic issues are also explained.

9.7.1.1 Distributed Generation and Cogeneration

With the increasing costs of transmission and distribution of electrical power, the utility companies are looking toward the options of onsite power generation and distributed generation. One of the most suitable options available involves the use of fuel cells for this purpose. Continued research has shown that the usage of fuel cells increases the overall efficiency of the system. Moreover, there are no emissions and, hence, no pollution involved. In this section, various fuel cell systems proposed for the above mentioned applications are addressed. Their specific advantages and disadvantages are also discussed.

Fuel cells are connected to the utility grids via interface power electronic converters. These systems usually work on natural gas, which is readily available at most of the load sites. This is one of the most promising technologies for distributed generation. In these generation units, the waste heat may be used after recovery for the facility heating purposes. In addition, as is shown in Figure 9.24, additional efficiency is gained from the usage of the thermal energy from the fuel cell exhaust to power a non-combusting gas turbine. In these hybrid systems, the fuel cell exhaust supplies a turbine-generator unit, the capacity of which maybe as high as 15MW. Since there is no combustion in the turbine, there is no production of pollutants [27].

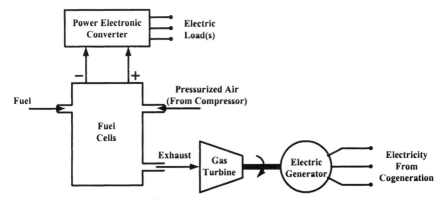

Figure 9.24 Block diagram of a typical fuel cell hybrid system.

As is shown in Figure 9.24, pressurized air and fuel are the inputs to the fuel cell. In fact, the efficiency of the system with pressurized air and fuel is much higher compared to the system in which air and fuel are fed at normal pressure to the fuel cell. For these cogeneration applications, pressurized SOFC is preferred. The main reason behind the selection of the SOFC is its high operating temperatures. Hence, the exhaust quality obtained will be comparatively higher.

The hybrid fuel cell system using a pressurized fuel cell combined with the use of gas turbines provides high efficiency and low emissions. As discussed earlier, an additional gain in efficiency is acquired by using the thermal exhaust from the fuel cell to power a non-combusting gas turbine. The typical applications for this technology is mainly in distributed generation, as mentioned earlier. The utility companies can suitably assign such units for specific load demands, thus reducing costs incurred in transmission.

SOFC, PAFC, MCFC, and PEM fuel cells are the most popular fuel cells that are being considered for the utility applications. As discussed, SOFC is considered to be the best option thus far for cogeneration applications. This is mainly due to the fact that this fuel cell operates at comparatively much higher temperatures and, hence, the fuel processing system is much simpler [28].

The gas industry has used the PAFC for smaller cogeneration applications, whereas the electric industry has used this fuel cell for large capacity generation. The stack life of this fuel cell is far higher than an engine based system [28]. In Japan, a project involving usage of PAFC has been developed, which produces enough power to supply electricity to approximately 4000 homes [29]. They are also considering applying this technology for heating and cooling applications for an entire district.

The usage of MCFC gives an added advantage of higher efficiency compared to the PAFC. However, this technology cannot be commercialized as easily since many changes are yet to be made in the MCFC. Once the costs of these fuel cells reduce, they will be considered over the PAFC for larger applications. This is because the exhaust of the MCFC is of much higher quality, when compared to that of the PAFC [28].

The PEM fuel cell is also being viewed as a future contender for distributed generation applications. It is considered to be the best option for supplying power to residential buildings. With the use of the PEMFC for distributed generation, the utility supply can be used to supply back-up power and peak power [28]. Extensive efforts are being concentrated into reducing the entry-level costs of these systems, thus making them the favorites to replace existing technologies for distributed generation.

9.7.1.2 Applications for Buildings

Energy saving is the main advantage of utilizing fuel cell based generation systems in residential and commercial buildings. These applications are generally considered to meet the electrical and thermal load demands in buildings. By supplying electric and thermal energy using these systems, energy usage in buildings can be improved considerably. Thus, the primary energy consumption in the building is reduced. The conventional equipments used for heating and cooling purposes in buildings will form an efficient cogeneration system, when used in conjunction with fuel cells [30].

Generally, the PEM fuel cell is considered to be the best option for supplying the electric and thermal loads of buildings. At the moment, most of the building electricity program has its efforts concentrated on commercial buildings rather than on residential buildings. Even so, there are efforts being made and projects are being undertaken to power homes using this technology.

When designing a fuel cell system for buildings, the important aspects to be kept in mind are the thermal-to-electric ratio and the electricity capacity factor. Thermal-to-electric ratio is the ratio of the annual thermal load of the building to the annual electric load. If this ratio is low, it indicates little usage of waste heat for heating purposes. Electric capacity factor is basically the ratio of annual energy usage in the building to the energy that would have been used if the building operated continuously at peak load [30]. These parameters are important from the point of view of cogeneration applications since they affect the total costs incurred. In addition, these systems should be compact so that the entire system space is small and can be easily accommodated inside buildings.

The PEM fuel cell is considered favorable for these power applications insides buildings due to the fact that it is much cheaper and can be suitably sized to fit buildings. However, in addition to the fuel cell, a battery pack is also included in the system. This battery pack generally consists of lead acid batteries, which supply the peak power demand.

From the above discussions, it is clear that various types of fuel cells have different properties and characteristics, when being considered for building applications. Hence, the best option must be selected based on the issues discussed. It is feasible that the needs of loads can be met satisfactorily by the usage of a particular fuel cell system.

9.7.1.3 Typical Power Electronic Converter Configurations

Design of fuel cell based generation systems for utility applications is different from the design of these systems for portable devices or vehicular applications. This is mainly because, in utility applications, the output stage has an additional DC/AC inverter, which converts the DC voltage from DC link to a suitable AC voltage of magnitude and frequency matching those required by the utility or the AC load. Figure 9.25 shows a typical power electronic interface for fuel cell utility applications. The output stage of this system is similar to that of an uninterruptible power supply (UPS), with the only major difference being the built-in generating capability in the case of fuel cells [31].

When designing an inverter for fuel cells used in utility applications, the interconnection issues must be considered. The fuel cell system is designed so that it can be used in grid-paralleled operation or in stand-alone mode. Furthermore, protection issues must be considered and suitable static relays must be used for fault detection and clearing. Electric power quality issues such as total harmonic distortion (THD) must be considered as well.

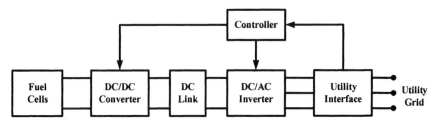

Figure 9.25 Block diagram of a typical fuel cell based generation system for utility applications.

The main purpose of power electronic converters, in Figure 9.25, is to convert the DC power output from the fuel cell to a suitable AC voltage, which can be connected to the utility grid or electric loads directly. In Figure 9.25, the fuel cell output is connected to the DC/DC converter, which regulates the DC link voltage. DC/DC converter is usually a PWM Boost chopper. Additional filters may be used to further improve the quality. DC/AC inverter is connected to the DC link and converts the DC voltage to AC. The control of the inverter is generally achieved by PWM techniques. Digital methods using DSPs or microcontrollers are mostly used for the control purposes [32].

The final stage of the system of Figure 9.25 may utilize a transformer, which brings the voltage to a suitable level for interconnecting the fuel cell system to the utility grid. A feedback command from the utility grid is given to the controller section, which, in turn, controls the switching of the inverter switches. Thus, the power output from the inverter is controlled to a suitable level in accordance with the utility grid.

As is shown in Figure 9.26, a multi-level inverter may be used to increase the voltage to the level of the utility grid. These inverters have a unique structure, which allows them to produce high levels of voltages with low production of harmonics [33]. By utilizing these inverters, there is no need for the output stage transformer. This is the main advantage of the system of Figure 9.26, since transformers are relatively more expensive and require more space, thus increasing the volume and weight of the system. However, in order to provide electric isolation, a high frequency transformer, with higher efficiency and lower weight and volume, may be used in the DC/DC converter or DC/AC inverter.

A multilevel inverter using cascaded H-bridges is presented in [33]. Each H-bridge needs a separate DC source. These DC sources are readily available, when utility is interfaced with the fuel cells. Thus, the fuel cells act as the DC sources for the individual bridges. In this topology, the conduction angles can be chosen such that the THD of voltage is reduced. Generally, these angles are selected so as to eliminate the lower order harmonics [33]. In the system proposed in [33], each inverter can generate three different levels of output voltages. They are $+V_{dc}$, 0, and $-V_{dc}$. By suitable switching combinations, these

Fuel Cell Based Vehicles

output voltages are easily attainable. Redundant voltage levels are also possible, thus improving the reliability of the system. This is due to the fact that, even upon loss of one level, the inverter still remains operational.

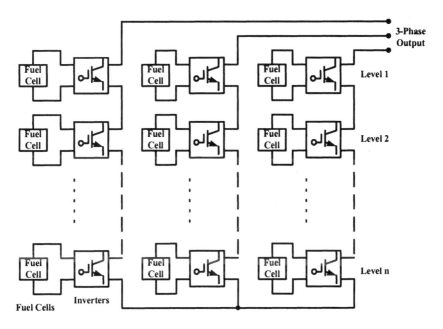

Figure 9.26 A typical multi-level inverter for fuel cell utility applications.

9.7.2 Portable Applications

Major applications of fuel cells have been presented in previous sections. In the following sections, other applications of fuel cells including portable applications, power supplies for telecommunication equipment, underwater vehicles, medical applications, and micro power sources are introduced. These applications are discussed and types of fuel cell used, typical requirements to be satisfied for these particular applications, and power electronic interface are explained.

There are some distinct differences between power sources for portable applications and stationary power sources. A portable power source needs to have a small volume and a low weight so that it is easy to maneuver. In this regard, it must have a high value of energy density. Moreover, the portable power units having output of several hundred watts must satisfy other requirements such as easy handling, long life, quick start up, portability, low cost, and safety [34].

When being considered to be used in conjunction with outdoor equipment, the power source must be able to work in varying environmental conditions. In addition, it must be less noisy and must not emit harmful gases such as NO_x and SO_x. The power source must have a rugged structure and must have a high value of efficiency as well.

As was mentioned, the fuel cells are being considered as one of the better options for fulfilling the above-mentioned requirements. In the following sections, discussions on various types of fuel cells and their properties are presented.

In recent years, there have been many efforts to commercialize portable power supplies using fuel cells. Based on the requirements for a portable power source, there exist three fuel cell options: phosphoric acid fuel cell (PAFC), direct methanol fuel cell (DMFC), and proton exchange membrane fuel cell (PEMFC). Each of these fuel cell technologies is discussed here for portable applications.

The phosphoric acid fuel cell (PAFC) is used for portable applications in the range of 200-300W. It operates at a temperature of about 200°C and uses metal hydride (MH) as a fuel supplier. A typical PAFC based power unit is shown in Figure 9.27. As is shown in Figure 9.27, the unit consists of an air cooled PAFC stack, a fuel supplier, and a DC/DC converter [34]. The MH cylinder supplies pure hydrogen fuel to the anode of the PAFC. This portable power unit can generate typically up to 200W for more than an hour. The DC/DC converter keeps the operating voltage regulated and protects the fuel cell from load variations [34].

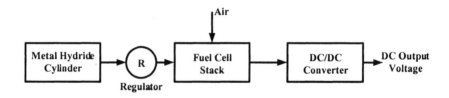

Figure 9.27 Block diagram of a PAFC based portable power unit.

The operation begins by opening a valve connecting the MH cylinder and the fuel cell. It takes approximately 10 minutes for the fuel cell to complete the heating up stage. The amount of fuel supply can be regulated using the regulator, as is shown in Figure 9.27.

Upon manufacture of such a unit, many tests need to be performed on it to ensure proper operation under outdoor conditions. Operational limits can be determined by testing the unit for various environmental and working conditions. Tests carried out by manufacturers include different start-up and shut-down tests and induced tests such as high/low operating temperature, voltage variations, and the lack of fuel. Furthermore, manufacturers usually test the fuel cell performance with varying flow rates of the primary fuel [35].

Upon development and testing of the PAFC based power units for various portable devices, it is found that the PAFC has an undesirably long start-up time: about 10 minutes. Moreover, the treatment of waste heat is comparatively much more complicated than other fuel cells.

As was explained in previous sections, the direct methanol fuel cell (DMFC) has a much simpler construction compared to other fuel cells. For portable applications, the liquid feed version of the DMFC is widely considered a good option. The operation is based on the oxidation of an aqueous solution of methanol to CO_2 at the anode and reduction of O_2 to H_2O at the cathode [36]. Thus, by using the DMFC, many issues concerned with the PAFC are solved. Furthermore, with the lack of an acidic solution, there is no concern of corrosion. This fuel cell can be operated at comparatively lower temperatures in the range of 60-100°C. The stack performance of such fuel cells is generally governed by the fuel concentration [36].

Portable power sources using DMFC, in the range of 50-150W, are being considered for various military applications [37]. In addition, the DMFC performance has been improved recently by using better techniques for membrane assemblies, higher activity catalysts, and improved electrode structures [37]. Therefore, there has been considerable improvement in power density of this fuel cell, which lies between 100-300 mW/sq. cm.

A typical DMFC system for portable applications is depicted in Figure 9.28. As is shown in Figure 9.28, a dilute methanol solution is fed to the anode by passing the concentrated methanol through a mixer stage. This is done because operating the fuel cell at optimum performance level depends on concentration of the methanol solution. Air is introduced in the stack via an air blower or a compressor [37]. Again, tests are conducted on the unit for power density, efficiency, performance under transient loads, and long-term stability during operation and storage [37].

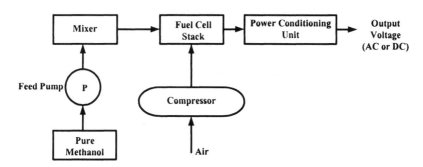

Figure 9.28 A typical DMFC system for portable applications.

As mentioned earlier, the operating temperature of the PAFC is approximately 200°C leading to longer warm-up time. This is a highly undesirable characteristic for a portable power source. As opposed to this, the PEMFC operates at a much lower temperature of approximately 100°C. This lower operating temperature allows a comparatively shorter start-up time. Furthermore, the PEMFC has a higher energy density; hence, system volume and weight is also low [38].

A typical PEMFC portable power system can supply nearly 1kW of power at a desirable AC or DC voltage. Moreover, the warm-up time, as mentioned earlier, is nearly 1 minute, which is acceptable for most portable applications. Schematic representation of a PEMFC power supply is shown in Figure 9.29.

As is depicted in Figure 9.29, it mainly consists of a fuel system, a power system, and a control system. The hydrogen gas is fed to the anode via hydrogen cylinders, while air is supplied to the cathode through a blower or fan [38]. The DC voltage generated from the cell is converted to the required DC or AC voltage by the power electronic converter. Batteries are also used for the warm-up purpose and are charged with a charger [38]. The control unit controls the operation of the system such that, in case of occurrence of a problem such as over voltage and over current, it shuts off the hydrogen supply; hence, power output is stopped.

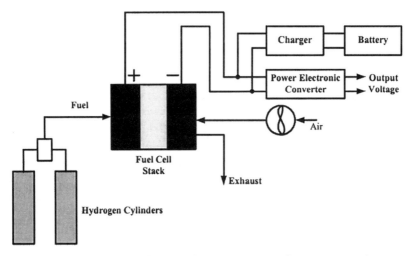

Figure 9.29 Schematic diagram of a portable PEMFC unit.

On comparing the volume, weight, and warm-up time of the PEMFC with the PAFC system, it has been found that the PEMFC system is better. Warm-up time required is only about 1 minute for the PEMFC; this is only 10% of the warm-up time required by the PAFC system. A pictorial representation of the above-mentioned differences between the two systems is shown in Figure 9.30.

Fuel Cell Based Vehicles

Furthermore, the operating time has been found to be inversely related to the output power [38].

Researchers are investigating the usage of butane as a fuel for the PEMFC based portable power units. They are considering manufacturing reformers connected with the PEMFC and working on butane as a fuel [39]. The main reason is that butane is readily available at homes or in the field as compared to hydrogen gas. Moreover, butane can be easily acquired in the form of compressed gas cylinders, which are easily transportable.

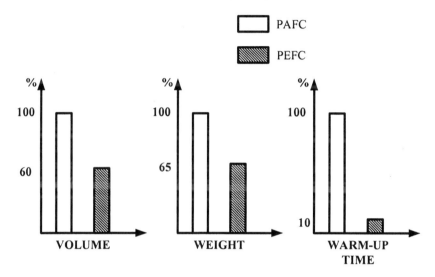

Figure 9.30 Relative volume, weight, and warm-up time of PAFC and PEMFC portable power systems.

In the past few years, there has been a considerable demand for portable power sources due to the increased production of outdoor devices. Fuel cells offer a profitable solution for the outdoor and portable needs. Some of the major portable applications for fuel cell power systems include long term back-up power supplies, portable devices such as radios and cellular phones, and portable military equipment and systems.

9.7.2.1 Back-Up Power Supply

Conventional back-up power supplies use engine generators and supply power to the load through an uninterruptible power supply (UPS) system. The output waveform of an engine generator is generally distorted; therefore, through the UPS system, the quality of the supply voltage for the load is improved.

In a system with fuel cell back-up power, the engine generator is replaced with a portable fuel cell system. A typical layout of a fuel cell based back-up power system used in conjunction with a UPS is shown in Figure 9.31. The efficiency is higher and the output voltage of the fuel cell system can be connected to the load directly. Thus, a great amount of power can be saved since the load power is not provided through the UPS double conversions.

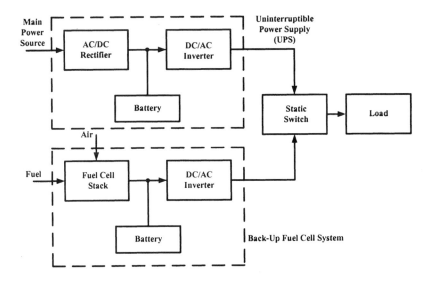

Figure 9.31 Block diagram of a fuel cell back-up power supply.

In the case of a power failure in the system of Figure 9.31, the UPS system supplies the load with the required AC power through the inverter. During this time period, the battery voltage reduces, and upon reaching a predetermined value, a signal is given to the control unit, and the portable fuel cell system begins to warm up [38]. After completion of the warm-up period, the static switch connects the load to the fuel cell system. Therefore, load is fed by the suitable AC power from the fuel cell system.

A fuel cell back-up power supply can be used without the UPS. In fact, there is no need for the UPS when the fuel cell system provides the back-up power for the load. However, in the normal operating mode of the system, when the fuel cell back-up power supply is disconnected from the load by the static switch, the UPS system can be used to provide an uninterrupted voltage for the load.

As is shown in Figure 9.32, the UPS and back-up fuel cell based power system of Figure 9.31 can be integrated to form a UPS with back-up fuel cell system. As the main advantage, the system of Figure 9.32 has only one DC/AC inverter compared to the system of Figure 9.31, which has two DC/AC inverters.

Fuel Cell Based Vehicles

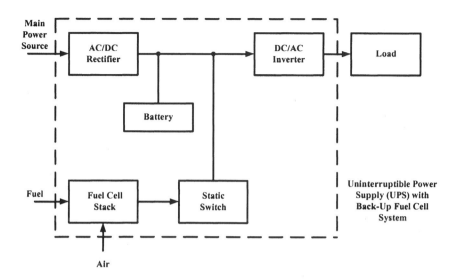

Figure 9.32 Block diagram of a UPS with back-up fuel cell system.

9.7.2.2 Small Portable Applications

DMFC systems are being considered to replace batteries in small portable applications such as radios and cellular phones. The batteries are troublesome to replace from time to time and they are expensive in the long run [40]. In order that the fuel cell systems can be an able replacement for rechargeable batteries, they must be compact in size and must be able to run in low power ranges. Hence, the best option amongst fuel cells is the DMFC.

The physical set-up of the DMFC system for a portable radio is fairly simple. In this case, instead of an infeasible pump, gravitational force is used to supply methanol to the fuel cell [40]. The main compartment of the radio consists of a container wherein a methanol storage bag is located, which is easily detachable. Turning the radio on or off is done by regulating a valve connected to this methanol bag. Air is acquired from the back of the radio through a series of small inlets [40].

Battery charging equipments for cellular phones are being considered to be fuel cell based. In fact, micro fuel cells for cellular phones and other portable devices could be seen in the market in 3-4 years. However, issues related to the technical aspects such as fuel cell efficiency, distribution of fuel cells, and cost of materials need to be considered and worked upon before commercialization of fuel cell systems for small portable devices [40].

Fuel cell systems have also been considered to replace rechargeable batteries for a long time now for military applications. The army applications demand high power and high energy density from the power sources. The rechargeable batteries, which have been used in the past, have been found to be too heavy. They also have a lower life [41]. Advanced soldier power systems demand low cost, low weight, small volume and size, ability to withstand fog, chemicals, and vibrations, and grater reliability and maintainability. Furthermore, they must take into account human factors during usage [41], [42].

Few PAFC systems ranging from 2 to 5 kW have been used in the past. In addition, PEMFC systems ranging from 200 to 300W have been used to power military portable devices [41], [42]. The major challenge is to provide hydrogen in a safe manner through storage technologies such as pressurized gas, liquid hydrogen, and hydrides. Oxygen must also be treated specially before supplying it to the fuel cell cathode electrode due to the presence of fog and chemicals in the ambient air.

The main advantage of fuel cells over rechargeable batteries is that they have higher energy density per load of fuel and long system life. Furthermore, the cost per unit of the system life is comparatively lower [43]. Apart from these merits, the PEMFC systems have rugged structure and, hence, can withstand the severe operating conditions posed to them during army missions.

Some of the military applications of fuel cells include tactical applications, such as in the operations center, lighting, radios, and other equipment [44]. Cooling equipment could also be powered by fuel cells ranging from 100 to 200W. Other applications include land warrior systems, power sensor suites, small unit operations, mobile power systems, and generators [44].

9.7.3 Fuel Cell Systems for Telecommunication Applications

PAFC has found applications in the telecommunication industry for powering telecommunication equipment as well as for cooling them. This idea has been put to practice in Japan and has been proved to be feasible. These systems can also be used for the purpose of co-generation in telecommunication buildings. They can be operated in parallel with the commercially available power supplies as well.

In past years, an engine generator was used to provide power to the telecommunication equipment and cooling devices such as air conditioners in case of an emergency [45]. Figure 9.33 depicts a typical block diagram of these conventional telecommunication power systems.

Approximately 40-50% of the energy consumed in a telecommunication building is by the main equipment, whereas about 30% is utilized by the cooling equipment [46]. The PAFC system, thus, helps in providing the power to both the main as well as the cooling equipment. A typical schematic diagram illustrating a telecommunication energy system based on fuel cells is shown in Figure 9.34.

As is shown in Figure 9.34, the fuel cell operates in parallel with the commercial AC power supply. Hence, output can be held constant even under power stoppages due to the maintenance or failure of equipment since the commercial power supply provides the necessary power in absence of the fuel cell [46].

Fuel Cell Based Vehicles

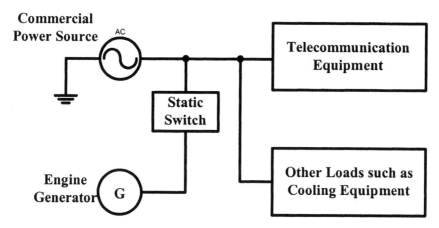

Figure 9.33 Typical block diagram of a conventional telecommunication power system.

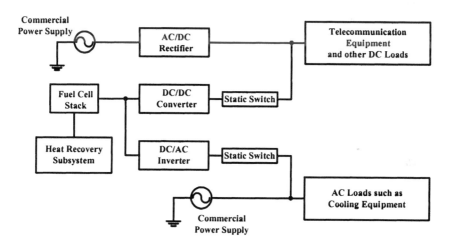

Figure 9.34 Typical block diagram of a fuel cell based telecommunication power system.

In addition, there are two methods of parallel operation for DC and AC loads. The DC power is supplied to the telecommunication equipment whereas the AC power output from the DC/AC inverter is fed to the AC loads and cooling devices that, in turn, cool the main equipment. The fuel cell output is kept constant by inverter voltage phase control and, hence, good interconnection characteristics can be observed during load changes [46].

Recently, pipeline fuels have been investigated for use as a fuel for the fuel cell stack due to the easy availability. Furthermore, fuels such as liquefied natural gas, hydrogen, methanol, propane, kerosene, and naphtha have been considered for use in future as the primary fuel [46]. Fuel cells are also being considered to power the switching equipment for telephones. Again, the PAFC is used in this case and is operated in the 30-100% range of its ratings. This is because between 0 and 30% of its rated power, the PAFC requires approximately 2 hours to warm up. However, from 30-100% power, it only takes a fraction of a second [47]. Typical power ratings of such systems range between 30-45kW.

Another application of fuel cells is for mobile power sources, which typically require quicker starts and have a larger number of start-ups, as compared to other commercial power sources [48]. Moreover, they must satisfy high efficiency, low noise, and low emissions requirements. These power sources can be used for on-site power generation, thus cutting down the losses due to distribution and helping in improving overall system efficiency. In addition, there has been a considerable demand for such mobile power sources in on-board shelters, trailers, and trains. They may serve as a power source for the on-board electronic equipment as well [48]. Much research has been done in recent years on mobile fuel processors and, hence, these power sources have become virtually practicable.

9.7.4 Fuel Cell Based Power Systems for Undersea Vehicles

Fuel cells offer an attractive solution over batteries for underwater vehicles. This is because they are highly efficient and quiet, and can be easily refueled. The PEMFC systems have been developed for such applications ranging from 20-30kW. These power units consist of equipment for fluid handling, heat management, and control systems [49].

Some of the major issues to be considered are safety, efficiency, reliability, durability, and capability to withstand shocks and vibrations [49]. Furthermore, tests are conducted on the units to ensure approximately 5000 hours of operation. Also, these units operate at about 200°F. It is made sure that the PEMFC membrane is absolutely saturated with water since it is hazardous to keep it dry for long durations [49].

As mentioned earlier, safe reactant storage, such as of hydrogen gas, is a major challenge. One of the methods practiced is increasing the pressure of storage. However, it is a highly unsafe procedure and, moreover, a huge amount of auxiliary equipments are required [49]. Apart from this, cryogenic storage and reversible metal hydrides (MH) are used for the purpose. The MH technology is similar to the cryogenic storage method, but it is much simpler and has lower system limits [49]. Furthermore, chemical hydride methods are being considered for future usage. If its development is successful, then it could offer up to 8-10 times the energy density offered by batteries [49].

9.7.5 Various Other Applications

Another interesting application for fuel cells involves usage in medical devices and gadgets. Scientists have developed a micro direct methanol fuel cell (μDMFC) for medical devices to be implanted into the human body, as in the cerebrospinal fluid shunt pump and micro-insulin pump [50]. Some of the factors in favor of fuel cells for medical applications are the compact size, manufacturing simplicity, clean operating features, not producing hazardous materials, and comparatively high efficiency. They are also much more powerful than other power sources.

In this case, the oxygen inhaled by the patient acts as the air to be supplied to the cathode of the μDMFC. On the other hand, the methanol fuel is supplied to the anode of the μDMFC via an external injection at regular intervals [50]. This μDMFC consists of a PEM and 2 silicon substrates with channels. Because of the danger that methanol poses to the human body, scientists are considering usage of ethanol as a fuel to micro power sources [50].

Various other interesting applications include airplane propulsion, navigation balloons, and in navy ships. The fuel cell is an ideal energy source for solar powered airplane. The problem addressed here is the power to be supplied for flights at night. During the daytime, the solar cells located on the airplane wings power the propulsion motors and charge the regenerative fuel cells. During the night, it is the fuel cell system that provides the necessary propulsion power [51]. The airplane basically must fly at altitudes of approximately 50,000 feet. Hence, the fuel cell must be extremely light in weight and must have a high value of electrical efficiency of the order of 60%.

Another application of fuel cells is for powering balloons, which are launched in the atmosphere to gather scientific information about the earth, sun, and other such phenomena [51]. Earlier, batteries provided the power to such air-balloons. Their power ratings were between 100-200W. However, future missions may require higher rated power sources of the order of up to 5kW. At these power ratings, batteries become extremely heavy, which is an undesirable characteristic for this application. Moreover, they require a vast amount of ancillary equipment for their operation at these power ratings. Thus, nowadays, PEMFC stacks are being considered for such applications. They are designed so as to withstand the large variation in temperature in the atmosphere, which ranges between $-50°C$ and $100°C$ [51].

Fuel cells have also solved many problems in the U.S. Navy, which has traditionally used diesel and turbine generators in the past. This is because the engine generators, as mentioned earlier, have a low efficiency and are noisy in operation. Furthermore, they emit gases, which are unacceptable for certain missions and at various harbors around the world [51]. Hence, the U.S. Navy has turned its attention towards alternate energy sources, one of them being the fuel cell. Typical power range lies between 2-3MW for these applications. Also, these fuel cells need to run on distillate fuel at efficiencies of about 40-50%. Moreover, cost for generating power needs is limited to $1500/kW. Again, the system weight, reliability, maintainability, and water management systems are key issues to be considered [51].

9.8 Conclusion

Fuel cells have emerged as one of the most promising technologies for meeting new energy demands. They are environmentally clean, quiet in operation, and highly efficient for generating electricity. This new technology provides a cost-effective system with greater flexibility to meet the deregulated power generation market criteria, especially for distributed power generation. Furthermore, fuel cells show a great promise to enhance automotive power systems in electric, hybrid electric, and more electric vehicles. Therefore, they find applications in a variety of transportation systems as well.

Distributed generation and cogeneration are providing the impetus toward the applications of fuel cells in utility systems. In fact, fuel cells offer an attractive solution with improved efficiency and flexibility as distributed generation systems. In this chapter, applications of fuel cells related to these issues were discussed. Furthermore, various types of fuel cells being considered were described and their potential opportunities to penetrate the market were addressed. Applications of fuel cells for powering buildings and homes were also discussed wherein their advantages were highlighted. In addition, power electronic converter topologies for usage with fuel cells for utility applications were addressed.

In recent years, much attention has been given to public health issues, with direct relation to vehicular emissions. As seen in this chapter, considerable R&D is being carried out in the development of fuel cell based light-duty passenger vehicles and heavy-duty transit buses. This chapter has discussed various topologies being considered by the automotive industry for the hybrid FC/battery vehicles and DHFCVs. Furthermore, similar technologies related to transit buses has been explained. In addition, all of these technologies have been compared and the latest trends in the automotive industry have been highlighted.

Over the past few years, several prototype fuel cell passenger cars and transit buses have been demonstrated in the U.S., Canada, Europe, and Japan. Thus, it is clear that the interest in fuel cells for transportation is growing rapidly, with an added boost from their high efficiency and low emissions. Through this chapter, it can also be concluded that transit buses form a perfect platform for fuel cell vehicular technology. This is due to the fact that public transportation presents an excellent opportunity to advertise the benefits of this new technology. With due progress in the development techniques and with reduction in costs, fuel cell vehicular systems will gain the rightful popularity.

9.9 References

[1] R. Anahara, S. Yokokawa, and M. Sakurai, "Present status and future prospects of fuel cell power systems," *Proceedings of IEEE*, vol. 81, no. 3, pp. 399-408, March 1993.

[2] J. Hirschenhofer, "How the fuel cell produces power?" *IEEE Aerospace and Electronic Systems Magazine*, vol. 7, no. 11, pp. 24-25, Nov. 1992.

[3] P. G. Grimes, "Historical pathways for fuel cells," *IEEE Aerospace and Electronic Systems Magazine*, vol. 15, no. 12, pp. 7-10, Dec. 2000.

[4] A. T. Raissi, A. Banerjee, and K. G. Sheinkopf, "Current technology of fuel cell systems," in *Proc. 32nd Intersociety Energy Conversion Engineering Conf.*, vol. 3, Honolulu, Hawaii, Aug. 1997, pp. 1953-1957.

[5] R. H. Goldstein, "EPRI 1990 Fuel Cell Status," in *Proc. 25th Intersociety Energy Conversion Engineering Conf.*, vol. 3, Reno, Nevada, Aug. 1990, pp. 170-175.

[6] T. Rehg, R. Loda, and N. Minh, "Development of a 50kW, high efficiency, high power density, co-tolerant PEM fuel cell stack system," in *Proc. 15th Annual Battery Conf. on Applications and Advances*, Long Beach, California, Jan. 2000, pp. 47-49.

[7] J. T. Brown, "Solid oxide fuel cell technology," *IEEE Trans. on Energy Conversions*, vol. 3, no. 2, pp. 193-198, June 1988.

[8] S. R. Narayanan, T. I. Valdez, A. Kindler, C. Witham, S. Surampudi, and H. Frank, "Direct methanol fuel cells – status, challenges and prospects," in *Proc. 15th Annual Battery Conf. on Applications and Advances*, Long Beach, California, Jan. 2000, pp. 33-36.

[9] K. Scott, "The direct methanol fuel cell," *IEE Colloquium on Compact Power Sources*, London, UK, May 1996, pp. 6/1-6/3.

[10] A. R. Eggert, D. Friedman, P. Badrinarayanan, S. Ramaswamy, and K. Heinz-Hauer, "Characteristics of an indirect-methanol fuel cell system," in *Proc. 35th Intersociety Energy Conversion Engineering Conf.*, vol. 2, Las Vegas, Nevada, July 2000, pp. 1326-1332.

[11] S. Smedley, "Zinc air fuel cell for industrial and specialty vehicles," *IEEE Aerospace and Electronic Systems Magazine*, vol. 15, no. 12, pp. 19-22, Dec. 2000.

[12] J. F. Contadini, "Social Cost Comparison among Fuel Cell Vehicle Alternatives," in *Proc. 35th Intersociety Energy Conversion Engineering Conf.*, Las Vegas, NV, July 2000, vol. 2, pp. 1341-1351.

[13] D. MacArther, "Fuel Cells for Electric Vehicles: Issues and Progress," in *Proc. Fourteenth Annual Battery Conf. on Applications and Advances*, California, Jan. 1999, pp. 1-4.

[14] R. M. Moore, "Indirect Methanol and Direct Methanol Fuel Cell Vehicles," in *Proc. 35th Intersociety Energy Conversion Engineering Conf.*, Las Vegas, NV, July 2000, pp. 1306-1316.

[15] T. E. Lipman, "Manufacturing and Lifecycle Costs of Battery Electric Vehicles, Direct-Hydrogen Fuel Cell Vehicles, and Direct-Methanol Fuel

Cell Vehicles," in *Proc. 35th Intersociety Energy Conversion Engineering Conf.*, Las Vegas, NV, July 2000, pp. 1352-1358.
[16] L. C. Iwan and R. F. Stengel, "The Application of Neural Networks to Fuel Processors for Fuel Cell Vehicles," *IEEE Trans. on Vehicular Technology*, vol. 50, no. 1, pp. 125-143, Jan. 2001.
[17] A. F. Burke and M. Miller, "Fuel Efficiency Comparisons of Advanced Transit Buses using Fuel Cell and Engine Hybrid Electric Drivelines," in *Proc. 35th Intersociety Energy Conversion Engineering Conf.*, Las Vegas, NV, July 2000, pp. 1333-1340.
[18] J. F. Miller, C. E. Webster, A. F. Tummillo, and W. H. DeLuca, "Testing and Evaluation of Batteries for a Fuel Cell Powered Hybrid Bus," in *Proc. 32nd Intersociety Energy Conversion Engineering Conf.*, Honolulu, Hawaii, July 1997, vol. 2, pp. 894-898.
[19] S. Romano, "The DOE/DOT Fuel Cell Bus Program and its Application to Transit Missions," in *Proc. 25th Intersociety Energy Conversion Engineering Conf.*, Reno, NV, August 1990, vol. 3, pp. 293-296.
[20] A Federal Transit Administration Project, "Clean, Quiet Transit Buses Are Here Today," See http://fuelcellbus.georgetown.edu/ (December 2001).
[21] L. Eudy, R. Parish, and J. Leonard, "Hydrogen Fuel Cell Bus Evaluation," in *Proc. 2001 DOE Hydrogen Program Review*, Baltimore, Maryland, April 2001.
[22] F. R. Kalhammer, P. R. Prokopius, V. P. Roan, and G. E. Voecks, "Fuel Cells for Future Electric Vehicles," in *Proc. Fourteenth Annual Battery Conf. on Applications and Advances*, Long Beach, California, Jan. 1999, pp. 5-10.
[23] K. Scott, "Direct Methanol Fuel Cells for Transportation," in *Proc. IEE Seminar on Electric, Hybrid, and Fuel Cell Vehicles*, London, UK, April 2000, pp. 3/1-3/3.
[24] F. A. Wyczalek, "Hybrid Electric Vehicles – Year 2000," in *Proc. 35th Intersociety Energy Conversion Engineering Conf.*, Las Vegas, NV, July 2000, pp. 349-355.
[25] W. Tillmetz, G. Dietrich, and U. Benz, "Regenerative fuel cells for space and terrestrial use," in *Proc. 25th Intersociety Energy Conversion Engineering Conf.*, Las Vegas, NV, Aug. 1990, pp. 154-158.
[26] M. Warshay, P. Prokopius, M. Le, and G. Voecks, "The NASA fuel cell upgrade program for the space shuttle orbiter," in *Proc. 32nd Intersociety Energy Conversion Engineering Conf.*, Hawaii, Aug. 1997, pp. 228-231.
[27] S. Hamilton, "Fuel cell-MTG hybrid – the most exciting innovation in power in the next 10 years," in *Proc. of IEEE Power Engineering Society Summer Meeting,* Edmonton, Alberta, Canada, vol. 1, July 1999, pp. 581-586.

[28] J. B. O'Sullivan, "Fuel cells in distributed generation," in *Proc. of IEEE Power Engineering Society Summer Meeting*, Edmonton, Alberta, Canada, vol. 1, July 1999, pp. 568-572.

[29] A. Hagiwara, "Fuel cell technologies in search of new directions: based on the experiences over 20 years," in *Proc. of IEEE Power Engineering Society Summer Meeting*, Edmonton, Alberta, Canada, vol. 1, July 1999, pp. 576-580.

[30] R. J. Fiskum, "Fuel cells for buildings program," in *Proc. of 32nd Intersociety Energy Conversion Engineering Conf.*, Honolulu, Hawaii, USA, vol. 2, Aug. 1997, pp. 796-799.

[31] M. J. Kornblit and K. A. Spitznagel, "Installing a fuel cell on an existing power system," in *Proc. of IEEE Industry Applications Society Annual Meeting*, Toronto, Canada, vol. 2, 1993, pp. 1379-1383.

[32] K. M. Salim, Z. Salam, F. Taha, and A. H. M. Yatim, "Development of a fuel cell power conditioner system," in *Proc. of IEEE International Conf. On Power Electronics and Drives Systems*, Hong Kong, vol. 2, July 1999, pp. 1153-1156.

[33] L. M. Tolbert and F. Z. Peng, "Multilevel converters as a utility interface for renewable energy systems," in *Proc. of IEEE Power Engineering Society Summer Meeting*, Seattle, USA, vol. 2, July 2000, pp. 1271-1274.

[34] T. Sakai, T. Ito, F. Takesue, M. Tsutsumi, N. Nishizawa, and A. Hamada, "Portable power source with low temperature operated PAFC," in *Proc. 35th International Power Sources Symposium*, New York, June 1992, pp. 49-52.

[35] V. Recupero, V. Alderruci, R. Di Leonardo, M. Lagana, and N. Giordano, "Performance of a portable 100W phosphoric acid fuel cell under different operating conditions," in *Proc. 25th Intersociety Energy Conversion Engineering Conf.*, Las Vegas, NV, Aug. 1990, pp. 269-274.

[36] T. I. Valdez, S. R. Narayanan, H. Frank, and W. Chun, "Direct methanol fuel cells for portable applications," in *Proc. 12th Annual Battery Conf. on Applications and Advances*, California, Jan. 1997, pp. 239-244.

[37] S. R. Narayanan, T. I. Valdez, N. Rohatgi, W. Chun, G. Hoover, and G. Halpert, "Recent advances in direct methanol fuel cells," in *Proc. 14th Annual Battery Conf. on Applications and Advances*, California, Feb. 1999, pp. 73-77.

[38] N. Kato, T. Murao, K. Fujii, T. Aoki, and S. Muroyama, "1-kW portable fuel cell system based on PEFCs," in *Proc. 3rd International Conf. on Telecommunications Energy Special*, Dresden, Germany, May 2000, pp. 209-213.

[39] N. Hashimoto, H. Kudo, J. Adachi, M. Shinagawa, N. Yamaga, and A. Igarashi, "Development of reforming technology for a portable generator," in *Proc. 32nd Intersociety Energy Conversion Engineering Conf.*, Hawaii, Aug. 1997, pp. 847-850.

[40] A. Jansen, S. V. Leeuwen, and A. Stevels, "Design of a fuel cell powered radio, a feasibility study into alternative power sources for portable products," in *Proc. IEEE International Symposium on Electronics and Environment*, California, May 2000, pp. 155-160.

[41] W. G. Taschek and R. Jacobs, "Silent energy source for tactical applications," *IEEE Aerospace and Electronic Systems Magazine*, vol. 8, no. 5, pp. 25-28, May 1993.

[42] A. Patil and R. Jacobs, "U.S. army PEM fuel cell programs," in *Proc. 32nd Intersociety Energy Conversion Engineering Conf.*, Hawaii, Aug. 1997, pp. 793-795.

[43] R. Jacobs, H. Christopher, R. Hamlen, R. Rizzo, R. Paur, and S. Gilman, "Portable power source needs of the future army – batteries and fuel cells," *IEEE Aerospace and Electronic Systems Magazine*, vol. 11, no. 6, pp. 19-25, June 1996.

[44] A. Patil and R. Jacobs, "U.S. army small fuel cell development program," *IEEE Aerospace and Electronic Systems Magazine*, vol. 15, no. 3, pp. 35-37, March 2000.

[45] Y. Kuwata, T. Take, T. Aoki, and T. Ogata, "Multi-fuel cell energy system for telecommunications co-generation system," in *Proc. 18th International Telecommunications Energy Conf.*, Texas, Oct. 1996, pp. 676-683.

[46] T. Koyashiki and K. Yotsumoto, "Advanced fuel cell energy system for telecommunications use," in *Proc. 14th International Telecommunications Energy Conf.*, Washington, DC, Oct. 1992, pp. 4-11.

[47] A. Ascoli, "Fuel cell generator for telephone switching equipment," in *Proc. 11th International Telecommunications Energy Conf.*, Pisa, Italy, Oct. 1989, pp. 18.2/1-18.2/2.

[48] W. G. Taschek, E. G. Starkovich, and R. Jacobs, "Fuel cells for mobile electric power," in *Proc. 9th Annual Battery Conf. on Applications and Advances*, California, Jan. 1994, pp. 50-53.

[49] H. J. DeRonck, "Fuel cell power systems for submersibles," in *Proc. Oceans Engineering for Today's Technology and Tomorrow's Preservation*, Brest, France, Sep. 1994, pp. 449-452.

[50] W. Y. Sim, G. Y. Kim, and S. S. Yang, "Fabrication of micro power source (MPS) using a micro direct methanol fuel cell (µDMFC) for medical applications," in *Proc. 14th IEEE International Conf. on Micro Electro-Mechanical Systems*, Interlaken, Switzerland, Jan. 2001, pp. 341-344.

[51] H. Oman, "News from the 34th intersociety energy conversion engineering conference – new applications for fuel cells," *IEEE Aerospace and Electronic Systems Magazine*, vol. 14, no. 12, pp. 15-22, Dec. 1999.

10

Electrical Modeling Techniques for Energy Storage Devices

This chapter focuses on the electrical modeling techniques of batteries, fuel cells (FCs), photovoltaic cells (PVs), and ultracapacitors (UCs). All of these devices have been investigated recently for their typical storage and supply capabilities. Hence, these devices must be modeled precisely, taking into account the concerned practical issues. Initially, this chapter will review several types of suitable models for each of the above-mentioned devices and the most appropriate model amongst them will be presented. Furthermore, a few important applications of these devices will be highlighted.

Through prior research results, it is well known that energy storage devices provide additional advantages to improve stability, power-quality, and reliability of the power supply source. The major types of storage devices being considered nowadays, namely, batteries and ultracapacitors, will be presented in this chapter. It is empirical that precise storage device models are created and simulated for several applications, such as hybrid electric vehicles (HEVs) and various power system applications.

The performances of batteries and ultracapacitors have, over the years, been predicted through many different mathematical models. Some of the important factors that need to be considered while modeling these energy storage devices include storage capacity, rate of charge/discharge, temperature, and shelf life.

The electrical models of two of the most promising renewable energy sources, namely, fuel cells and PV cells, are also described in this chapter. They have recently been studied widely, as they do not produce much emission, and are considered to be environmentally friendly. However, these renewable energy sources are large, complex, and expensive at the same time. Hence, designing and building new prototypes is a difficult and expensive affair. A suitable solution to overcome this problem is to carry out detailed simulations on accurately modeled devices. Through this chapter, the various types of equivalent electrical models for fuel cells and PV cells will be looked at and analyzed for suitability for operation at system levels.

10.1 Battery Modeling

As mentioned earlier, precise battery models are required for several applications such as for the simulation of energy consumption of electric vehicles and portable devices, or for power system applications. The major challenge in modeling a battery source is dealing with the non-linear characteristics of the equivalent circuit parameters, which require lengthy experimental and numerical procedures. The battery itself has some internal parameters, which need to be taken care of [1].

In this section, three basic types of battery models will be presented, namely, the ideal model, linear model, and the Thevenin model. Lastly, a simple lead-acid battery model for traction applications that can be simulated using PSpice software will also be presented.

10.1.1 Ideal Model

The ideal model of a battery basically ignores the internal parameters and, hence, is very simple. Figure 10.1(a) depicts an ideal model of a battery wherein it is clear that the model is primarily made up of only a voltage source [1].

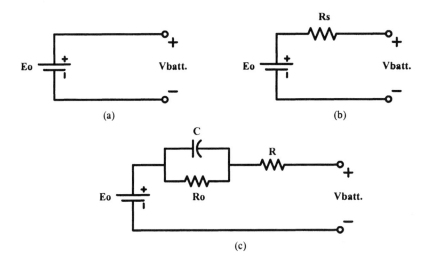

Figure 10.1 Battery models (a) ideal model, (b) linear model, and (c) Thevenin model.

10.1.2 Linear Model

This is, by far, the most commonly used battery model. As is clear from Figure 10.1(b), this model consists of an ideal battery with open-circuit voltage E_o and an equivalent series resistance R_s [1], [2]. The terminal voltage of the battery is represented by "V_{batt}." This terminal voltage can be obtained from the open-circuit tests as well as from load tests conducted on a fully charged battery [2].

Although this model is quite widely used, it still does not consider the varying characteristics of the internal impedance of the battery with the varying state of charge (SOC) and electrolyte concentration [2].

10.1.3 Thevenin Model

This model consists of electrical values of the open-circuit voltage (E_o), internal resistance (R), capacitance (C), and the overvoltage resistance (R_o) [1], [2]. As observed in Figure 10.1(c), capacitor C depicts the capacitance of the parallel plates and resistor R_o depicts the non-linear resistance offered by the plate to the electrolyte [2].

In this model, all the elements are assumed to be constants. However, in actuality, they depend on the battery conditions. Thus, this model is not the most accurate, yet it is the most widely used. In this view, a new approach to evaluate batteries is introduced. The modified model is based on operation over a range of load combinations [3]. The electrical equivalent of the proposed model is as depicted in Figure 10.2.

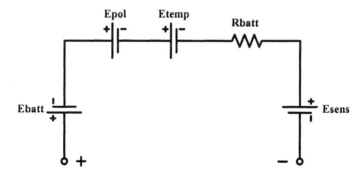

Figure 10.2 Main circuit representation of modified battery model.

As is clear from Figure 10.2, the main circuit model consists of the following five sub-circuits:

(a) E_{batt}: This is a simple DC voltage source designating the voltage in the battery cells.

(b) E_{pol}: It represents the polarization effects due to the availability of active materials in the battery.
(c) E_{temp}: It represents the effect of temperature on the battery terminal voltage.
(d) R_{batt}: This is the battery's internal impedance, the value of which depends primarily on the relation between cell voltage and state of charge (SOC) of the battery [3].
(e) V_{sens}: This is basically a voltage source with a value of 0V. It is used to record the value of battery current.

Thus, this simulation model is capable of dealing with various modes of charge/discharge. It is comparatively more precise and can be extended for use with Ni-Cd and Li-ion batteries, which could be applied to hybrid electric vehicles and other traction applications. Only a few modifications need to be carried out in order to vary the parameters, such as load state, current density, and temperature [3].

10.2 Modeling of Fuel Cells

Over the past few years, there have been great environmental concerns shown with respect to emissions from vehicles. These concerns along with the recent developments in fuel cell technology have made room for the hugely anticipated fuel cell market [4]. Fuel cells are nowadays being considered for applications in hybrid electric vehicles, portable applications, renewable power sources for distributed generation applications, and other similar areas where the emission levels need to be kept at a minimum.

The proton exchange membrane (PEM) fuel cell has the potential of becoming the primary power source for HEVs utilizing fuel cells. However, such fuel cell systems are large and complex and, hence, need accurate models to estimate the auxiliary power systems required for use in the HEV. In this section, fuel cell modeling techniques will be highlighted, thus avoiding the need to build huge and expensive prototypes. To have a clearer picture refer to Figure 10.3, showing the schematic representation of a fuel cell/battery hybrid power system. The battery pack in Figure 10.3 is used to compensate for the slow start-up and transient response of the fuel processor [4]. Furthermore, the battery can be used for the purpose of regenerative braking in the HEV.

As mentioned previously, since fuel cell systems are large, complex, and expensive, designing and building new prototypes is difficult [5], [6]. Therefore, the feasible alternative is to model the system and examine it through simulations. The fuel cell power system consists of a reformer, a fuel cell stack, and a DC/DC (Buck/Boost) or DC/AC power converter. The final output from the power electronic converter is in the required DC or AC form acquired from the low-voltage DC output from the fuel cell stack. An electrical equivalent model of a fuel cell power system is discussed here, which can be easily simulated using a computer simulation software.

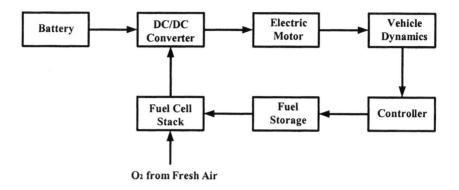

Figure 10.3 Schematic representation of a fuel cell/battery hybrid power system.

In the electrical equivalent model, a first-order time-delay circuit with a relatively long time-constant can represent the fuel reformer. Similarly, the fuel cell stack can also be represented by a first-order time-delay circuit, but with a shorter time-constant [5]. Thus, the mathematical model of the reformer and stack are represented as:

$$\frac{Vcr}{Vin} = \frac{\frac{1}{Cr.S}}{Rr + \frac{1}{Cr.S}} = \frac{1}{1 + RrCr.S} \qquad (10.1)$$

$$\frac{Vcs}{Vcr} = \frac{\frac{1}{Cs.S}}{Rs + \frac{1}{Cs.S}} = \frac{1}{1 + RsCs.S} \qquad (10.2)$$

Here, $RrCr = \tau r$ is the time-constant of the reformer and $RsCs = \tau s$ is the time-constant of the fuel cell stack [5]. The equivalent circuit is as shown in Figure 10.4.

By simulating the above equivalent circuit of Figure 10.4, the system operation characteristics can be investigated. In order to achieve a fast system response, the DC/DC or DC/AC converter can utilize its short time-constant for control purposes. But, eventually, the fuel has to be controlled, despite its long time-delay [5]. The inputs to the chemical model of a fuel cell include mass flows of air (O_2) and hydrogen (H_2), cooling water, relative humidity of oxygen and hydrogen, and the load resistance. The outputs from the chemical model include temperature of the cell, power loss, internal resistance, heat output, efficiency, voltage, and total power output [6]. Generally, in case of excess of hydrogen supply, it is re-circulated in order to avoid any wastage.

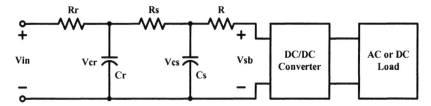

Figure 10.4 Equivalent circuit model of a fuel cell power system.

10.3 Modeling of Photovoltaic (PV) Cells

As mentioned earlier, photovoltaic systems have been studied widely as a renewable energy source because they are not only environmentally friendly, but also have infinite energy available from the sun. Although the PV system has the above-mentioned advantages, its study involves precise management of factors such as solar irradiation and surface temperature of the PV cell [7]. The PV cells typically show varying v-i characteristics depending on the factors mentioned above. Figure 10.5 shows the output characteristics of a PV cell with changing levels of illumination.

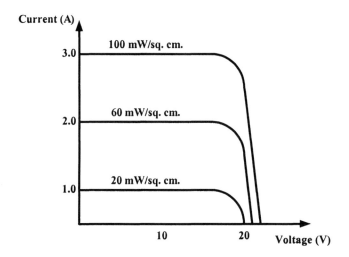

Figure 10.5 Typical v-i characteristics of PV cell with varying illumination levels.

As is clear from Figure 10.5, the current level increases with increase in the irradiation level. Figure 10.6 shows the v-i curves with varying cell temperatures.

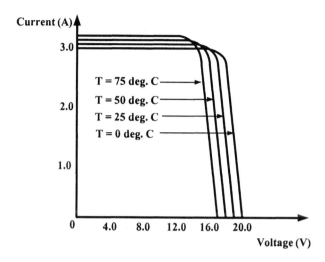

Figure 10.6 Typical *v-i* characteristics of PV cell with varying cell temperatures.

As depicted in Figure 10.6, the output curves for varying cell temperatures show higher voltage level as the cell temperature increases. Therefore, while modeling the PV cell, adequate consideration must be given to these two characteristics in particular.

Keeping the above-mentioned factors in mind, the electrical equivalent circuit modeling approach is proposed here. This model is basically a non-linear distributed circuit in which the circuit elements consist of the familiar semiconductor device parameters. Eventually, running a suitable computer simulation can easily simulate this model.

The PV cell can basically be considered as a current source with the output voltage primarily dependent on the load connected to its terminals [8]. The equivalent circuit model of a typical PV cell is as shown in Figure 10.7.

As is clear from Figure 10.7, there are various parameters involved in the modeling of a typical PV cell. These parameters are:

I_L – light generated current (A),

I_{D1} – diode saturation current (A),

I_{D2} – additional current due to diode quality constant (A),

I_{sh} – shunt current (A),

R_s – cell series resistance (Ω),

R_{sh} – cell shunt resistance (Ω), and

I – cell generated current (A).

The model depicted in Figure 10.7 examines all the characteristic measurements of the p-n junction cell type. From the above circuit, the following equation for cell current can be obtained:

$$J = JL - Jo1\left\{\exp\left[\frac{q(V + J.R_s)}{kT}\right] - 1\right\} \\ - Jon\left\{\exp\left[\frac{q(V + J.R_s)}{A.kT}\right] - 1\right\} - G_{sh}(V + J.R_s) \quad (10.3)$$

Here, q and k are electron charge and Boltzman's constant, respectively.

The voltage at the terminals of the diodes in Figure 10.7 can be expressed as follows:

$$V = V_{oc} - IR_s \\ + \frac{1}{\Delta}\log_n\left\{\frac{\beta(I_{sc} - I) - V/R_{sh}}{\beta.I_{sc} - V_{oc}/R_{sh}} + \exp[\Delta(I_{sc}.R_s - V_{oc})]\right\} \quad (10.4)$$

Here, β is the voltage change temperature coefficient (V/°C).

For the PV cell model of Figure 10.7, R_s and R_{sh} are usually estimated when the cell is not illuminated. Thus, these values can be approximated from the dark characteristic curve of the cell. The generated light current (I_L) is calculated by the collective probability of free electrons and holes. It can be expressed as follows:

$$I_L = q.N\left[\sum f_c(x_N) + \sum f_c(x_P) + 2I\right] \quad (10.5)$$

Here, $f(x)$ is the probability distribution function and N is the rhythm of generated electrons and holes.

Once the equations of the cell model are formulated, the efficiency of the PV cell can be obtained as:

$$Efficiency = \frac{P_{out}}{P_{in}} = \frac{f.I_{sc}.V_{oc}}{P_{in}} \quad (10.6)$$

A distinct advantage of such a computer model is the fact that, with a very few number of changes, it can receive data from different kinds of PV cells, maintaining satisfactory results [8].

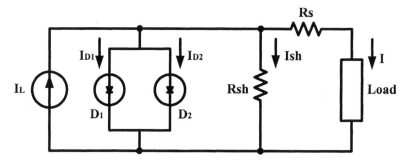

Figure 10.7 Schematic of equivalent circuit model of a PV cell.

10.4 Modeling of Ultracapacitors

Ultracapacitors (also known as super capacitors and double-layer capacitors) work on the electro-chemical phenomenon of a very high capacitance/unit area using an interface between electrode and electrolyte [9]. Typical values of such capacitors range from 400 F - 800 F and have low values of resistivity (approximately 10^{-3} Ω-cm) [9], [10]. These UCs operate at high energy densities, which are commonly required for applications such as space communications, digital cellular phones, electric vehicles, and hybrid electric vehicles. In some cases, usage of a hybridized system employing a battery alongside the UC provides an attractive energy storage system, which offers numerous advantages. This is particularly due to the fact that the UC provides the necessary high power density whereas the battery provides the desired high energy density. Such a hybridized model will be covered in this section.

10.4.1 Double-Layer Model

A simple electrical equivalent circuit of a double-layer UC is as shown in Figure 10.8. Its parameters include equivalent series resistance (ESR), equivalent parallel resistance (EPR), and the overall capacitance.

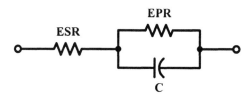

Figure 10.8 Electrical equivalent circuit of an ultracapacitor.

The ESR in Figure 10.8 is important during charging/discharging since it is a lossy parameter, which, in turn, causes the capacitor to heat up. On the other hand, the EPR has a leakage effect and, hence, it only affects the long-term storage performance [9]. For the purpose of simplification in calculations, the EPR parameter is dropped. Furthermore, the dropping of the EPR parameter does not have any significant impact on the results. The circuit for analysis is, thus, simply an ideal capacitor in series with a resistance and the corresponding load [9]. Hence, the value of resistance can be written as:

$$R = n_s \frac{ESR}{n_p} \tag{10.7}$$

Here, R is the overall resistance, n_s is the number of series capacitors in each string, and n_p is the number of parallel strings of capacitors. Furthermore, the value for the total capacitance can be expressed as:

$$C = n_p \frac{C_{rated}}{n_s} \tag{10.8}$$

Here, C is the overall value of capacitance and C_{rated} is the capacitance of individual capacitor. This model can be used in conjunction with a DC/DC converter, which, in turn, acts as a constant power load, as shown in Figure 10.9.

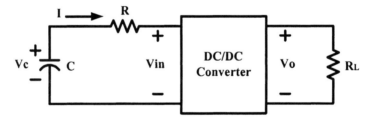

Figure 10.9 Circuit showing UC connected to a constant power load.

The capacitor bank can be used in stand-alone mode or can be operated in parallel with a battery of suitable size for the applications mentioned earlier. A brief description of such a hybrid model is described in the following section.

10.4.2 Battery/Ultracapacitor Hybrid Model

As stated earlier, combining a battery and a UC to operate in parallel is an attractive energy storage system with many advantages. Such a hybrid systems uses both the high power density of the UC as well as the high energy density of the battery. In this section, an electrical equivalent model of such a system will be presented, which can be used to evaluate its voltage behavior. The model is as depicted in Figure 10.10.

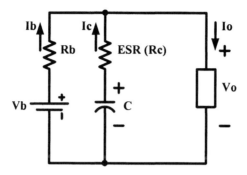

Figure 10.10 Equivalent circuit model of a battery/UC hybrid system.

The equivalent circuit of Figure 10.10 shows an equivalent series resistance (R_c) and a capacitor (C) as a model of the UC, whereas the Li-ion battery can be modeled simply by using a series resistance (R_b) and a battery. The values of R_c and C depend on the frequency due to the porous nature of the electrodes of the UC [10]. When the pulse-width (T) is varied, the discharge rate of the UC can be varied and can be shown to be equal to a frequency of $f = \frac{1}{T}$. The following equations can be written for I_o and V_o from the equivalent circuit model of Figure 10.10:

$$I_o = I_c + I_b \tag{10.9}$$

$$V_o = V_b - I_b R_b = \left[V_b - \frac{1}{C} \int_0^T I_c \, dt \right] - I_c R_c \tag{10.10}$$

Here, I_o and V_o are the output current and voltage delivered to the load, respectively.

From the above two equations, it is possible to achieve a voltage drop $\Delta V = V_b - V_o$ due to a pulse current of I_o. This voltage drop can be finally expressed as:

$$\Delta V = \frac{I_o R_b R_c}{R_b + R_c + \frac{T}{C}} + \frac{I_o R_b \frac{T}{C}}{R_b + R_c + \frac{T}{C}} \tag{10.11}$$

The currents delivered by the battery (I_b) and capacitor (I_c) can also be derived and their ratio can be expressed as:

$$\frac{I_c}{I_b} = \frac{R_b}{R_c + \dfrac{T}{C}} \qquad (10.12)$$

It can be seen that for a long pulse, I_c can be limited by the value of C. Furthermore, it can be concluded that during the pulsed discharge, about 40-50% of the total current is delivered by C. Upon computer simulation of the equivalent circuit model, it is possible to study the fact that, during peak power demand, UC delivers energy to assist the battery whereas, during low power demand, UC receives energy from the battery [10].

Due to the advanced energy storage capabilities of the UC, it can be used for applications requiring repeated short bursts of power such as in vehicular propulsion systems. In a typical scenario, both the battery and the UC provide power to the motor and power electronic DC/AC inverter during acceleration and overtaking, whereas they receive power via regenerative braking during slow-down/deceleration [11]. The two most popular topologies for inserting batteries and UCs into drivetrains are as shown in Figures 11(a) and 11(b).

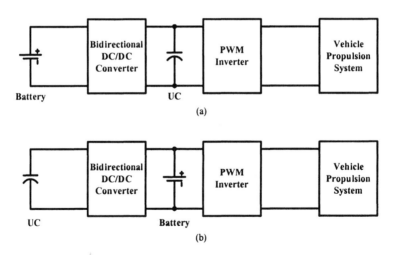

Figure 10.11 Typical topologies of batteries and UCs in drivetrains.

As is clear in the topology of Figure 10.11(a), the UC bank is placed on the DC bus whereas, in Figure 10.11(b), the DC bus houses the battery. Amongst the two topologies, the one of Figure 10.11(a) has a much more degraded energy efficiency since the whole of the battery energy has to go through the DC/DC converter. Another drawback worth highlighting is that a very high voltage UC bank is required, which is extremely expensive. Hence, more often than not, the topology of Figure 10.11(b) is generally considered for HEV applications [11]. In a typical brushless DC (BLDC) motor driven electric vehicle (EV) propulsion system, a UC bank could be used to achieve a wider drive range and good acceleration/deceleration performance, and to lower the costs.

Future projections with regards to performance of UCs show that energy densities of as high as 10-20 Wh/kg are easily achievable using carbon electrode materials with specific capacitance values of nearly 150-200 F/gm [12]. Currently, extensive R&D on UCs is being carried out in the U.S., Canada, Europe, and Japan. As mentioned earlier, most of the research on UCs focuses on EV and HEV applications as well as medical and power system applications.

10.5 Conclusion

This chapter has dealt with the modeling of four of the major types of energy storage devices, namely, batteries, fuel cells, PV cells, and ultracapacitors. These models can easily be simulated and validated for their performance by running a simple computer simulation. In addition, this chapter presents the various applications where these energy storage devices are employed, thus making their modeling and simulation studies worthwhile.

10.6 References

[1] Y-H. Kim and H-D. Ha, "Design of interface circuits with electrical battery models," *IEEE Trans. on Industrial Electronics*, vol. 44, no. 1, pp. 81-86, Feb. 1997.

[2] H. L. Chan and D. Sutanto, "A new battery model for use with battery energy storage systems and electric vehicles power systems," in *Proc. 2000 IEEE Power Engineering Society Winter Meeting*, Singapore, vol. 1, Jan. 2000, pp. 470-475.

[3] J. Marcos, A. Lago, C. M. Penalver, J. Doval, A. Nogueira, C. Castro, and J. Chamadoira, "An approach to real behavior modeling for traction lead-acid batteries," in *Proc. 32nd Annual Power Electronic Specialists Conf.*, Vancouver, BC, Canada, vol. 2, June 2001, pp. 620-624.

[4] H-G. Jeong, B-M. Jung, S-B. Han, S. Park, and S-H. Choi, "Modeling and performance simulation of power systems in fuel cell vehicles," in *Proc. 3rd International Power Electronics and Motion Control Conf.*, Beijing, China, vol. 2, Aug. 2000, pp. 671-675.

[5] Y-H. Kim and S-S. Kim, "An electrical modeling and Fuzzy logic control of a fuel cell generation system," *IEEE Trans. on Energy Conversion*, vol. 14, no. 2, pp. 239-244, June 1999.

[6] W. Turner, M. Parten, D. Vines, J. Jones, and T. Maxwell, "Modeling of a PEM fuel cell for use in a hybrid electric vehicle," in *Proc. 49th IEEE*

Vehicular Technology Conf., Amsterdam, Netherlands, vol. 2, Sep. 1999, pp. 1385-1388.

[7] J-H. Yoo, J-S. Gho, and G-H. Choe, "Analysis and control of PWM converter with V-I output characteristics of solar cell," in *Proc. 2001 IEEE International Symposium on Industrial Electronics*, Pusan, Korea, vol. 2, June 2001, pp. 1049-1054.

[8] G. A. Vokas, A. V. Machias, and J. L. Souflis, "Computer modeling and parameters estimation for solar cells," in *Proc. 6th Mediterranean Electrotechnical Conf.*, Ljubljana, Slovenia, vol. 1, May 1991, pp. 206-209.

[9] R. L. Spyker and R. M. Nelms, "Double layer capacitor/DC-DC converter system applied to constant power loads," in *Proc. 31st Intersociety Energy Conversion Engineering Conf.*, Washington, DC, vol. 1, Aug. 1996, pp. 255-259.

[10] J. P. Zheng, T. R. Jow, and M. S. Ding, "Hybrid power sources for pulsed current applications," *IEEE Trans. on Aerospace and Electronic Systems*, vol. 37, no. 1, pp. 288-292, Jan. 2001.

[11] X. Yan and D. Patterson, "Improvement of drive range, acceleration, and deceleration performance in an electric vehicle propulsion system," in *Proc. 30th Power Electronic Specialists Conf.*, Charleston, South Carolina, vol. 2, June 1999, pp. 638-643.

[12] A. F. Burke, "Prospects for ultracapacitors in electric and hybrid vehicles," in *Proc. 11th Annual Battery Conf. on Applications and Advances*, Long Beach, CA, Jan. 1996, pp. 183-188.

11

Advanced Motor Drives for Vehicular Applications

11.1 Brushless DC Motor Drives

New advancements in fast semiconductor switches and cost-effective DSP processors have revolutionized the adjustable speed motor drives. These new opportunities have contributed to novel configurations for electric machines in which the burden is shifted from complicated structures to software and control algorithms. This in turn has resulted in considerable reduction in cost, while improving the performance of the overall drive system. Brushless DC (BLDC) is one example of this trend. Very compact geometry, high efficiency, and a simple control are among the main incentives for satisfying many adjustable speed applications with this emerging technology. Figure 11.1 shows the DC motor and brushless DC motor configurations. The DC motor has mechanical components, such as commutators and brushes, whereas the brushless DC motor replaces the mechanical commutators with electronic switches.

11.1.1 Machine Structure

A BLDC machine can be categorized according to the way the magnets are mounted on the rotor. The magnets can either be surface mounted or interior mounted.

11.1.1.1 Surface-Mounted BLDC Machines

Figure 11.2 (a) shows the surface mounted permanent magnet rotor. Each permanent magnet is mounted on the surface of the rotor. This rotor is simple to build. The magnets are skewed on the rotor surface to minimize cogging torque. However, there is a possibility of the magnets flying apart at very high-speed operation.

11.1.1.2 Interior-Mounted BLDC Machines

Figure 11.2 (b) shows the interior-mounted permanent magnet rotor. Each permanent magnet is mounted inside of the rotor. This configuration is not as common as the surface-mounted BLDC, but it is a better candidate for high-speed operation. Note that there is rotor saliency caused inductance variation on the stator windings in this type of motor due to the permanent magnets being buried in the rotor. This adds some salience torque to the permanent magnet torque and is beneficial for other reasons, such as field weakening and sensorless controls.

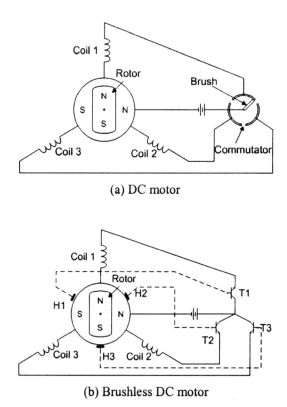

(a) DC motor

(b) Brushless DC motor

Figure 11.1 Schematics of DC and BLDC motors.

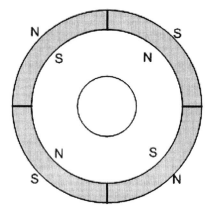

(a) Surface-mounted PM rotor

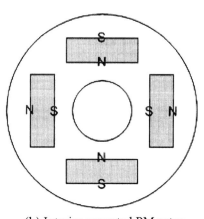

(b) Interior-mounted PM rotor

Figure 11.2 Cross sectional view of the permanent magnet rotor.

11.1.1.3 Machine Classification

There are two major classes of BLDC motor drives, which can be characterized by the shapes of their respective back-EMF waveforms: trapezoidal and sinusoidal.

11.1.1.3.1 Trapezoidal Back-EMF

The trapezoidal back-EMF BLDC motor is designed to develop trapezoidal back-EMF waveforms. Ideally, it has the following characteristics:

- Rectangular distribution of magnetic flux in the airgap
- Rectangular current waveform
- Concentrated stator windings

Excitation of the stator poles is by 120° wide quasi-square current waveforms, with a 60° electrical interval of zero current each side, per cycle. The square stator excitation current waveforms, along with the trapezoidal back-EMF, permits important control system simplifications, compared to the sinusoidal back-EMF machines. In particular, the rotor position sensor requirements are much simpler since only six commutation instants are necessary per electrical cycle. Figure 11.3 shows the winding configuration of the trapezoidal back-EMF BLDC machine.

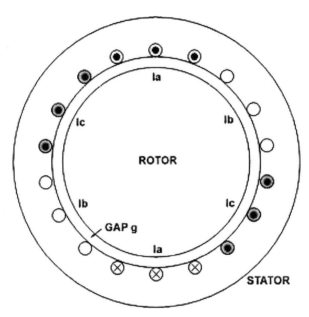

Figure 11.3 Trapezoidal back-EMF BLDC machine winding.

Figure 11.4 (a) shows equivalent circuit and (b) shows trapezoidal back-EMF, stator winding current profiles, and Hall sensor signals of a 3-phase BLDC motor drive. The voltages seen in this figure (e_a, e_b, and e_c) are the line-to-neutral back-EMF voltages, which are the result of the permanent-magnet flux crossing the airgap in the radial direction and linking the windings of the stator with the rotor speed. The windings of the stator are positioned in the standard 3-phase full-pitch, concentrated arrangement, and thus the phase trapezoidal back-EMF waveforms are displaced by 120° electrical degrees. The stator phase current is on for 120° and off for 60°, resulting in each phase current flowing for 2/3 of the electrical 360° period, 120° positively and 120° negatively. To drive the motor with maximum torque/ampere, it is necessary for the phase current pulses to be synchronized with the line-neutral back-EMF voltages of that phase. This is different from the operation of the conventional synchronous motors in which no rotor position information is required for its operation. The BLDC is therefore called a self-synchronous motor.

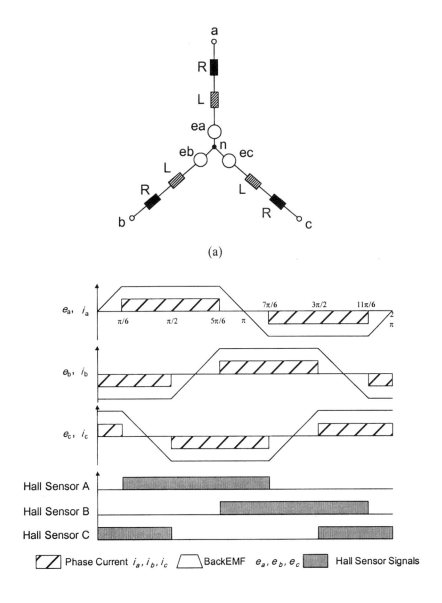

Figure 11.4 (a) Three-phase equivalent circuit, (b) back-EMFs, currents, and hall sensor signals.

11.1.1.3.2 Sinusoidal Back-EMF

The sinusoidal back-EMF BLDC motor is designed to develop sinusoidal back-EMF waveforms. It has the following ideal characteristics:

- Sinusoidal distribution of magnetic flux in the airgap
- Sinusoidal current waveforms
- Sinusoidal distribution of stator conductors

The most fundamental aspect of this motor is that the back-EMF generated in each phase winding by the rotation of the rotor magnet is a sinusoidal function of the rotor angle. The stator phase current of a sinusoidal back-EMF BLDC machine is similar to that of the AC synchronous motor. The rotating stator MMF is similar to that of the synchronous motor and, therefore, can be analyzed with a phasor diagram. Figure 11.5 shows the winding configuration of the sinusoidal shape back-EMF BLDC machine.

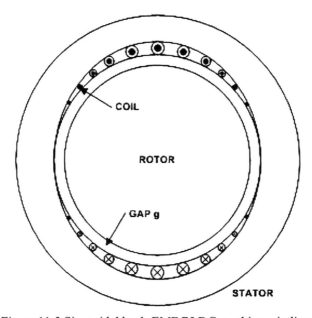

Figure 11.5 Sinusoidal back-EMF BLDC machine winding.

11.1.2 Control of the Brushless DC Machine

11.1.2.1 Analysis of BLDC Motor Drive

The analysis of the BLDC drive is based on the following simplifying assumptions:

- The motor is not saturated.

- Stator resistances of the windings are equal and self- and mutual inductances are constant.
- Power semiconductor devices in the inverter are ideal.
- Iron losses are negligible.

Under the above assumptions, a BLDC motor can be represented by

$$\begin{bmatrix} v_a \\ v_b \\ v_c \end{bmatrix} = \begin{bmatrix} R_s & 0 & 0 \\ 0 & R_s & 0 \\ 0 & 0 & R_s \end{bmatrix} \begin{bmatrix} i_a \\ i_b \\ i_c \end{bmatrix} + \begin{bmatrix} L-M & 0 & 0 \\ 0 & L-M & 0 \\ 0 & 0 & L-M \end{bmatrix} \frac{d}{dt}\begin{bmatrix} i_a \\ i_b \\ i_c \end{bmatrix} + \begin{bmatrix} e_a \\ e_b \\ e_c \end{bmatrix} \quad (11.1)$$

where e_a, e_b, and e_c are back-EMFs and $L - M = L_s$ is the stator self-inductance per phase. The electromagnetic torque is expressed by

$$T_e = \frac{1}{\omega_r}(e_a i_a + e_b i_b + e_c i_c) \quad (11.2)$$

The interaction of T_e with the load torque determines the motor speed dynamics:

$$T_e = T_L + J\frac{d\omega_r}{dt} + B\omega_r \quad (11.3)$$

where T_L is load torque, J is the total rotating inertia, and B is friction. Based on the equivalent circuit of Figure 11.6, the system equations can be expressed, using the Laplace transform, as

$$V_t(s) = E_s(s) + (R_s + sL_s)I_s(s) \quad (11.4)$$

$$E_s(s) = k_E \omega_r(s) \quad (11.5)$$

$$T_e(s) = k_T I_s(s) \quad (11.6)$$

$$T_e(s) = T_L(s) + (B + sJ)\omega_r(s) \quad (11.7)$$

From the above equations, the transfer function of the drive system is

$$\omega_r(s) = \frac{k_T}{(R_s + sL_s)(sJ + B) + k_T k_E} V_t(s) \\ - \frac{R_s + sL_s}{(R_s + sL_s)(sJ + B) + k_T k_E} T_L(s) \quad (11.8)$$

The physical significance of the electrical and mechanical time constants in this transfer function is as follows:

- The electrical time constant determines how quickly the armature current changes in response to a step change in the terminal voltage, when the rotor speed is assumed to be constant.
- The mechanical time constant determines how quickly the speed changes in response to a step change in the motor torque.

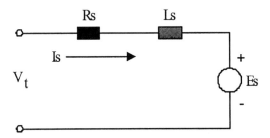

Figure 11.6 Simplified equivalent circuit of BLDC motor.

11.1.2.2 Control of BLDC Motor Drive

Figure 11.7 shows the block diagram of the classical position and speed control scheme for a BLDC motor drive. If only speed control is desired, the position controller and position feedback circuitry may be eliminated. Usually, both position and velocity feedback transducers are required in high-performance position controllers. A position sensor without a velocity sensor would necessitate differentiating the position signal to obtain the velocity, which would tend to amplify noise in an analog system. However, in digital systems, this usually is not a problem. The position sensor, or some other means of obtaining position information, is required, however, in position and speed control of BLDC motor.

Many high-performance applications include current feedback for torque control. At the minimum, a DC bus current feedback is required to protect the drive and the machine from over-currents. The controller blocks, "position controller", and "velocity controller" may be any type of classical controller such as a proportional-integral (PI) controller or a more advanced controller such as an artificial intelligence based controller. The "current controller and commutation sequencer" provides the properly sequenced gating signals to the 3-phase inverter. By comparing the sensed currents with a reference current, current control is achieved by hysteresis control method or some other method. Using position information, the commutation sequencer causes the inverter to electronically commutate, as in the mechanical commutator in a conventional DC machine. The commutation angle of each stator winding is selected to maximize the torque per stator current ampere.

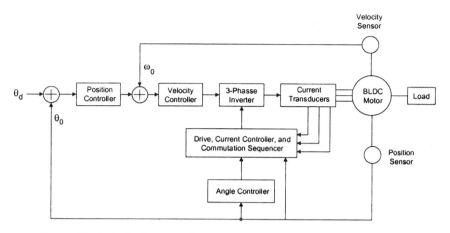

Figure 11.7 Block diagram of a classical speed and position control for a BLDC motor.

The "position sensor" is usually either a 3-element Hall-effect sensor or an optical encoder. The resolver is another option, which allows phase shifting (advancing) of the current pulses with respect to rotor position. Advancing of the angle is required due to the electrical time constant of the windings. A given amount of time is needed for the current to build up. At higher speeds, there is less time available for this current buildup. Furthermore, the growing back-EMF, due to increasing speed, is opposing the rapid rise of the phase current. This is called the field weakening operation of the BLDC. A problem with this type of operation is that the drive will produce lower torque per ampere, as in field-weakening operations of DC machines.

11.1.3 Sensorless Techniques

Many methods of sensorless operation of BLDC motors have been described in the literature [1]-[47]. The sensorless techniques can be primarily grouped into the following categories:

- those using measured currents, voltages, fundamental machine equations, and algebraic manipulations
- those using observers
- those using back-EMF methods

11.1.3.1 Methods Using Measurables and Math

This consists of two sub-categories: (1) those which calculate the flux linkages using measured voltages and currents, and (2) those which utilize a model's prediction of a measurable voltage or current, compare the model's value with the actual measured voltage or current, and calculate the change in position which is proportional to the difference between the measured and actual voltage or current. The first sub-category is described in [1]-[8]. The fundamental idea is to start with the voltage equation of the machine:

$$V = Ri + \frac{d}{dt}\psi \qquad (11.9)$$

where

V is the voltage vector,

I is the current vector,

R is the resistance matrix, and

φ is the flux linkage vector.

This equation is then manipulated to obtain,

$$\psi = \int_0^t (V - Ri)\, d\tau \qquad (11.10)$$

With the knowledge of initial position, machine parameters, and the flux linkages' relationship to rotor position, the rotor position can be estimated. By determining the rate of change of the flux linkage from the integration results, the speed can also be determined. An advantage of the flux-calculating method is that line-line voltages may be used in the calculations and, thus, no motor neutral connection is required. This is desirable because the most common BLDC configuration is the Y-connected stator windings with no neutral connection.

The second sub-category is described in [9]-[12]. This method consists of first developing an accurate d-q model of the machine. For background in the d-q model of the machine, see the above references. Utilizing the measured currents and their d-q transformation, the output voltages of the model are compared to the measured and transformed voltages. The difference is proportional to the difference in angular reference between the model's coordinate system and the actual coordinate system. This is the rotor position with reference to the actual coordinate system's reference. Conversely, the measured voltages are used and the differences in the currents are found. In either case, the difference between the measured and the calculated is used as the multiplier in an update equation for the rotor position.

11.1.3.2 Methods Using Observers

These methods determine the rotor position and/or speed using observers. The first of these are those utilizing the well-known Kalman filter as a position estimator [20]-[25]. One of the first to appear in the literature was by M.

Advanced Motor Drives for Vehicular Applications

Schroedl in 1988. In his many publications, Schroedl utilized various methods of measuring system voltages and currents, which could produce rough estimates of the angular rotor position. The Kalman filtering added additional refinement to the estimates of position and speed. Other observer-based systems include those utilizing non-linear [27]-[29], full-order [2], [30], [31], and sliding-mode observers [1], [10], [32].

Observers are like models of a system, which take as inputs the output and inputs of the actual system. Typically, the state of the system is produced as the output. The observer-produced state, or a part of it, can then be fed back into the system as if it was the actually measured state to be used as in any closed-loop system. The full-order observer is designed using the designer's choice of eigenvalues. The eigenvalues are usually chosen to be slightly faster (of a larger negative value) than those of the actual system so the state estimation error approaches zero as time approaches infinity. If the observer eigenvalues are chosen to be an order of magnitude larger (or more) than the actual system's, then the estimated values will converge to the actual values within a sufficiently short time [24].

In [1], [32], and [34], reduced-order observers are utilized to estimate the back-EMF, which is used to derive position and speed. In [29], the rotor position and speed are directly estimated, as the outputs of a non-linear reduced-order observer. Sliding-mode observers, an offspring of the sliding-mode control, utilize a varying switching function in order to confine the state estimation error to a trajectory which leads toward zero on a phase-plane sliding surface [35]. In [36] and [37], the sliding-mode observers were based on the observation of the d-q transformed stator currents. From the resulting current estimations, the back-EMFs were estimated, producing speed and position estimates.

11.1.3.3 Methods Using Back-EMF Sensing

A) Terminal Voltage Sensing

In normal operation of the BLDC motors, the flat top part of a phase back-EMF is aligned with the phase current. The switching instants of the converter can be obtained by knowing the zero-crossings of the back-EMF and a speed-dependent period of time delay [39]. The method utilizes this concept to estimate commutation instants. The BLDC motor has a rectangular flux density distribution in the airgap. For this reason the induced back-EMF in the stator windings is trapezoidal. By monitoring the phase back-EMF during the silent phase intervals, the zero crossing can be detected. To produce a proper switching pattern for the motor, the terminal voltages are filtered by low pass filters. The low pass filters are used to eliminate higher harmonics in the terminal voltages. Low pass filters can be designed to introduce a near 90-degree phase delay. At the point where a filtered terminal voltage crosses the neutral point voltage, back-EMF of that phase becomes zero and that point corresponds to the transition at the output of a comparator. Comparator outputs can be decoded to provide the gating signals for the inverter switches.

Since back-EMF is zero at rotor stand-still and proportional to speed after that, it is not possible to use the terminal voltage sensing method to obtain a switching pattern at low speeds. As the speed increases, the average terminal voltage increases and the frequency of excitation increases. The capacitive reactance in the filters varies with the frequency of excitation, introducing a speed-dependent delay in switching instants. This speed-dependent reactance disturbs current alignment with the back-EMF and stator field orientation in the airgap, which causes problems at higher speeds. With this method, a reduced speed operating range is normally used, typically around 1000-6000 rpm.

In conclusion, the zero-crossing is a good method for steady-state operation; however, phase differences in the circuits used due to speed variations do not allow optimal torque/amp over a wide speed range.

B) Third Harmonic Back-EMF Sensing

Rather than using the fundamental of the phase back-EMF waveform as in the previous technique, the third harmonic of the back-EMF can be used in the determination of the switching instants in the wye connected 120-degree current conduction operating mode [40]. This method is not as sensitive to phase delay as the zero voltage crossing method, as the frequency to be filtered is three times as high. The reactance of the filter capacitor in this case dominates the phase angle output of the filter.

The terminal voltage equation of the BLDC motor for phase A can be expressed as

$$v_a = Ri_a + L_s \frac{di_a}{dt} + e_a \tag{11.11}$$

Back-EMF voltage, e_a, is made up of many voltage harmonic components,

$$\begin{aligned} e_a = E(\cos \omega_e t + k_3 \cos 3\omega_e t + k_5 \cos 5\omega_e t \\ + k_7 \cos 7\omega_e t + ...) \end{aligned} \tag{11.12}$$

The third harmonic of the terminal voltages is acquired by the summation of the terminal voltages.

$$\begin{aligned} v_{as} + v_{bs} + v_{cs} &= 3Ek_3 \cos 3\omega_e t + v_{high\,freq.} \\ &= v_3 + v_{high\,freq.} \end{aligned} \tag{11.13}$$

The summed terminal voltage contains only the triplens due to the fact that only zero sequence current components can flow through the motor neutral. This voltage is dominated by the third harmonic, which does not require much filtering. To obtain switching instants, the filtered voltage signal, which provides the third harmonic voltage component, is integrated to find the third harmonic flux linkage.

$$\lambda_{r3} = \int v_3 dt \tag{11.14}$$

The third harmonic flux linkage lags the third harmonic of the phase back-EMF voltages by 30 degree. The zero crossings of the third harmonic of the flux linkage correspond to the commutation instants of the BLDC motor. The third harmonic method provides a wider speed range than the zero-crossing method, does not introduce as much phase delay as the zero-crossing method, and requires less filtering.

C) Freewheeling Diode Conduction

This method uses indirect sensing of the zero-crossing of the phase back-EMF to obtain the switching instants of the BLDC motor [41]. In the 120 degree conducting Y-connected BLDC motor, one of the phases is always open-circuited. For a short period after opening the phase, there remains phase current flowing, via a free-wheeling diode, due to the inductance of the windings. This open phase current becomes zero in the middle of the commutation interval, which corresponds to the point where back-EMF of the open phase crosses zero. The biggest downfall of this method is the requirement of six additional isolated power supplies for the comparator circuitry for each free-wheeling diode.

D) Back-EMF Integration

In this method, position information is extracted by integrating the back-EMF of the unexcited phase [42]-[45]. The integration is based on the absolute value of the open phase's back-EMF. Integration of the voltage divider scaled-down back-EMF starts when the open phase's back-EMF crosses zero. A threshold is set to stop the integration which corresponds to a commutation instant. As the back-EMF is assumed to vary linearly from positive to negative (trapezoidal back-EMF assumed), and this linear slope is assumed speed-insensitive, the threshold voltage is kept constant throughout the speed range. If desired, current advance can be implemented by the change of the threshold. Once the integrated value reaches the threshold voltage, a reset signal is asserted to zero the integrator output.

In conclusion, the integration approach is less sensitive to switching noise, and it automatically adjusts to speed changes; but its low speed operation is poor. With this type of sensorless operation scheme, up to 3600 rpm has been reported [45].

11.1.4 Advantages and Disadvantages of BLDC Motor Drives for Vehicular Applications

Advantages of BLDC motors are:

- High efficiency: BLDC motors are the most efficient of all electric motors. This is due to the use of permanent magnets for the excitation, which consumes no power. The absence of a mechanical commutator and brushes means low mechanical friction losses and, therefore, higher efficiency.
- Compactness: The recent introduction of high-energy density magnets (rare-earth magnets) has allowed achieving very high flux densities in the BLDC motor. This allows achieving accordingly high torques, which in turn allows making the motor small and light.
- Ease of control: The BLDC motor can be controlled as easily as a DC motor because the control variables are easily accessible and constant throughout the operation of the motor.
- Ease of cooling: There is no current circulation in the rotor. Therefore, the rotor of a BLDC motor does not heat up. The only heat production is on the stator, which is easier to cool than the rotor because it is static and on the periphery of the motor.
- Low maintenance, great longevity, and reliability: The absence of brushes and mechanical commutator suppresses the need for associated regular maintenance and suppresses the risk of failure associated with these elements. The longevity is therefore only a function of the winding insulation, bearings, and magnet life-length.
- Low noise emissions: There is no noise associated with the commutation because it is electronic and not mechanical. The driving converter switching frequency is high enough so that the harmonics are not audible.

Disadvantages of BLDC motors are:

- Cost: Rare-earth magnets are much more expensive than other magnets and result in an increased motor cost.
- Limited constant power range: A large constant power range is crucial to achieving high vehicle efficiencies. The permanent magnet BLDC motor is incapable of achieving a maximum speed greater than twice the base speed.
- Safety: Large rare-earth permanent magnets are dangerous during the construction of the motor because of flying metallic objects attracted towards them. There is also a danger in case of vehicle wreck if the wheel is spinning freely; the motor is still excited by its magnets and high voltage is present at the motor terminals that can possibly endanger the passengers or rescuers.

Advanced Motor Drives for Vehicular Applications

- Magnet demagnetization: Magnets can be demagnetized by large opposing magneto motive forces and high temperatures. The critical demagnetization force is different for each magnetic material. Great care must be brought to cooling the motor, especially if it is compactly built.
- High-speed capability: The surface mounted permanent magnet motors cannot reach high speeds because of the limited mechanical strength of the assembly between the rotor yoke and the permanent magnets.
- Inverter failures in BLDC motor drives: Because of the permanent magnets on the rotor, BLDC motors present major risks in case of short circuit failures of the inverter. Indeed, the rotating rotor is always energized and constantly induces an electromotive force in the short-circuited windings. A very large current circulates in those windings and an accordingly large torque tends to block the rotor. The dangers of blocking one or several wheels of a vehicle are not negligible. If the rear wheels are blocked while the front wheels are spinning, the vehicle will spin uncontrollably. If the front wheels are blocked, the driver has no directional control over the vehicle. If only one wheel is blocked, it will induce a yaw torque that will tend to spin the vehicle, which will be difficult to control. In addition to the dangers to the vehicle, it should be noted that the large current resulting from an inverter short circuit poses a risk to demagnetize and destroy the permanent magnets.

Open circuit faults in BLDC motor drives are no direct threat to the vehicle stability. The impossibility of controlling a motor due to an open circuit may however pose problems in terms of controlling the vehicle. Because the magnets are always energized and cannot be controlled, it is difficult to control a BLDC motor in order to minimize the fault. This is a particularly important issue when the BLDC motor is operated in its constant power region. Indeed, in this region, a flux is generated by the stator to oppose the magnet flux and allow the motor to rotate at higher speeds. If the stator flux disappears, the magnet flux will induce a large electromotive force in the windings, which can be harmful to the electronics or passengers.

11.2 Switched Reluctance Motor Drives

11.2.1 Basic Operation

The structure of the switched reluctance motor (SRM) is simple, robust, and very reliable in operation. The machine has a salient pole stator with concentrated excitation windings and a salient pole rotor with no conductors or permanent magnets. A typical three phase 6/4 switched reluctance motor is shown in Figure 11.8. The motor has six stator poles and four rotor poles. The coil is wound around each stator pole and is connected in series with the coil on the diametrically opposite stator pole to form a phase winding. The reluctance of the flux path between the two diametrically opposite stator poles varies as a pair of the rotor poles rotates into and out of alignment. The inductance of a phase winding is maximum when the rotor is in aligned position and minimum when the rotor is in unaligned position.

Figure 11.8 Cross sectional view of a 6/4 switched reluctance machine.

Figure 11.9 shows a typical change in the stator self-inductance as a function of the rotor position. The flat portion La of the inductance profile is caused by a difference in the width between the stator and rotor poles. This flat portion is normally provided to avoid or reduce negative torques during demagnetization. Usually, stator poles are made narrow, thus providing more space for the winding at the same time. A pulse of positive torque is produced if current flows in a phase winding as inductance of that phase winding is increasing. A positive voltage is applied to a phase winding in the unaligned rotor position, where the inductance is lowest, and thus, the rate of rise of current is high. In the region where the rotor poles overlap the excited stator teeth (region of rising inductance), the current level is usually maintained constant with the aid of a current chopping regulator. In this region, shear forces are generated, which tend to align the rotor poles. As the rotor poles approach alignment position, the phase current is commutated. In the aligned position, the magnetic forces will tend to close the air gap by pulling opposite members together. The stator, under such conditions, is subjected to compressive forces

Advanced Motor Drives for Vehicular Applications

while the rotor is under tension. A negative torque contribution is avoided if the current is reduced to zero before the inductance starts to decrease again. Rotation is maintained by switching on and off the current in the stator phase winding in synchronism with the rotor position. The ideal inductance profile, ideal phase current, and ideal torque are shown in Figure 11.10.

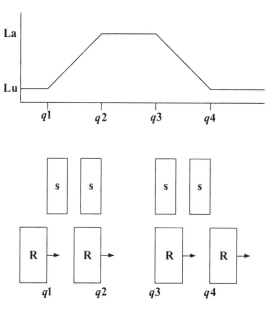

Figure 11.9 Ideal inductance profile.

The flux versus current curve is shown in Figure 11.11. The ϖ_f is the stored field energy, where as ϖ' is the co-energy.

$$co-energy = \varpi' = \int_0^i \phi \, di \qquad (11.15)$$

The torque produced by one phase at any rotor position is given by the equations below.

$$T = \left[\frac{\partial \varpi'}{\partial \theta} \right] \qquad (11.16)$$

$$T = \frac{1}{2} i^2 \frac{dL}{d\theta} \qquad (11.17)$$

The positive or negative sign of current does not affect the torque, since torque is proportional to the square of current. Torque is totally related to the slop of the inductance. Therefore, phase excited during positive slop gives motoring mode and phase excited during negative slope gives generating mode. This means that a single machine can be operated as motor as well as generator.

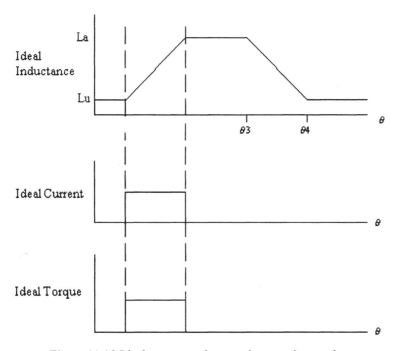

Figure 11.10 Ideal current and torque in motoring mode.

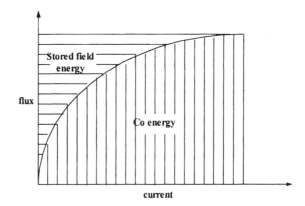

Figure 11.11 Flux versus current characteristic.

11.2.2 Torque-Speed Characteristics and Control Techniques

The torque-speed curve is shown in Figure 11.12. The torque speed plane can be divided into five regions. Below the base speed, the torque remains constant. Base speed (ω_b) is the lowest speed at which maximum power can be extracted. This region offers the flexibility of current control or hysteresis control to obtain the desired performance from the motor. The control parameters are I_{max}, θ_{on}, and θ_{off}. In hysteresis control, we define I_{max} and I_{min}, or ΔI. If we decrease the ΔI, the switching frequency increases. It must be noted that, at very low speeds (region 1 and 2), motional back-emf is much smaller than the DC bus voltage and can be neglected. As the speed increases, the motional back-emf is quite considerable (region 3), necessitating the advancement of the current turn-off angle to obtain a maximum average torque. Hysteresis current control is simple and has variable switching frequency and residual current ripples as shown in Figure 11.13.

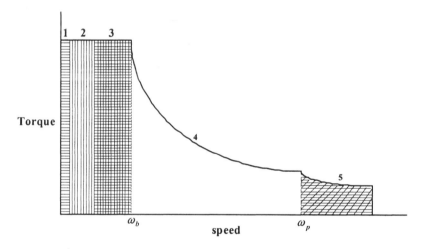

Figure 11.12 Torque-speed characteristics of SRM.

When the speed increases further, the motional back-emf exceeds the DC bus voltage and the machine operates in the single pulse mode (region 4). In this mode, the current is limited by the motional back-emf and never reaches the rated current. Hence, current control is not possible and the torque is maintained at the optimal value by controlling the θ_{on} and θ_{off} angles. Usually, high-speed operation requires advancement of the turn on angle well before the unaligned position (Figure 11.14). In this region, torque is inversely proportional to the speed; therefore, it is called the "constant power region." By further increasing the speed in the constant power region, the available time for online computations of the rotor position will be limited. In addition, simultaneous conduction of stator phases in this region will result in an unusual magnetic distribution in the back iron of the stator with major contribution from mutual fluxes. This mode of operation is called "ultra-high speed" (region 5). It must be noted that this mode of operation will end once continuous conduction in all phases occurs.

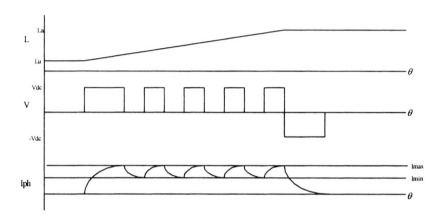

Figure 11.13 Hysteresis current control of SRM.

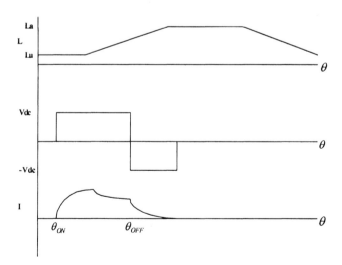

Figure 11.14 Single pulse operation of SRM.

Advanced Motor Drives for Vehicular Applications

11.2.3 Power Electronic Driver

The SRM relies upon reluctance torque rather than the more conventional reactive torque of wound field synchronous, surface magnet PM, and induction machines. The SRM is a doubly salient reluctance machine with independent phase windings on the stator. The rotor does not have any winding, and is usually made of steel lamination. The stator and rotor have an unequal number of poles, with three phase 6/4, four phase 8/6, and three phase 12/8 being common configurations. Due to the absence of rotor windings this motor is very simple to construct, has low inertia, and allows an extremely high-speed operation.

The absence of rotor copper loss eliminates the problem, as in an induction motor, associated with rotor cooling, which has a poor thermal path. SRMs normally are low cost machines for their extremely simple construction. Moreover, due to the unipolar excitation, it is possible to design SRM converters which require a minimum of one switch per phase. SRM operation is safe. This motor is particularly suitable for hazardous environments. The SRM drive produces zero or small open circuit voltage and no short circuit current. Furthermore, most SRM converters are immune to shoot through faults, unlike the inverters of induction and brushless DC motors. A circuit diagram of the most commonly SRM converter, the classical converter, is shown in Figure 11.15.

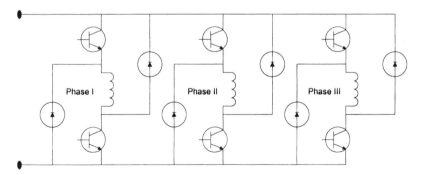

Figure 11.15 Classical SRM converter.

Rapid acceleration and extremely high speed operation in SRM is possible for its low rotor inertia and simple construction. As was explained, SRM operates in constant torque from zero speed up to the rated speed. Above rated speed up to a certain speed, the operation is in constant power. The range of this constant power operation depends on the motor design and its control. Beyond constant power operation and up to the maximum speed, the motor operates in the natural mode, where the torque reduces as the square of the speed. Because of this wide speed range operation, SRM is particularly suitable for traction applications in propulsion systems.

11.3 References

[1] T. Senjyu and K. Uezato, "Adjustable speed control of brushless DC motors without position and speed sensors," in *Proc. IEEE/IAS Conf. on Industrial Automation and Control: Emerging Technologies*, 1995, pp. 160-164.

[2] A. Consoli, S. Musumeci, A. Raciti, and A. Testa, "Sensorless vector and speed control of brushless motor drives," *IEEE Trans. on Industrial Electronics*, vol. 41, pp. 91-96, Feb. 1994.

[3] P. Acarnley, "Sensorless position detection in permanent magnet drives," *IEE Colloquium on Permanent Magnet Machines and Drives*, pp. 10/1-10/4, 1993.

[4] T. Liu and C. Cheng, "Adaptive control for a sensorless permanent-magnet synchronous motor drive," *IEEE Trans. on Aerospace and Electronic Systems*, vol. 30, pp. 900-909, July 1994.

[5] R. Wu and G. R. Slemon, "A permanent magnet motor drive without a shaft sensor," *IEEE Trans. on Industry Applications*, vol. 27, pp. 1005-1011, Sep./Oct. 1991.

[6] T. Liu and C. Cheng, "Controller design for a sensorless permanent magnet synchronous drive system," *IEE Proc.-B*, vol. 140, pp. 369-378, Nov. 1993.

[7] N. Ertugrul, P. P. Acarnley, and C. D. French, "Real-time estimation of rotor position in PM motors during transient operation," in *Proc. IEE Fifth European Conf. on Power Electronics and Applications*, 1993, pp. 311-316.

[8] N. Ertugrul and P. Acarnley, "A new algorithm for sensorless operation of permanent magnet motors," *IEEE Trans. on Industry Applications*, vol. 30, pp. 126-133, Jan./Feb. 1994.

[9] T. Takeshita and N. Matsui, "Sensorless brushless DC motor drive with EMF constant identifier," in *Proc. IEEE Conf. on Industrial Electronics, Control, and Instrumentation*, vol. 1, 1994, pp. 14-19.

[10] N. Matsui and M. Shigyo, "Brushless DC motor control without position and speed sensors," *IEEE Trans. on Industry Applications*, vol. 28, pp. 120-127, Jan./Feb. 1992.

[11] N. Matsui, "Sensorless operation of brushless DC motor drives," in *Proc. IEEE Conf. on Industrial Electronics, Control, and Instrumentation*, vol. 2, Nov. 1993, pp. 739-744.

[12] N. Matsui, "Sensorless PM brushless DC motor drives," *IEEE Trans. on Industrial Electronics*, vol. 43, pp. 300-308, April 1996.

[13] M. A. Hoque and M. A. Rahman, "Speed and position sensorless permanent magnet synchronous motor drives," in *Proc. 1994 Canadian*

Conf. on Electrical and Computer Engineering, vol. 2, Sep. 1994, pp. 689-692.
[14] H. Watanabe, H. Katsushima, and T. Fujii, "An improved measuring system of rotor position angles of the sensorless direct drive servomotor," in *Proc. IEEE 1991 Conf. on Industrial Electronics, Control, and Instrumentation*, 1991, pp. 165-170.
[15] J. S. Kim and S. K. Sul, "New approach for the low speed operation of the PMSM drives without rotational position sensors," *IEEE Trans. on Power Electronics*, vol. 11, pp. 512-519, 1996.
[16] A. H. Wijenayake, J. M. Bailey, and M. Naidu, "A DSP-based position sensor elimination method with on-line parameter identification scheme for permanent magnet synchronous motor drives," *IEEE Conf. Record of the 13th Annual IAS Meeting*, vol. 1, 1995, pp. 207-215.
[17] H. Watanabe, S. Miyazaki, and T. Fujii, "Improved variable speed sensorless servo system by disturbance observer," 16th *Annual Conf. of IEEE Industrial Electronics Society*, vol. 1, pp. 40-45, 1990.
[18] J. Oyama, T. Abe, T. Higuchi, E. Yamada, and K. Shibahara, "Sensorless control of a half-wave rectified brushless synchronous motor," *Conf. Record of the 1995 IEEE Industry Applications Conf.*, vol. 1, pp. 69-74, 1995.
[19] J. S. Kim and S. K. Sul, "New approach for high performance PMSM drives without rotational position sensors," *IEEE Conf. Proc. 1995, Applied Power Electronics Conf. and Exposition*, vol. 1, pp. 381-386, 1995.
[20] M. Schrodl, "Digital implementation of a sensorless control algorithm for permanent magnet synchronous motors," *Proc. Int'l. Conf. "SM 100", ETH Zurich (Switzerland)*, pp. 430-435, 1991.
[21] M. Schrodl, "Operation of the permanent magnet synchronous machine without a mechanical sensor," *IEE Proc. Int'l. Conf. on Power electronics and Variable Speed Drives*, pp. 51-56, July 1990.
[22] M. Schrodl, "Sensorless control of permanent magnet synchronous motors," *Electric Machines and Power Systems*, vol. 22, pp. 173-185, 1994.
[23] B. J. Brunsbach, G. Henneberger, and T. Klepsch, "Position controlled permanent magnet excited synchronous motor without mechanical sensors," *IEE Conf. on Power Electronics and Applications*, vol. 6, pp. 38-43, 1993
[24] R. Dhaouadi, N. Mohan, and L. Norum, "Design and implementation of an extended Kalman filter for the state estimation of a permanent magnet synchronous motor," *IEEE Trans. on Power Electronics*, vol. 6, pp. 491-497, July 1991.
[25] A. Bado, S. Bolognani, and M. Zigliotto, "Effective estimation of speed and rotor position of a PM synchronous motor drive by a Kalman filtering

technique," *PESC'92 Record, 23rd Annual IEEE Power Electronics Specialist Conf.*, vol. 2, pp. 951-957, 1992.

[26] M. S. Santini, A. R. Stubberud, and G. H. Hostetter, *Digital Control System Design*, New York: Saunders College Publishing, 1994.

[27] K. R. Shouse and D. G. Taylor, "Sensorless velocity control of permanent-magnet synchronous motors," *Proc. of the 33rd Conf. on Decision and Control*, pp. 1844-1849, Dec. 1994.

[28] J. Hu, D. M. Dawson, and K. Anderson, "Position control of a brushless DC motor without velocity measurements," *IEE Proc. on Electronic Power Applications*, vol. 142, pp. 113-119, March 1995.

[29] J. Solsona, M. I. Valla, and C. Muravchik, "A nonlinear reduced order observer for permanent magnet synchronous motors," *IEEE Trans. on Industrial Electronics*, vol. 43, pp. 38-43, Aug. 1996.

[30] R. B. Sepe and J. H. Lang, "Real-time observer-based (adaptive) control of a permanent-magnet synchronous motor without mechanical sensors," *IEEE Trans. on Industry Applications*, vol. 28, pp. 1345-1352, Nov./Dec. 1992.

[31] L. Sicot, S. Siala, K. Debusschere, and C. Bergmann, "Brushless DC motor control without mechanical sensors," *IEEE Power Electronics Specialist Conf.*, pp. 375-381, 1996.

[32] T. Senjyu, M. Tomita, S. Doki, and S. Okuma, "Sensorless vector control of brushless DC motors using disturbance observer," *PESC'95 Record, 26th Annual IEEE Power Electronics Specialists Conf.*, vol. 2, pp. 772-777, 1995.

[33] R. Bronson, *Matrix Methods - An Introduction*, New York: Academic Press, Inc., 1991.

[34] J. Kim and S. Sul, "High performance PMSM drives without rotational position sensors using reduced order observer," *Record of the 1995 IEEE Industry Applications Conf.*, vol. 1, pp. 75-82, 1995.

[35] Y. Kim, J. Ahn, W. You, and K. Cho, "A speed sensorless vector control for brushless DC motor using binary observer," *Proc. of the 1996 IEEE IECON 22nd Int'l. Conf. on Industrial Electronics, Control, and Instrumentation*, vol. 3, pp. 1746-1751, 1996.

[36] T. Furuhashi, S. Sangwongwanich, and S. Okuma, "A position-and-velocity sensorless control for brushless DC motors using an adaptive sliding mode observer," *IEEE Trans. on Industrial Electronics*, vol. 39, pp. 89-95, April 1992.

[37] Z. Peixoto and P. Seixas, "Application of sliding mode observer for induced e.m.f., position and speed estimation of permanent magnet motors," *IEEE Conf. Proc. of IECON*, pp. 599-604, 1995.

[38] J. Hu, D. Zhu, Y. Li, and J. Gao, "Application of sliding observer to sensorless permanent magnet synchronous motor drive system," *IEEE Power Electronics Specialist Conf.*, vol. 1, pp. 532-536, 1994.

[39] K. Iizuka, H. Uzuhashi, and M. Kano, "Microcomputer control for sensorless brushless motor," *IEEE Trans. on Industry Applications*, vol. IA-27, pp. 595-601, May/June 1985.
[40] J. Moreira, "Indirect sensing for rotor flux position of permanent magnet AC motors operating in a wide speed range," *IEEE Trans. on Industry Applications Society*, vol. 32, pp. 401-407, Nov./Dec. 1996.
[41] S. Ogasawara and H. Akagi, "An approach to position sensorless drive for brushless DC motors," *IEEE Trans. on Industry Applications*, vol. 27, pp. 928-933, Sep./Oct. 1991.
[42] T. M. Jahns, R. C. Becerra, and M. Ehsani, "Integrated current regulation for a brushless ECM drive," *IEEE Trans. on Power Electronics*, vol. 6, pp. 118-126, Jan. 1991.
[43] R. C. Becerra, T. M. Jahns, and M. Ehsani, "Four-quadrant sensorless brushless ECM drive," *IEEE Applied Power Electronics Conf. and Exposition*, pp. 202-209, March 1991.
[44] D. Regnier, C. Oudet, and D. Prudham, "Starting brushless DC motors utilizing velocity sensors," *Proc. of the 14th Annual Symposium on Incremental Motion Control Systems and Devices*, Champaign, Illinois: Incremental Motion Control Systems Society, pp. 99-107, June 1985.
[45] D. Peters and J. Harth, "I.C.s provide control for sensorless DC motors," *EDN*, pp. 85-94, April 1993.
[46] J. P. Johnson, "Synchronous-misalignment detection/correction technique of sensorless BLDC control," Ph.D. Dissertation, Texas A&M University, 1998.
[47] J. R. Hendershot and T. J. E. Miller, *Design of Brushless Permanent-Magnet Motors*, Oxford, 1994.
[48] T. J. E. Miller and J. R. Hendershot, *Switched Reluctance Motors & Their Controls*, Magna Physics Publishing, Madison, WI, 1993.
[49] R. Krishnan, *Switched Reluctance Motor Drives: Modeling, Simulation, Analysis, Design, and Applications*, CRC Press, 2001.
[50] M. Barnes and C. Pollock, "Power electronic converters for switched reluctance drives," *IEEE Transactions on Power Electronics*, vol. 13, pp. 1100-1111, Nov. 1998.
[51] V. R. Stefanovic and S. Vukosavic, "SRM inverter topologies: a comparative evaluation," *IEEE Transactions on Industry Applications*, vol. 27, pp. 1034-1047, Nov./Dec. 1991.
[52] K. W. E. Cheng, D. Sutanto, C. Y. Tang, X. D. Xue, and Y. P. B. Yeung, "Topology analysis of switched reluctance drives for electric vehicle," in *Proc. 8th International Power Electronics and Variable Speed Drives Conf.*, 2000, pp. 512-517.
[53] B. Fahimi, G. Suresh, J. Mahdavi, and M. Ehsani, "A new approach to model switched reluctance motor drive: application to analysis, design and control," in *Proc. IEEE Power Electronics Specialists Conference*, Fukuoka, 1998.

[54] J. Mahdavi, G. Suresh, B. Fahimi, and M. Ehsani, "Dynamic modeling of non-linear SRM using Pspice," in *Proc. IEEE Industry Applications Society Annual Meeting,* New Orleans, 1997.

[55] S. Filizadeh, L. S. Safavian, and A. Emadi, "Control of variable reluctance motors: A comparison between classical and Lyapunov-based fuzzy schemes," *Journal of Power Electronics*, pp. 305-311, Oct. 2002.

12
Multi-Converter Vehicular Dynamics and Control

In general, there are two kinds of loads in multi-converter power electronic systems. One group is conventional constant voltage loads that require fixed voltage for their operation. The other group is constant power loads sinking constant power from their input buses [1]. Power electronic converters, when tightly regulated, behave as constant power loads [1]-[8]. An example is an electric motor drive, which tightly regulates the speed when the rotating load has one-to-one torque-speed characteristic. Another example is a voltage regulator, which tightly regulates the voltage for an electric load that has one-to-one voltage-current characteristic.

Constant power loads have a negative impedance characteristic. This means that although, in constant power loads, the instantaneous value of impedance is positive, the incremental impedance is always negative. In fact, the current through a constant power load decreases/increases when the voltage across it increases/decreases. This is a destabilizing effect for the system and it is known as negative impedance instability.

Because of the non-linearity and time-dependency of DC/DC converters, and because of the negative impedance destabilizing characteristics of constant power loads, classical linear control methods, which are often used to design controllers for DC/DC converters, have stability limitations around the operating points. Therefore, stabilizing control methods must be used to ensure large-signal stability.

The purpose of this chapter is to present an assessment of the negative impedance instability concept of the constant power loads in the multi-converter vehicular power electronic systems. We address the fundamental problems faced in the stability studies of these systems. In addition, necessary and sufficient conditions of stability for DC vehicular distribution systems with constant power and conventional constant voltage loads are expressed.

Negative impedance stabilizing controllers for PWM DC/DC converters are designed and simulated under large changes in the loads and for different operations. Furthermore, the stability is verified using the second theorem of Lyapunov for a large range of variations in the constant power and constant voltage loads. Therefore, the presented controllers have the large-signal capabilities. Moreover, dynamic responses are improved.

12.1 Multi-Converter Vehicular Power Electronic Systems

International space station (ISS), spacecraft and aircraft power systems, more electric vehicles (MEVs) such as more electric cars (MECs) and hybrid electric vehicles (HEVs), advanced industrial electrical systems, telecommunications, and terrestrial computer systems are the best examples of multi-converter power electronic systems. In these systems, multiple power electronic converters are used as source, load, and distribution converters to supply needed power at different voltage levels and both DC and AC forms. Most of the loads are also in the form of electric motor drives, DC/DC choppers, DC/AC inverters, and AC/DC rectifiers. Of prime concern in such systems is the way that the system responds to dynamics caused by interconnection between converters. In addition, changes in power demands, faulty operations of converters, and so forth disturb dynamics of the system. However, the dynamics associated with these multi-converter systems and related issues such as modeling, design, control, and stability assessment are not widely known.

Examples of multi-converter vehicular power electronic systems, which are used in land, sea, air, and space vehicles, have been reviewed in previous chapters. In addition, a brief description of the conventional and advanced architectures, role of power electronics, and present trends have been given. The purpose of this chapter is to present a conceptual definition of these unconventional power systems. Furthermore, this chapter addresses the fundamental issues faced in these systems.

12.1.1 System Definition

In this section, we present a representative system that is consistent with the practical systems explained in previous chapters. Then we will study the dynamics associated with the multi-converter environment in this system. Figure 12.1 shows the concept of a representative system. The primary source of power is the starter/generator system, which includes a bidirectional power electronic converter. The storage system through the battery charge/discharge unit is also connected to the main bus. The primary source of power may be arrays of solar panels as in the ISS. The main switching DC/DC converters #1 and #2 supply power to buses #2 and #3, respectively, which then supply power to several smaller buses. These, in turn, power a variety of loads.

Multi-Converter Vehicular Dynamics and Control

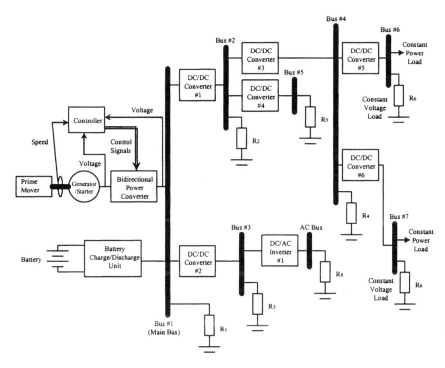

Figure 12.1 A representative multi-converter power electronic system.

The system of Figure 12.1 is a hybrid system with a main DC distribution system. AC loads are feeding from the main AC bus, which is provided by DC/AC inverter #1. This inverter along with some resistive loads are connected to the main DC/DC converter #2. Two different constant power loads are also connected to buses #6 and #7. Tables 12.1 and 12.2 show the bus voltages and the type of power electronic converters for the representative system of Figure 12.1, respectively.

12.1.2 Simulation Considerations

Conventionally, each power electronic converter in a multi-converter system is modeled, designed, and controlled by considering its stand-alone operation. However, because of the interactions between converters in an integrated multi-converter system, dynamic performances of the converters and, as a result, the system may be degraded. These interactions may even cause system instability.

To show that dynamics at the output of each converter affects other converters and bus voltages throughout the system, we have simulated the representative system of Figure 12.1 considering load step changes at different buses. The simulation results are presented in chapter 7. There are also some severe transients associated with the multi-converter environment. These transients may even destabilize the system. As an example, we have simulated the representative system of Figure 12.1 when there is a tightly regulated Buck-Boost converter connected to bus #7. This tightly regulated converter, as is explained in the next section, behaves as a constant power load. Due to the negative impedance characteristic of this constant power load at bus #7, the DC/DC converter #6 is unstable. Figure 12.2 shows the voltage at bus #7.

Table 12.1 Bus voltages of the representative system.

Battery Bus	Bus #1	Bus #2	Bus #3	Bus #4	Bus #5	Bus #6	Bus #7	AC Bus
120v	120v	60v	60v	30v	60v	100v	10v	54v, 60Hz

Table 12.2 Power electronic converters of the representative system.

Battery Charge/Discharge Unit	Buck-Boost Converter
DC/DC Converter #1	Buck Converter
DC/DC Converter #2	Buck Converter
DC/DC Converter #3	Buck Converter
DC/DC Converter #4	Buck-Boost Converter
DC/DC Converter #5	Boost Converter
DC/DC Converter #6	Buck Converter
DC/DC Load Converter	Buck-Boost Converter with Hysteresis Control
DC/AC Load Inverter	Sliding-Mode Buck Converter & Square Wave Inverter
DC/AC Inverter #1	Square Wave Inverter with RLC Load

12.2 Constant Power Loads and Their Characteristics

Most of the power electronic converters in the multi-converter vehicular power electronic systems, when they are tightly regulated, behave as constant power loads. In fact, there is a trend in the loads of multi-converter systems to be constant power. On the other hand, several loads such as electric motors, actuators, and power electronic converters have to be controlled such that constant output power is maintained for them. Output power is equal to the input power, if we neglect the losses, that is, assuming 100% efficiency for the drive system. Furthermore, the output power is constant; therefore, the input power is constant. As a result, these loads present constant power characteristics to the system.

Multi-Converter Vehicular Dynamics and Control

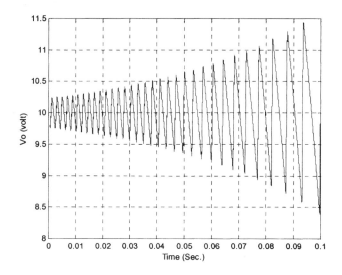

Figure 12.2 Output voltage of DC/DC converter #6 when there is a tightly regulated converter connected to bus #7.

An example of constant power loads, as is shown in Figure 12.3, is a DC/AC inverter which drives an electric motor and tightly regulates the speed when the rotating load has a one-to-one torque-speed characteristic. The simplest form of one-to-one torque-speed characteristics is linear relation between torque and speed. In Figure 12.3, the controller tightly regulates the speed; therefore, the speed (ω) is almost constant. Since the rotating load has a one-to-one torque-speed characteristic, for every speed there is one and only one torque. As a result, for a constant speed (ω), torque (T) is constant and power, which is the multiplication of speed and torque, is constant. If we assume a constant efficiency for the drive system, considering the constant power of the rotating load, the input power of the DC/AC inverter will be constant. Therefore, the DC/AC inverter presents a constant power load characteristic to the power system.

Figure 12.4 shows another example of constant power loads. It is a DC/DC converter which feeds an electric load and tightly regulates the voltage when the electric load has a one-to-one voltage-current characteristic. The simplest form of these loads is a resistor, which has a linear relation between voltage and current.

In constant power loads, although the instantaneous value of impedance is positive ($V/I>0$), the incremental impedance is always negative ($dV/dI<0$). In fact constant power loads have negative impedance characteristics, which might impact the power quality and stability of the multi-converter systems. Figure 12.5 depicts the negative impedance behavior of constant power loads. In the next sections, we will assess the effects of constant power loads in the multi-converter power electronic systems.

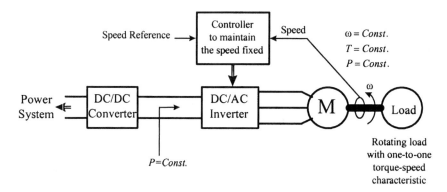

Figure 12.3 A DC/AC inverter that presents a constant power load characteristic to the system.

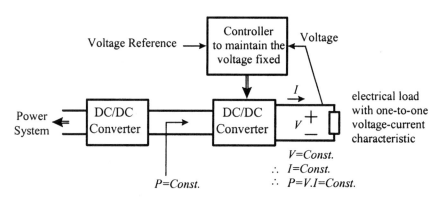

Figure 12.4 A DC/DC voltage regulator that presents a constant power load characteristic to the system.

12.3 Concept of Negative Impedance Instability

In a constant power load, power, which is the product of voltage and current of the load $(V.I)$, is constant. Therefore, as is shown in Figure 12.5, if the voltage across a constant power load increases/decreases, the current through it decreases/increases. This is a destabilizing effect to the system that the constant power load is connected to.

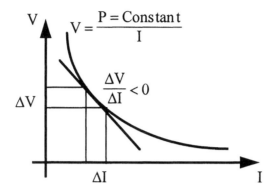

Figure 12.5 The negative impedance behavior of constant power loads.

Figure 12.6 depicts a simple circuit in which a constant power load with power P is connected in parallel to the capacitor C with initial voltage V_o. The state equations are given by

$$\begin{cases} v_o = \dfrac{P}{i_o} \\ -i_o = C\dfrac{dv_o}{dt} \end{cases} \qquad (12.1)$$

Solving these equations gives:

$$\begin{aligned} v_o^2 &= V_o^2 - \dfrac{2P}{C}t \\ i_o &= \dfrac{P}{\sqrt{V_o^2 - \dfrac{2P}{C}t}} \end{aligned} \qquad (12.2)$$

where t represents time. From equations (12.2), as is depicted in Figure 12.7, i_o and v_o go to infinity and zero, respectively, in $t = V_o^2 C/2P$. Therefore, the simple circuit of Figure 12.4, with only a capacitor and a constant power load, is unstable.

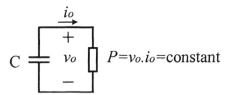

Figure 12.6 A constant power load parallel to a capacitor.

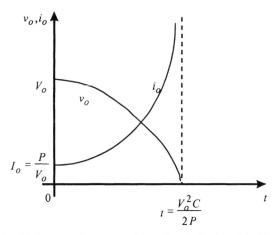

Figure 12.7 Voltage and current of the circuit depicted in Figure 12.6.

Figure 12.8 depicts another simple circuit in which a constant power load with power P is connected in series to the inductor L and voltage source v. The equilibrium operating point of the circuit is obtained when source voltage equals the constant power load voltage. Circuit will operate in steady-state at this point. Stability of an equilibrium point from the steady-state voltage-current curves of the source and load is the concept of steady-state stability of the circuit without solving the differential equations of the circuit valid for transient operation. Figure 12.9 depicts v-i characteristics of typical voltage sources and constant power loads.

An equilibrium point will be regarded as stable when the operation will be restored to it after a small departure from it due to a disturbance in the source or load. Let us examine the steady-state stability of equilibrium point A in Figure 12.9. Suppose the disturbance causes a reduction of Δi in the current. At the new current, the source voltage is less than the load voltage; consequently, the current decreases again and the operating point moves away from A. In fact, the circuit acts like a positive feedback. Finally current and load voltage go to zero and infinity, respectively. Similarly, an increase of Δi in current caused by a disturbance will make the load voltage less than the source voltage, resulting in more increase in current, which will move the operating point away from A. Thus, A is an unstable point of equilibrium.

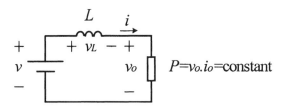

Figure 12.8 A constant power load series with an inductor.

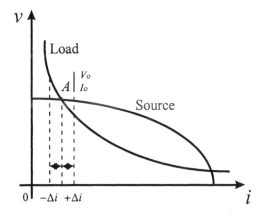

Figure 12.9 v-i characteristics of typical voltage sources and constant power loads.

Figure 12.10 depicts voltage-current characteristics of typical voltage sources and resistive loads. Let us examine the steady-state stability of equilibrium point B in Figure 12.10, which is obtained when the same source feeds another load. The load of Figure 12.10 has a positive incremental resistance instead of the negative incremental resistance characteristic of constant power loads. Suppose the disturbance causes an increase of Δi in current. It causes the source voltage to become less than the load voltage; consequently, v_L becomes negative and current decreases. As a result, operation will be restored to B. Similarly, when working at B, a reduction in current will make the source voltage greater than the load voltage, which will make the voltage v_L positive and, therefore, the current increases making restoration of operation to point B. Therefore, the circuit is steady-state stable at point B. In fact, the circuit acts like a negative feedback.

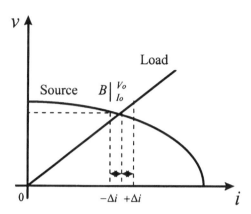

Figure 12.10 *v-i* characteristics of typical voltage sources and resistive loads.

The preceding discussion suggests that an equilibrium point will be stable when an increase in current causes load voltage to exceed the source voltage and a decrease in current causes load voltage to become less than the source voltage. This can be derived mathematically, for the circuit of Figure 12.8, as the following. In the circuit of Figure 12.8, the current equation is given by:

$$v - v_o = L\frac{di}{dt} \tag{12.3}$$

Assume that a small perturbation in current, Δi, causes Δv and Δv_o perturbations in v and v_o, respectively.

$$(v + \Delta v) - (v_o + \Delta v_o) = L\frac{d}{dt}(i + \Delta i) \tag{12.4}$$

Subtracting (12.3) from (12.4) gives

$$\Delta v - \Delta v_o = L\frac{d}{dt}(\Delta i) \tag{12.5}$$

Assuming small signal variations at operating point (OP) results in the following:

$$\Delta v \cong \left(\frac{dv}{di}\right)_{OP} \Delta i$$

$$\Delta v_o \cong \left(\frac{dv_o}{di}\right)_{OP} \Delta i \tag{12.6}$$

By inserting equations (12.6) in (12.5):

$$\left[\left(\frac{dv}{di}\right)_{OP} - \left(\frac{dv_o}{di}\right)_{OP}\right]\Delta i = L\frac{d}{dt}(\Delta i) \quad (12.7)$$

Let $(\Delta i)_o$ be the initial current perturbation. The solution of (12.7) is as follows:

$$\Delta i = (\Delta i)_o e^{-\frac{1}{L}\left[\left(\frac{dv_o}{di}\right)_{OP} - \left(\frac{dv}{di}\right)_{OP}\right]t} \quad (12.8)$$

The operating point is stable if Δi goes to zero as t goes to infinity. It means:

$$\left(\frac{dv_o}{di}\right)_{OP} > \left(\frac{dv}{di}\right)_{OP} \quad (12.9)$$

12.4 Negative Impedance Instability in the Single PWM DC/DC Converters

The DC/DC PWM Buck converter of Figure 12.11, which is operating with the switching period of T and duty cycle d, with a constant power load is considered.

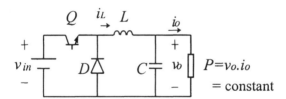

Figure 12.11 DC/DC Buck converter with constant power load.

Based on the analysis of the circuits depicted in Figures 26 and 28, we expect that the Buck converter of Figure 12.11 is unstable. However, to examine the stability, we consider small-signal analysis as the following.

12.4.1 Small-Signal Modeling

During the continuous conduction mode of operation, the state space equations when the switch is ON are given by:

$$0 \langle t \langle dT \begin{cases} \dfrac{di_L}{dt} = \dfrac{1}{L}[v_{in} - v_o] \\ \dfrac{dv_o}{dt} = \dfrac{1}{C}\left[i_L - \dfrac{P}{v_o}\right] \end{cases} \quad (12.10\text{-a})$$

and the equations when the switch is OFF are represented by:

$$dT \langle t \langle T \begin{cases} \dfrac{di_L}{dt} = \dfrac{1}{L}[-v_o] \\ \dfrac{dv_o}{dt} = \dfrac{1}{C}\left[i_L - \dfrac{P}{v_o}\right] \end{cases} \quad (12.10\text{-b})$$

Using the state space averaging method [9]-[13], these sets of equations can be shown by:

$$\begin{cases} \dfrac{di_L}{dt} = \dfrac{1}{L}[dv_{in} - v_o] \\ \dfrac{dv_o}{dt} = \dfrac{1}{C}\left[i_L - \dfrac{P}{v_o}\right] \end{cases} \quad (12.11)$$

Equations (12.11) are non-linear. For studying the small-signal stability of the Buck converter of Figure 12.11, small perturbations in the state variables due to the small disturbances in the input voltage and duty cycle are assumed as the following.

$$v_{in} = V_{in} + \tilde{v}_{in}$$
$$d = D + \tilde{d} \quad (12.12)$$
$$v_o = V_o + \tilde{v}_o$$
$$i_L = I_L + \tilde{i}_L$$

As is shown in Figure 12.12, $\tilde{v}_{in}$ and $\tilde{d}$ are the inputs, $\tilde{v}_o$ is the output, and $\tilde{i}_L$ and $\tilde{v}_o$ are the state variables of the small-signal model of the Buck converter. The stability of this small-signal model can be determined by calculating the transfer functions and their pole locations.

$$\tilde{v}_o(s) = H_1(s) * \tilde{d}(s) + H_2(s) * \tilde{v}_{in}(s) \quad (12.13)$$

$$H_1(s) = \frac{\tilde{v}_o(s)}{\tilde{d}(s)} = \frac{\dfrac{V_{in}}{LC}}{s^2 - \left(\dfrac{P}{CV_o^2}\right)s + \left(\dfrac{1}{LC}\right)} \quad (12.14)$$

$$H_2(s) = \frac{\tilde{v}_o(s)}{\tilde{v}_{in}(s)} = \frac{\dfrac{D}{LC}}{s^2 - \left(\dfrac{P}{CV_o^2}\right)s + \left(\dfrac{1}{LC}\right)} \quad (12.15)$$

The poles of the transfer functions $H_1(s)$ and $H_2(s)$ have positive real parts. Therefore, the system is unstable as the effect of the constant power load. Figure 12.13 depicts the simulation of the open-loop Buck converter with the parameters given in Table 12.3 and duty cycle of 0.5.

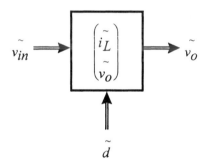

Figure 12.12 Small-signal inputs, output, and state variables of the Buck converter.

Table 12.3 Parameters of the Buck converter.

V_{in}	L	C	P	f
20 V	1 mH	10 mF	0.01 W	10 kHz

In reference [3], the dynamic properties of the Buck converter with constant power load are studied, and the line-to-output and control-to-output transfer functions are derived. Reference [3] also considers both voltage mode control and current mode control, in continuous conduction mode and discontinuous conduction mode.

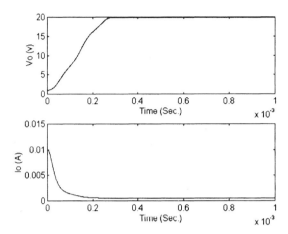

Figure 12.13 Simulation of open-loop Buck converter with duty cycle 0.5.

However, there is another type of instability based on the system parameters, which is of great importance in stabilizing the DC/DC converters in the next section. Based on the P, C, L, V_o, I_o values, the converter can go to instability very fast. It is like the instability explained in Figure 12.7. In this case, the converter doesn't have enough time even to change the status of the switch. In the other words, this type of instability happens in a tiny time interval. Suppose the inductor current is almost constant (I_{in}) in this short period of time. Figure 12.14 shows the equivalent circuit for this very fast instability occurring in the Buck converter of Figure 12.11. The state equations are given by

$$\begin{cases} v_o = \dfrac{P}{i_o} \\ I_{in} - i_o = C\dfrac{dv_o}{dt} \end{cases} \quad (12.16)$$

Solving these equations gives:

$$t = \dfrac{C}{I_{in}^2}\left[P*\ln\left(\dfrac{P - I_{in}*v_o}{P - I_{in}*V_o}\right) - I_{in}(v_o - V_o)\right] \quad (12.17)$$

where t represents time. From equations (12.16) and (12.17), as it is depicted in Figure 12.15, i_o and v_o go to infinity and zero, respectively, in $t = t_1$.

$$t_1 = \frac{C}{I_{in}^2}\left[P*\ln\left(\frac{P}{P-I_{in}*V_o}\right)+I_{in}*V_o\right] \tag{12.18}$$

Therefore, the system with constant power load is unstable. The stabilizing controller of the next section cannot stabilize the system in this case, because this instability happens in a time interval less than the switching period. Therefore, we have to increase the switching frequency of *C* and *L* and then we should apply the sliding-mode controller of the next section.

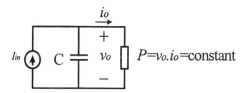

Figure 12.14 Equivalent circuit for very fast instability occurring in the Buck converter.

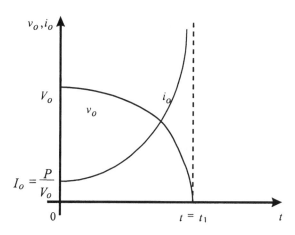

Figure 12.15 Voltage and current of the circuit depicted in Figure 12.14.

12.4.2 Designing Stabilizing Controller

As was explained, because of the negative impedance characteristics of constant power loads, conventional linear control methods have stability limitations around the operating points. To overcome the instability problem, reference [4] presents a nonlinear *PI* stabilizing controller. However, the main disadvantage of the method proposed in [4] is variable switching frequency. In this section, a new approach to the design of stabilizing controllers for PWM DC/DC converters with constant power loads using sliding-mode control is presented. The proposed controllers improve large-signal stability and dynamic responses.

12.4.2.1 State Space Averaging Model

In this section, state space averaging methods [9]-[13] are used to model PWM DC/DC converters. The Buck converter of Figure 12.11, which is operating with the switching period T and duty cycle d, is considered. During continuous conduction mode of operation, the state space equations using the state space averaging method can be shown by

$$\begin{cases} \dot{x}_1 = -\frac{1}{L}x_2 + \frac{d}{L}V_{in} \\ \dot{x}_2 = \frac{1}{C}x_1 - \frac{1}{RC}x_2 \end{cases} \tag{12.19}$$

where x_1 and x_2 are the moving averages of i_L and v_o, respectively.

12.4.2.2 Sliding-Mode Control

The objective of the control system in DC/DC converters with constant power loads is to control the output power. P_{out} is output power and K is output power reference. To design this type of controller, the moving averages of their state variables are used. This will significantly simplify the design. The sliding surface in the state space is described by $P_{out}=K$. According to the sliding-mode control [14]-[16] there is:

$$\begin{cases} \dot{P}_{out} < 0 \quad \text{if} \quad P_{out} > K \\ \dot{P}_{out} > 0 \quad \text{if} \quad P_{out} < K \end{cases} \tag{12.20}$$

For this, a first order path is chosen according to the following equation and the convergence speed is under control.

$$\dot{P}_{out} = -\lambda \left(P_{out} - K \right) \tag{12.21}$$

where λ is a positive real number and is called the convergence factor. Figure 12.16 shows the convergence relation for control of DC/DC converters with constant power loads.

In order to design the controller, it is necessary to combine (12.21) with (12.19). This will result in an equation for duty cycle d in terms of the state

Multi-Converter Vehicular Dynamics and Control

variables and the system parameters. This equation is important, since it would control the output power. A main contribution of this section is the fact that the relationship which is obtained for the duty cycle d has fewer state variables in it. The fewer times state variables appear in the duty cycle relation, the fewer times feedback of the state variables will be needed. By inserting the convergence equation (12.21) into equations (12.19), there is:

$$d(t) = \frac{\frac{1}{L}x_2^2 - \lambda\left(1 - \frac{\lambda RC}{2}\right)\left(\frac{x_2^2}{R} - K\right) - \frac{1}{C}\left(\frac{\lambda RC}{2}\right)^2 \frac{1}{x_2^2}\left(\frac{x_2^2}{R} - K\right)^2}{\frac{1}{L}x_2 * v_{in}} \qquad (12.22)$$

At the steady state where $P_{out} = K$, the duty cycle is given by

$$d^* = \frac{K}{V_{in}} = \frac{V_0}{V_{in}} \qquad (12.23)$$

where V_0 is constant and it is the steady state value of v_0. Equation (12.23) is the Buck converter input and output relation in the steady state.

Based on (12.21), the larger the convergence factor the faster the converter follows its control commands. However, it is not possible to increase the convergence factor too much. The reason is that there are practical limits of the system. Besides, there is an upper limit on the value of the convergence factor in the theory. In other words, in the equation (12.22) the convergence factor must be so that in the different functions of converter, duty cycle d is real and in the reasonable range. For the Buck converter under study a convergence factor of 10000 is selected. The designed controller with the parameters given in Table 12.4 was simulated. Figures 37-40 illustrate the results provided by system simulation per $K=10$ W.

Based on the fact that the sliding-mode controller holds the system in the way that $(P_{out}-K)$ reduces, it is possible to apply the following approximation in the equation (12.22).

$$\frac{1}{C}\left(\frac{\lambda RC}{2}\right)^2 \frac{1}{x_2^2}\left(\frac{x_2^2}{R} - K\right)^2 \approx 0 \qquad (12.24)$$

So the more simple equation is obtained for the duty cycle d.

$$d(t) = \frac{\frac{1}{L}x_2^2 - \lambda\left(1 - \frac{\lambda RC}{2}\right)\left(\frac{x_2^2}{R} - K\right)}{\frac{1}{L}x_2 * v_{in}} \qquad (12.25)$$

A simulation result similar to that of the previous controller was obtained. The only problem in the simplified sliding-mode controller comparing to the complete controller is its sensitivity to large load changes. Therefore, there is a limitation on load changes. A better response was obtained by reducing the convergence factor at the expenses of increasing the settling time.

Table 12.4 Parameters of the PWM DC/DC Buck converter.

V_{in}	L	C	P	f
20 V	1 mH	10 mF	10 W	10 kHz

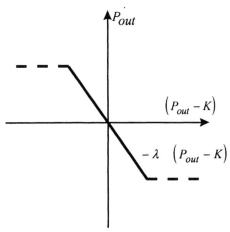

Figure 12.16 Convergence relation for control of DC/DC converters.

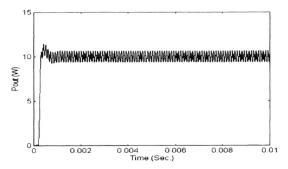

Figure 12.17 Start-up of sliding-mode controlled Buck converter.

Multi-Converter Vehicular Dynamics and Control 443

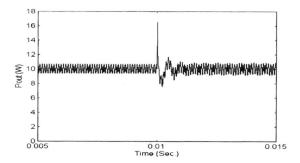

Figure 12.18 Dynamic response of sliding-mode controlled Buck converter to load step change from 10Ω to 5Ω at 0.01 sec.

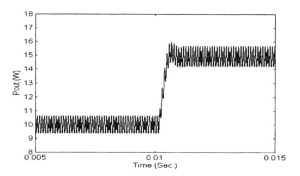

Figure 12.19 Dynamic response of sliding-mode controlled Buck converter to output power reference step change from 10 W to 15 W at 0.01 sec.

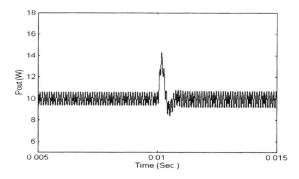

Figure 12.20 Dynamic response of sliding-mode controlled Buck converter to input voltage step change from 20V to 25V at 0.01 sec.

Designing stabilizing sliding-mode controllers for other PWM DC/DC converters such as Boost, Buck-Boost, Cuk, and Weinberg converters with constant power loads follows the same procedure.

12.4.2.3 Stability Analysis

The sliding surface for the DC/DC PWM Buck converter is defined as follows:

$$S = \{x | \ P_{out} - K = 0\} \tag{12.26}$$

The sliding-mode controller needs to keep the variables along the sliding surface; therefore,

$$P_{out} = K, \quad \dot{P}_{out} = 0 \tag{12.27}$$

The above conditions result in the following state space equations:

$$\begin{cases} \dot{x}_1 = -\frac{1}{L}x_2 + \frac{d}{L}V_{in} \\ 0 = \frac{1}{C}x_1 - \frac{1}{RC}x_2 \end{cases} \tag{12.28}$$

In order to prove the stability of the system, a continuously differentiable positive definite function, $V(x)$, needs to be defined. Let $V(x)$ be presented by the following quadratic function,

$$V(x) = \frac{1}{2}(x - x_e)^T P(x - x_e) \tag{12.29}$$

where x is the vector representation of the state variables given by

$$x = \begin{bmatrix} x_1 & x_2 \end{bmatrix}^T \tag{12.30}$$

P is a two-by-two identity matrix and x_e is the state variable equilibrium point shown by

$$x_e = \begin{bmatrix} x_{1e} & x_{2e} \end{bmatrix}^T \tag{12.31}$$

Derivation of (12.29) results as the following:

$$\dot{V}(x) = -\frac{R}{L}(x_1 - x_{1e})^2 \tag{12.32}$$

$\dot{V}(x)$ is a negative definite function and, consequently, $V(x)$ is a Lyapunov function. Therefore, the closed-loop system is uniformly asymptotically stable in the large [17] and the arbitrary operating point x_e is a stable equilibrium point. The stability analysis of other types of PWM DC/DC converters such as Boost, Buck-Boost, Cuk, and Weinberg converters with the proposed stabilizing sliding-mode controller follows the same procedure.

12.5 Stability of PWM DC/DC Converters Driving Several Loads

Figure 12.21 depicts the interconnecting converters in the multi-converter power electronic systems. In general, there are two kinds of loads in the system. One group is constant voltage loads, which require constant voltage for their operation. The other group is constant power loads sinking constant power from the source bus. The system has to provide constant power in a specified range of voltage (i.e., $V_{o,\min.} \leq v_o \leq V_{o,\max.}$) for them without going to instability.

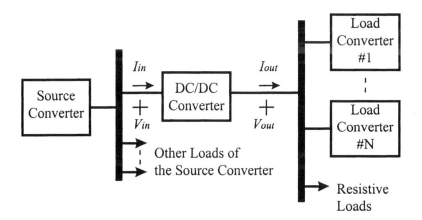

Figure 12.21 Interconnecting converters in the multi-converter power electronic systems.

Figure 12.22 shows a DC/DC converter with constant power and constant voltage loads in an example of advanced aircraft power systems of the future. In Figure 12.21, "other loads" means DC/AC inverters to provide AC supplies for AC loads or, even, other DC/DC converters to provide DC supplies with different voltage levels associated with different DC loads. The system depicted in Figure 12.22 will use solid-state bidirectional power electronic converters to condition variable-frequency power into a fixed voltage DC or a fixed frequency and voltage AC power to feed secondary power electronic converters.

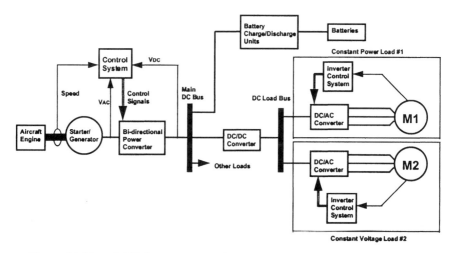

Figure 12.22 A DC/DC converter with constant power and constant voltage loads in an example of advanced aircraft power systems of the future.

Figure 12.23 shows the equivalent constant power and constant voltage loads, represented by P and R, respectively, of the DC/DC converter of Figure 12.21. Figure 12.24 depicts v-i characteristic of the equivalent load of the DC/DC converter. It also shows the stable region that conventional controllers can be applied for fixing the output voltage of the DC/DC converter.

If $v_o > V_o$, the slope of the v-i curve is positive and the equilibrium point is stable; that is, the operation will be restored to it after a small departure from it due to a disturbance in the source or load converter. If $v_o < V_o$, the incremental impedance is negative. In this case, the equilibrium point is unstable since the operating point moves away from it after a small departure due to a disturbance. Therefore, necessary and sufficient condition for stability can be expressed as follows.

$$P_{\text{Constant Power Loads}} < P_{\text{Constant Voltage Loads}} \tag{12.33}$$

In order to obtain (12.33) from the modeling, we suppose that the DC/DC converter of Figure 12.21 is a PWM Buck converter operating in continuous conduction mode, with switching period T and duty cycle d. It should be noticed that the procedure would be the same for other DC/DC PWM converters such as Boost, Buck-Boost, Cuk, and Weinberg converters.

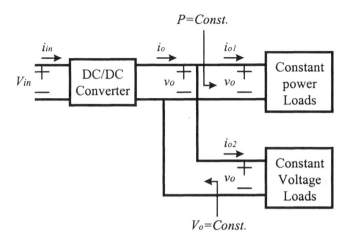

Figure 12.23 Equivalent constant power and constant voltage loads of the DC/DC converter.

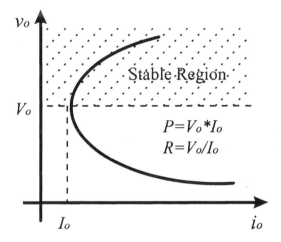

Figure 12.24 $v\text{-}i$ characteristic of the equivalent load of the DC/DC converter of Figure 12.23.

Constant power and constant voltage loads are all represented by a single dependent current source $i_{o1} = (P = Const.)/v_o$ and a resistance R, respectively. Using the state space averaging method the state equations of the DC/DC PWM Buck converter of Figure 12.21 can be shown by

$$\begin{cases} \dfrac{di_L}{dt} = \dfrac{1}{L}[dv_{in} - v_o] \\ \dfrac{dv_o}{dt} = \dfrac{1}{C}\left[i_L - \dfrac{v_o}{R} - \dfrac{P}{v_o}\right] \end{cases} \quad (12.34)$$

The small-signal transfer functions of the system of equations (12.34), assuming small perturbations in the state variables due to the small disturbances in the input voltage and duty cycle as (12.12), are as follows:

$$H_1(s) = \frac{\tilde{v}_o(s)}{\tilde{d}(s)} = \frac{\dfrac{V_{in}}{LC}}{s^2 + \left(\dfrac{1}{RC} - \dfrac{P}{CV_o^2}\right)s + \left(\dfrac{1}{LC}\right)} \quad (12.35)$$

$$H_2(s) = \frac{\tilde{v}_o(s)}{\tilde{v}_{in}(s)} = \frac{\dfrac{D}{LC}}{s^2 + \left(\dfrac{1}{RC} - \dfrac{P}{CV_o^2}\right)s + \left(\dfrac{1}{LC}\right)} \quad (12.36)$$

Necessary and sufficient condition for stability is determined to be the following; that is, poles of the transfer functions have negative real parts.

$$\frac{1}{RC} - \frac{P}{CV_o^2} > 0 \quad \Rightarrow \quad R < \frac{V_o^2}{P} \quad (12.37)$$

which is the same relation as (12.33). In the case of equations (12.33) and (12.37), we can control the DC/DC converter using conventional controllers such as the PI controller, as is shown in Figure 12.25, without going to instability. In other words, if equation (12.33) is satisfied, we can provide a constant voltage at the output of the DC/DC converter for constant voltage loads and, at the same time, supply constant power to the constant power loads.

However, if equation (12.33) is not satisfied, the system is unstable. In this case, the proposed method to provide constant power for constant power loads and supply constant voltage for constant voltage loads while ensuring stability is adding a DC/DC converter as a voltage regulator to feed constant voltage loads (Figure 12.26). The proposed stabilizing sliding-mode controller is used to provide constant power by the DC/DC converter in a predefined range of output voltage ($V_{o,\min.} \leq v_o \leq V_{o,\max.}$).

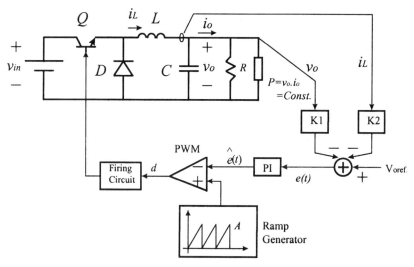

Figure 12.25 Conventional PI controller block diagram of a PWM DC/DC Buck converter with constant power and resistive loads.

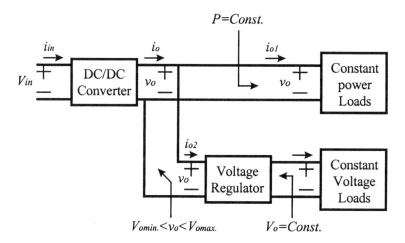

Figure 12.26 Adding a DC/DC converter as a voltage regulator to feed constant voltage loads.

In section 12.7, feedback linearization techniques are used to design stabilizing robust controllers for DC/DC PWM converters with constant power and constant voltage loads. It means that we provide a constant voltage at the output of the DC/DC converter for constant voltage loads, at the same time supplying constant power to constant power loads. For designing the controllers, the stability has been verified using the second theorem of Lyapunov for a large range of variations in the constant power and constant voltage loads ($R_{min} < R < \infty$, $0 < P < P_{max}$). Therefore, the proposed controllers have large-signal capabilities.

12.6 Stability Condition in a DC Vehicular Distribution System

As was explained, constant power loads have a negative impedance destabilizing effect on DC/DC converters. In order to guarantee small-signal stability, as was determined, the power of constant power loads must be less than the power of conventional constant voltage loads for a DC/DC converter. However, in a distribution system with the presence of output resistance of the source subsystem seen from the input side of the DC/DC converter, the stability margin is improved. Figure 12.27 depicts the equivalent circuit of a DC distribution system in a typical bus driving several constant power and constant voltage loads. P and R are representing the equivalent constant power and constant voltage loads, respectively. In Figure 12.27, the source subsystem includes the state space averaged circuit of the DC/DC converter as well as filters and distribution system at the input side of the DC/DC converter.

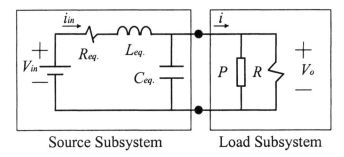

Figure 12.27 Equivalent circuit of a DC distribution system in a typical bus.

Voltages and currents of the circuit of Figure 12.27 are given by

$$\begin{cases} v_{in} = R_{eq}.i_{in} + L_{eq}.\dfrac{di_{in}}{dt} + v_o \\ i_{in} = C_{eq}.\dfrac{dv_o}{dt} + \dfrac{P}{v_o} + \dfrac{v_o}{R} \end{cases} \quad (12.38)$$

By linearizing the set of differential equations (12.38) around the operating point, the small-signal transfer function is expressed by

Multi-Converter Vehicular Dynamics and Control

$$H(s) = \frac{\tilde{v}_o(s)}{\tilde{v}_{in}(s)} = \frac{1/C_{eq.}L_{eq.}}{s^2 + \left(\dfrac{R_{eq.}}{L_{eq.}} + \left(\dfrac{1}{R} - \dfrac{P}{V_o^2}\right)\dfrac{1}{C_{eq.}}\right)s + \left(\dfrac{1+R_{eq.}}{C_{eq.}L_{eq.}}\right)} \quad (12.39)$$

V_o is the output nominal voltage. By considering the condition for which the poles of the system have negative real parts, necessary and sufficient condition for small-signal stability is determined by

$$P < \frac{V_o^2}{R} + \frac{R_{eq.}C_{eq.}}{L_{eq.}}V_o^2 \quad (12.40\text{-a})$$

$$\frac{V_o^2}{P} > (R \parallel R_{eq.}) \quad (12.40\text{-b})$$

This can be written as follows:

$$P_{CPL} < P_{CVL} + \frac{R_{eq.}C_{eq.}}{L_{eq.}}V_o^2 \quad (12.41\text{-a})$$

$$P_{CPL} < \frac{V_o^2}{(R \parallel R_{eq.})} \quad (12.41\text{-b})$$

CPL and CVL mean constant power loads and constant voltage loads, respectively. Since $R_{eq.}$ is very small, generally, the constraint (12.41-b) is satisfied. Therefore, (12.41-a) is the most important constraint. The effect of the distribution system and its resistance is shown in the second term at the right hand side of (12.41-a). This is an additional term compared to (12.33). It is related to the distribution system, filters, and the DC/DC converter parameters as well as the output nominal voltage. Therefore, in order to ensure small-signal stability for a converter in a distribution system, the power of constant power loads must be less than the power of constant voltage loads plus $V_o^2(R_{rq.}C_{eq.})/L_{eq.}$. If this second term is greater than the power of constant power loads itself, the converter is stable even without any resistive load. This is an improved stability condition.

In order to stabilize the system, based on relation (12.41-a), there are four different possible solutions. The first one is increasing $R_{eq.}$, which is not practical due to the power loss. The second one is decreasing $L_{eq.}$, which is not feasible also. However, in the design stage of the system, it should be considered a priority to have the minimum inductance possible. The third method is increasing $C_{eq.}$, which is easily possible by adding a filter. In fact, adding passive filters is the solution which has been proposed to maintain the stability of the system [4], [5]. The last method is increasing V_o. In most cases, there is no control over the nominal voltages of the system. However, systems with higher base voltages have better negative impedance stability than systems with low base voltages.

Another approach to stabilize the system, only if we have control over constant power loads, is manipulating the input impedances of the constant power loads to satisfy the stability condition (12.41-a) in all cases. Figure 12.28 shows a typical mechanical constant power load. In Figure 12.28, the DC/AC inverter drives an electric motor and tightly regulates the speed when the rotating load has a one-to-one torque-speed characteristic. The controller tightly regulates the speed; therefore, the speed is almost constant. Since the rotating load has a one-to-one torque-speed characteristic, for a constant speed, torque is constant; as a result, power, which is the multiplication of speed and torque, is also constant. If we assume a constant efficiency for the drive system, the input power of the inverter would be constant.

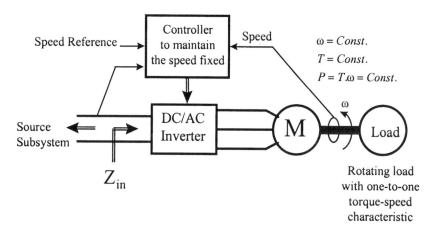

Figure 12.28 A DC/AC inverter presenting constant power load characteristic to the system.

The input impedance of the constant power load is expressed as follows:

$$Z_{in} = \frac{V_o^2}{P} \qquad (12.42)$$

Multi-Converter Vehicular Dynamics and Control

By considering this input impedance, the relation (12.40-a) can be shown by

$$\frac{1}{Z_{in}} < \frac{1}{R} + \frac{R_{eq.}C_{eq.}}{L_{eq.}} \tag{12.43}$$

In this method, the input impedance of constant power load is manipulated in such a way that the relation (12.43) is satisfied. In fact, instead of having a stable source control, changes in the load are made and the tightly regulated output is sacrificed. Based on this method, [4] and [5] propose a nonlinear stabilizing control to manipulate the input impedance of the converter/motor drive in an induction motor based electric propulsion system as well as a small distribution system consisting of a generation system, a transmission line, a DC/DC converter load, and a motor drive load. For their studies, since only constant power load is considered, the stability condition (12.43) becomes

$$Z_{in} > \frac{L_{eq.}}{R_{eq.}C_{eq.}} \tag{12.44}$$

In this chapter, we suppose that there is no control over the loads. Therefore, we design robust stabilizing control for the DC/DC converters to drive constant power and constant voltage loads without imposing any restriction for the loads or the system. As a result, there is no need for adding the filters or considering any additional restriction on the design of the distribution system. We use the same method for other converters in the multi-converter distribution system to stabilize them. However, for the original source of the system, the relations (12.41), with parameters defined at the source bus, should be satisfied. In fact, we must have a negative impedance stable source for the system.

12.7 Negative Impedance Stabilizing Control for PWM DC/DC Converters with Constant Power and Resistive Loads

The Buck converter of Figure 12.11, which is operating with the switching period T and duty cycle d, is considered. This converter is driving several different loads which require constant voltage for their operation. Some of these loads, such as tightly regulated power electronic converters, behave as constant power sinks. Equivalent constant power and constant voltage loads of the Buck converter, as is depicted in Figure 12.23, are represented by P and R, respectively. Upper and lower limits of P and R are given by

$$P_{min.} < P < P_{max.} (P_{CPL,min.} < P_{CPL} < P_{CPL,max.})$$
$$R_{min.} < R < R_{max.} (P_{CVL,max.} > P_{CVL} > P_{CVL,min.}) \tag{12.45}$$

For designing stabilizing controllers in this chapter, $P_{min.}$ and $R_{max.}$ are considered zero and infinity, respectively. Furthermore, we suppose $P_{max.} = P_{CPL,max.} = P_{CVL,max.}$. $P_{max.}$ occurs when all loads are on and behave as constant power sinks. It also occurs when all loads are on and have resistive characteristics as in constant voltage loads. $P_{min.}$ and $R_{max.}$ occur when there is no constant power load and no constant voltage load on, at the output of the DC/DC converter, respectively. They also happen when all loads are off.

Using the state space averaging method, during continuous conduction mode of operation, the state equations of the DC/DC converter of Figure 12.11 can be shown by

$$\begin{cases} \dfrac{di_L}{dt} = \dfrac{1}{L}[dv_{in} - v_o] \\ \dfrac{dv_o}{dt} = \dfrac{1}{C}\left[i_L - \dfrac{v_o}{R} - \dfrac{P}{v_o}\right] \end{cases} \quad (12.46)$$

The small-signal transfer functions of the system of equations (12.46), assuming small perturbations in the state variables due to small disturbances in the input voltage and duty cycle, are as follows.

$$H_1(s) = \frac{\tilde{v}_o(s)}{\tilde{d}(s)} = \frac{\dfrac{V_{in}}{LC}}{s^2 + \left(\dfrac{1}{RC} - \dfrac{P}{CV_o^2}\right)s + \left(\dfrac{1}{LC}\right)} \quad (12.47)$$

$$H_2(s) = \frac{\tilde{v}_o(s)}{\tilde{v}_{in}(s)} = \frac{\dfrac{D}{LC}}{s^2 + \left(\dfrac{1}{RC} - \dfrac{P}{CV_o^2}\right)s + \left(\dfrac{1}{LC}\right)} \quad (12.48)$$

Necessary and sufficient condition for stability is determined as follows; that is, poles of the transfer functions have negative real parts.

$$\frac{1}{RC} - \frac{P}{CV_o^2} \rangle 0 \quad \Rightarrow \quad R \langle \frac{V_o^2}{P} \quad (12.49)$$

This is the same stability condition as (12.33). It means that if the total power of constant power loads is less than the total power of constant voltage loads, we can control the DC/DC converter using conventional controllers such as *PI* controller without going to instability. In other words, if equation (12.49) is satisfied, we can provide constant voltage at the output of the DC/DC converter for constant voltage loads and, at the same time, supply constant power to the constant power loads. In the next section, using feedback linearization techniques, we design a negative impedance stabilizing controller for the DC/DC converter of Figure 12.11 with different loads in the range of (12.45).

12.7.1 Feedback Linearization Technique

In this section, we use the feedback linearization technique to control the converter [18], [19]. We look for a nonlinear feedback to cancel out the nonlinearity in the set of differential equations (12.46). In (12.46), there is no direct relation between the control input (dv_{in}) and the nonlinearity (P/v_o); therefore, the following change of variables is considered.

$$\begin{cases} x_1 = i_L - \dfrac{v_o}{R} - \dfrac{P}{v_o} \\ x_2 = v_o - V_{oRef.} \end{cases} \qquad (12.50)$$

where V_{oRef} is the output reference voltage for the Buck converter. Using change of variables (12.50), state equations (12.46) can be written as follows:

$$\begin{cases} \dot{x}_1 = \dfrac{dv_{in}}{L} - \dfrac{x_2 + V_{oRef.}}{L} - \dfrac{x_1}{RC} + \dfrac{Px_1}{C(x_2 + V_{oRef.})^2} \\ \dot{x}_2 = \dfrac{1}{C} x_1 \end{cases} \qquad (12.51)$$

In order to cancel out the nonlinearity in (12.51), the following nonlinear feedback is proposed:

$$\dfrac{dv_{in}}{L} = k_1 x_1 + k_2 x_2 - \dfrac{\hat{P} x_1}{C(x_2 + V_{oRef.})^2} + \omega \qquad (12.52)$$

$$\omega = \dfrac{V_{oRef.}}{L}$$

where k_1, k_2, and $\hat{P}$ are the parameters of the controller to be designed. With the nonlinear feedback, the state equations of (12.51) can be shown by

$$\begin{cases} \dot{x}_1 = (k_1 - \dfrac{1}{RC}) x_1 + (k_2 - \dfrac{1}{L}) x_2 + \dfrac{x_1}{C(x_2 + V_{oRef})^2} (P - \hat{P}) \\ \dot{x}_2 = \dfrac{1}{C} x_1 \end{cases} \qquad (12.53)$$

If $\hat{P} = P$, the system is linear. Other parameters of the controller (i.e., k_1 and k_2) can be designed such that the resulting system has poles at appropriate places. However, we don't have any control over the loads and they may change.

In fact, P and R vary according to (12.45). Therefore, in the next section, we design the controller to guarantee the stability of the system in the presence of load changes.

12.7.1.1 Stability Analysis

In order to evaluate the stability of the converter, a continuously differentiable positive definite function, $V(x)$, needs to be determined. We define $V(x)$ as follows:

$$V(x_1,x_2) = \frac{1}{2}Kx_1^2 + \frac{1}{2}KC(\frac{1}{L} - k_2)x_2^2 \tag{12.54}$$

where $K > 0$ and $k_2 < 1/L$; therefore, $V(x)$ is a positive definite function. The derivative of $V(x)$ is given by

$$\dot{V}(x_1,x_2) = K(k_1 - \frac{1}{RC})x_1^2 + K\frac{x_1^2}{C(x_2+V_{oRef.})^2}(P-\hat{P}) \tag{12.55}$$

In order to guarantee the stability of the converter, the derivative of $V(x)$ needs to be negative definite. Therefore, the parameters of the controller are chosen as follows.

$$\begin{cases} k_1 - \dfrac{1}{RC} < 0 \\ \hat{P} - P < 0 \\ k_2 < \dfrac{1}{L} \end{cases} \tag{12.56}$$

Considering upper and lower limits of P and R given in (12.45) as well as variations in the values of the inductor and capacitor of the converter, control parameters for a robust design are given by

$$\begin{cases} k_1 < \dfrac{1}{R_{max}.C_{max}} \\ \hat{P} > P_{max}. \\ k_2 < \dfrac{1}{L_{max}.} \end{cases} \tag{12.57}$$

Multi-Converter Vehicular Dynamics and Control

$L_{max.}$ and $C_{max.}$ are the maximum values that L and C can hold, respectively.

Therefore, $\dot{V}(x)$ is a negative definite function and, consequently, $V(x)$ is a Lyapunov function. Hence, the closed loop system is asymptotically stable and the operating point is a stable equilibrium point.

12.7.1.2 Simulation Results

A negative impedance stabilizing controller based on the proposed feedback linearization technique and *PI* controller has been designed and simulated. The parameters of the converter and designed robust controller are given in Table 12.5.

In order to study the converter dynamic performance under load variations, step changes in constant power and resistive loads have been investigated. Figures 49-51 depict the simulation results.

It must be mentioned that most of the practical constant power loads at the starting phase of their operations have positive incremental impedance characteristic. Therefore, their power is increasing until they reach the nominal power. After they hit the nominal power, the input power will be constant and they behave as constant power sinks, which have negative incremental impedance characteristic. In our simulations, as is depicted in Figure 12.32, we considered a linear *v-i* characteristic for the starting period of the constant power loads.

Table 12.5 Parameters of the Buck converter and stabilizing controller.

V_{in}	20 V
L	1 mH
C	10 mF
f	10 kHz
$V_{oRef.}$	10 V
R	20 ohms, 5 ohms < R < ∞
P	10 W, 0 < P < 20 W
PI controller	$K_p=1$, $K_I=100$
Feedback Linearization	$K_1=-195$, $K_2=900$, $\hat{P}=25$

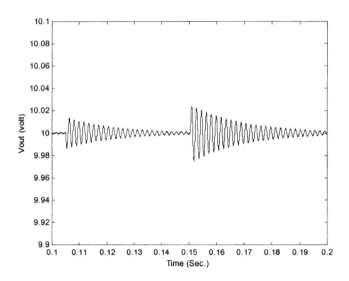

Figure 12.29 Dynamic response of the proposed controller for Buck converter to load step change from $P=10W$, $R=20\Omega$, to $P=10W$, $R=10\Omega$, and $P=10W$, $R=100\Omega$.

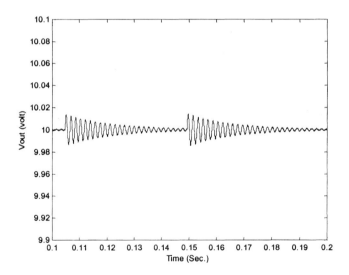

Figure 12.30 Dynamic response of the proposed controller for Buck converter to constant power load step change from $10W$ to $5W$ and 0 ($R=20\Omega$, i.e., $P_{Constant\ Voltage\ Load}=5W$).

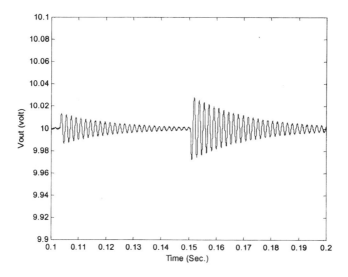

Figure 12.31 Dynamic response of the proposed controller for Buck converter to load step change from *P=10W, R=20Ω,* to *P=10W, R=1000Ω,* and *P=20W, R=1000Ω*.

As is shown, the controller designed in this section has large-signal capabilities. However, if the loads of the DC/DC converter are not changing much, we can use a simpler controller based on the linear state variable feedback method [19]. Figure 12.25 depicts the block diagram of a DC/DC PWM Buck converter with a linear state variable feedback controller. We have designed a stabilizing controller based on this method for a Buck converter with constant power loads. In addition, we have designed another stabilizing controller for the case that there are both constant power and constant voltage loads at the output of the Buck converter. Figures 53 and 54 show the simulation results. As is shown, the operations of these controllers are satisfactory in the presence of small variations in the loads. Therefore, these simple controllers are proposed for the converters operating in their nominal conditions without large changes in their loads.

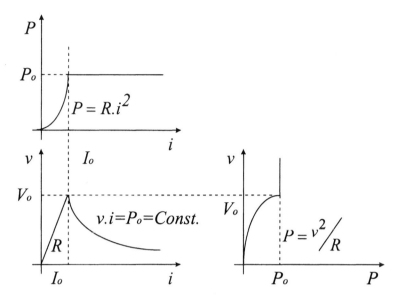

Figure 12.32 The v-i characteristic of a practical constant power load.

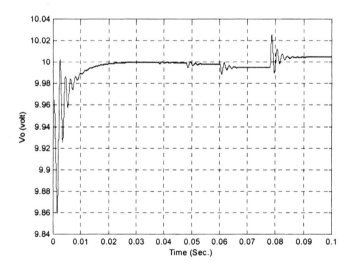

Figure 12.33 Dynamic response of the designed linear state variable feedback controller for the Buck converter with only constant power loads ($V_{in}=20v$, $L=1mH$, $C=10mF$, $P=10W$, $f=10kHz$, $V_{o,\,ref.}=10v$, $K_I=200$, $K_P=1$, $K_1=0.01$, $K_2=1$) to load step changes from $P=10W$ to $P=10.5W$ to $P=12W$ to $P=15W$ and to $P=5W$.

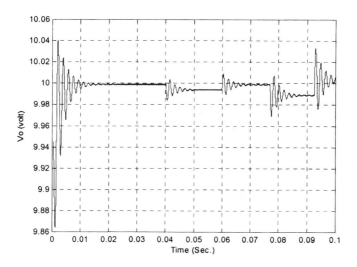

Figure 12.34 Dynamic response of the designed linear state variable feedback controller for the Buck converter with constant power and constant voltage loads ($V_{in}=20v$, $L=1mH$, $C=10mF$, $P=10W$, $R=20\Omega$, $f=10kHz$, $V_{o,\,ref.}=10v$, $K_1=500$, $K_P=1$, $K_I=0.01$, $K_2=1$) to load step changes from $P=10W$, $R=20\Omega$ to $P=10W$, $R=10\Omega$ to $P=5W$, $R=10\Omega$ to $P=15W$, $R=10\Omega$ and to $P=0W$, $R=10\Omega$.

12.8 Conclusion

In this chapter, the concept of negative impedance instability of the constant power loads in multi-converter vehicular power electronic systems was described. It was illustrated that if the power of constant power loads was greater than the power of constant voltage loads, the converter would be unstable. Furthermore, the necessary and sufficient conditions of stability for DC distribution systems containing DC/DC converters were explained. By applying the resultant models from the state space averaging method, different stabilizing controllers, such as sliding-mode and linear state variable feedback controllers, were designed. In addition, a nonlinear robust stabilizing controller based on the feedback linearization techniques for DC/DC PWM converters was proposed. The responses of the controllers under different operations and in the presence of significant variations in load and input voltage were studied and shown to be satisfactory. Large-signal control, simplicity of construction, and high reliability were described as the main advantages. To prove that the designed controllers were stable, Lyapunov's second theorem was used. Furthermore, recommendations on the design of multi-converter vehicular power electronic systems to avoid negative impedance instability were given.

12.9 References

[1] A. Emadi, "Modeling, analysis, and stability assessment of multi-converter power electronic systems," *Ph.D. Dissertation*, Texas A&M University, College Station, TX, Aug. 2000.

[2] A. Emadi, B. Fahimi, and M. Ehsani, "On the concept of negative impedance instability in advanced aircraft power systems with constant power loads," *Society of Automotive Engineers Journal*, Paper No. 1999-01-2545, 1999.

[3] V. Grigore, J. Hatonen, J. Kyyra, and T. Suntio, "Dynamics of a Buck converter with constant power load," in *Proc. IEEE 29th Power Electronics Specialist Conf.*, Fukuoka, Japan, May 1998, pp. 72-78.

[4] S. F. Glover and S. D. Sudhoff, "An experimentally validated nonlinear stabilizing control for power electronics based power systems," *Society of Automotive Engineers (SAE) Journal*, Paper No. 981255, 1998.

[5] S. D. Sudhoff, K. A. Corzine, S. F. Glover, H. J. Hegner, and H. N. Robey, "DC link stabilized field oriented control of electric propulsion systems," *IEEE Trans. on Energy Conversion*, vol. 13, no. 1, pp. 27-33, March 1998.

[6] A. S. Kislovski, "Optimizing the reliability of DC power plants with backup batteries and constant power loads," in *Proc. IEEE Applied Power Electronics Conf.*, Dallas, TX, March 1995, pp. 957-964.

[7] A. S. Kislovski and E. Olsson, "Constant-power rectifiers for constant-power telecom loads," in *Proc. 16th International Telecommunications Energy Conf.*, Nov. 1994, pp. 630-634.

[8] I. Gadoura, V. Grigore, J. Hatonen, and J. Kyyra, "Stabilizing a telecom power supply feeding a constant power load," in *Proc. 10th International Telecommunications Energy Conf.*, Nov. 1998, pp. 243-248.

[9] R. D. Middlebrook and S. Cuk, "A general unified approach to modeling switching converter power stages," in *Proc. IEEE Power Electronics Specialist Conf.*, June 1976, pp. 18-34.

[10] P. T. Krein, J. Bentsman, R. M. Bass and B. Lesieutre, "On the use of averaging for the analysis of power electronic systems," *IEEE Trans. on Power Electronics*, vol. 5, no. 2, pp. 182-190, 1990.

[11] J. Sun and H. Grotstollen, "Averaged modeling of switching power converters: reformulation and theoretical basis," in *Proc. IEEE Power Electronics Specialist Conf.*, June 1981, pp. 1165-1172.

[12] J. Mahdavi, A. Emadi, M. D. Bellar and M. Ehsani, "Analysis of power electronic converters using the generalized state space averaging approach," *IEEE Trans. on Circuits and Systems I: Fundamental Theory and Applications*, vol. 44, no. 8, pp. 767-770, Aug. 1997.

[13] M. O. Bilgic and M. Ehsani, "Analysis of inductor-converter bridge by means of state space averaging technique," in *Proc. IEEE Power Electronics Specialist Conference*, June 1988, pp. 116-121.

[14] V. I. Utkin, *Sliding Modes and their Application in Variable Structure Systems*, MIR, Moscow, Russia, 1974.
[15] H. Sira-Ramirez, "Sliding motions in bilinear switched networks," *IEEE Trans. on Circuits and Systems*, vol. CAS-34, no. 8, Aug. 1987.
[16] J. Mahdavi, A. Emadi, and H.A. Toliyat, "Application of state space averaging method to sliding mode control of PWM DC/DC converters," in *Proc. IEEE Industry Application Conf.*, New Orleans, LA, Oct. 1997, pp. 820-827.
[17] P. A. Ioannou and J. Sun, *Robust Adaptive Control*, PTR Prentice-Hall, Upper Saddle River, NJ, 1996.
[18] H. A. Khalil, *Nonlinear Systems*, Prentice-Hall, Upper Saddle River, NJ, 1996.
[19] J. E. Slotine and W. Li, *Applied Nonlinear Control*, Prentice-Hall, Upper Saddle River, NJ, 1991.

13

Effects of Constant Power Loads in AC Vehicular Systems

Conventional electrical power systems have a single-phase or three-phase AC distribution system. Most of the loads in these systems require constant voltage and frequency for their operation. Traditionally, these loads have a positive incremental impedance characteristic. However, with the expansion of solid-state devices in AC vehicular power systems to improve the performance and flexibility, power electronic converters and motor drives are extensively used in different vehicles. Consequently, power electronic converters are emergent loads in AC vehicular power systems.

As is explained in the previous chapter, power electronic loads, when tightly regulated, sink constant power from the distribution system. Therefore, they have a negative incremental impedance characteristic. Due to these unique characteristics, dynamics, and stability problems related to constant power behavior, conventional methods in AC power systems are not suitable for investigating the effects of power electronic loads in AC systems.

The purpose of this chapter is to present an assessment of the effects of constant power loads in AC vehicular distribution systems. Furthermore, recommendations for the design of AC power systems to avoid negative impedance instability are provided. Guidelines to design proper distribution architectures are also established.

13.1 Vehicular AC Distribution Systems

In this section, we consider a single-phase AC distribution system. As is depicted in Figure 13.1, there are different loads connected to each bus in the system. The duty of the source subsystem in Figure 13.1 is to maintain the amplitude and frequency of the bus voltage fixed. We assume that some of the loads, which are fed from the source subsystem, are tightly regulated power electronic converters. Therefore, these power electronic loads, which may be voltage regulators or motor drives, behave as constant power loads.

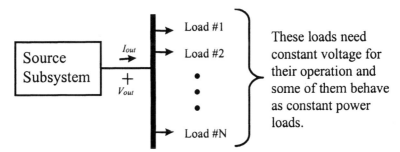

Figure 13.1 Conventional constant voltage and power electronic constant power loads connected to a typical bus.

Figure 13.2 shows the equivalent constant power and constant voltage loads connected to a typical bus. It also shows the equivalent circuit of the source subsystem at this typical bus. Since the stability of the system is defined as the absence of instability at each bus, we study the system stability at the typical bus with the equivalent circuit given in Figure 13.2. Because P and R are equivalent constant power and constant voltage loads, respectively, they can change according to the number and power of the loads. It means that if a load goes off or a new load comes on, P and R may change consequently. However, P and R instantaneously show the equivalent power of constant power loads and the equivalent resistance of constant voltage loads, respectively.

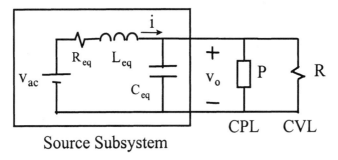

Figure 13.2 Equivalent circuit of the system of Figure 13.1.

In order to determine the necessary and sufficient conditions for stability of the circuit shown in Figure 13.2, constant power load P must be modeled. In

Effects of Constant Power Loads in AC Vehicular Systems

the next section, different AC constant power loads will be modeled. However, before modeling, some definitions related to average power are presented.

For an element with non-sinusoidal voltage and current in an AC distribution system, the average power can be expressed using the Fourier theorem. If $v(t)$ and $i(t)$ are the voltage across and current through the element, respectively, they can be described by a constant term plus an infinite series of cosine terms of frequency $n\omega$, where n is an integer and ω is the angular frequency.

$$v(t) = V_0 + \sum_{n=1}^{+\infty} V_n Cos(n\omega t + \psi_n) \tag{13.1}$$

$$i(t) = I_0 + \sum_{n=1}^{+\infty} I_n Cos(n\omega t + \psi'_n)$$

Average power can be written as

$$P_{average} = V_0 I_0 + \sum_{n=1}^{+\infty} \frac{1}{2} V_n I_n Cos(\psi_n - \psi'_n) \tag{13.2}$$

In terms of Fourier coefficients, the RMS values of voltage and current can be expressed as

$$V_{rms} = \sqrt{\frac{1}{2\pi} \int_0^{2\pi} v^2(t) d(\omega t)} = \sqrt{V_0^2 + \sum_{n=1}^{+\infty} \frac{1}{2} V_n^2}$$

$$I_{rms} = \sqrt{\frac{1}{2\pi} \int_0^{2\pi} i^2(t) d(\omega t)} = \sqrt{I_0^2 + \sum_{n=1}^{+\infty} \frac{1}{2} I_n^2} \tag{13.3}$$

In addition, power factor of this element is defined as the following:

$$Power\ Factor = \frac{P_{average}}{V_{rms} I_{rms}} \tag{13.4}$$

By assuming a sinusoidal voltage across the load, the current through a power electronic load, generally, is non-sinusoidal. Therefore, power factor can be written as the following:

$$Power\ Factor = \frac{I_{1,rms}}{I_{rms}} Cos(\psi_1 - \psi'_1) \tag{13.5}$$

Distortion and displacement factors are defined as

$$Distortion\ Factor = \frac{I_{1,rms}}{I_{rms}} \tag{13.6}$$

$$Displacement\ Factor = Cos(\psi_1 - \psi_1')$$

Most of the power electronic loads in AC distribution systems have a unity displacement factor. In the next section, we use this attribute to model AC constant power loads.

13.2 Modeling of AC Constant Power Loads

In this section, in order to study the effects of constant power loads in AC vehicular systems, a modeling approach considering small-signal variations around the operating points is presented [1]. Figure 13.3 shows a constant power load to be modeled with its voltage and current.

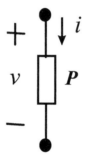

Figure 13.3 An AC constant power load.

Most of the AC power electronic loads have a rectifier at their front end. Therefore, generally, constant power loads are connecting to AC systems via a controlled or uncontrolled rectifier. Figure 13.4 shows a rectifier connecting two AC and DC subsystems. We assume that the DC subsystem behaves as a DC constant power load. These loads have been studied in detail in the previous chapter. However, in this chapter, we investigate the behavior of these loads from the AC subsystem point of view.

Figure 13.5 depicts typical input and output voltages and currents of the controlled rectifier of Figure 13.4 assuming the continuous conduction mode of operation. The firing angle of the rectifier is α. If the input inductance of the DC subsystem is large, the output current of the rectifier is almost constant, equal to its average.

Effects of Constant Power Loads in AC Vehicular Systems 469

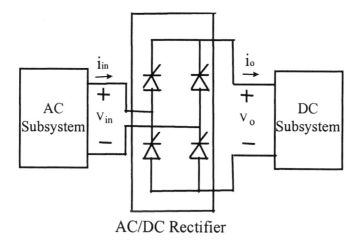

Figure 13.4 An AC/DC rectifier connecting a DC subsystem with constant power load to an AC subsystem.

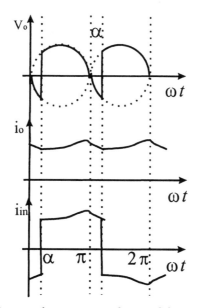

Figure 13.5 Voltage and current waveforms of the controlled rectifier of Figure 13.4.

13.2.1 AC Constant Power Loads with Diode Rectifier at the Front Stage

Assuming zero firing angle (i.e., $\alpha = 0$) for the rectifier of Figure 13.4, Figure 13.5 gives the voltage and current waveforms of an AC constant power load with diode rectifier at its front stage. The converter is in the continuous conduction mode of operation. The voltage of the constant power load is assumed to be sinusoidal. As is depicted in Figure 13.5, the current is not sinusoidal. By neglecting harmonics in this section, we consider only the fundamental component of the current. Therefore, voltage and current of the load can be written as

$$v(t) = V_{max}.Cos(\omega t)$$
$$i(t) = I_{max}.Cos(\omega t - \varphi) \tag{13.7}$$

Average power is

$$P = \frac{1}{2}V_{max}.I_{max}.Cos(\varphi) \tag{13.8}$$

For a diode rectifier, there is no phase difference between the voltage and the fundamental component of the current, assuming that the output current of the rectifier is constant. Therefore, average power is expressed as

$$\varphi = 0, \ Cos(\varphi) = 1$$
$$P = \frac{1}{2}V_{max}.I_{max}. = V_{rms}I_{rms} \tag{13.9}$$

In order to model this load, a small-signal perturbation in the amplitude of the voltage is considered. As a result, there will be a small-signal perturbation in the amplitude of the current through the load.

$$v(t) = \left(V_{max.} + \tilde{V}_{max.}\right)Cos(\omega t)$$
$$i(t) = \left(I_{max.} + \tilde{I}_{max.}\right)Cos(\omega t) \tag{13.10}$$

Since the average power is constant for the load, we can write

$$P = \frac{1}{2}\left(V_{max.} + \tilde{V}_{max.}\right)\left(I_{max.} + \tilde{I}_{max.}\right) = \frac{1}{2}V_{max.}I_{max.} \tag{13.11}$$

By neglecting the second order term, we can conclude

Effects of Constant Power Loads in AC Vehicular Systems

$$V_{max.} \tilde{I}_{max.} + I_{max.} \tilde{V}_{max.} = 0 \tag{13.12}$$

This relation, which specifies the small signal behavior of the load, can be rewritten as

$$\frac{\tilde{V}_{max.}}{\tilde{I}_{max.}} = -\frac{V_{max.}}{I_{max.}} = -R_{CPL} \tag{13.13}$$

R_{CPL} is defined as

$$R_{CPL} = \frac{V_{max.}}{I_{max.}} = \frac{V_{rms}^2}{P} \tag{13.14}$$

Therefore, the AC constant power load, considering small-signal variations, behaves as a negative resistance. The absolute value of this resistance is equal to the impedance of the constant power load at its operating point. Figure 13.6 shows the small-signal model of the load.

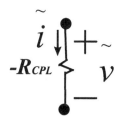

Figure 13.6 Small-signal model of AC constant power loads with diode rectifier at the front stage.

    This model is also valid for unity displacement converters. In these converters, having a unity displacement factor, as is defined in (13.6), means that the voltage across the load and the fundamental component of the current through the load are in phase. Therefore, the same model of Figure 13.6 is used for this category of AC power electronic loads. Generally, these converters have a Boost converter, as is shown in Figure 13.7, after the diode rectifier.
    In this section, we have assumed the continuous conduction mode of operation for the converters. However, some the power electronic converters are working is a discontinuous conduction mode or there is a Buck converter after the rectifier, as is shown in Figure 13.8. Therefore, the input current of the load is not continuous. Figure 13.9 shows the discontinuous current for the converter of Figure 13.8.

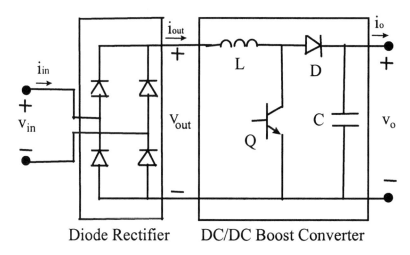

Figure 13.7 An AC power electronic load with a diode rectifier and a Boost converter at the front end.

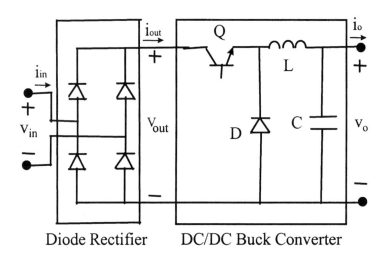

Figure 13.8 An AC power electronic load with a diode rectifier and a Buck converter at the front end.

If the voltage and fundamental component of the current are in phase, the small-signal model given in Figure 13.6 is valid. However, if there is a phase difference, the small-signal behavior of the load is inductive. We will explain this case in the next section.

Effects of Constant Power Loads in AC Vehicular Systems

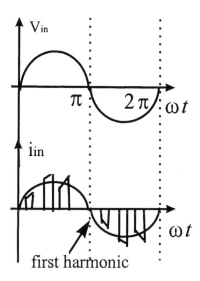

Figure 13.9 Input voltage and discontinuous current of the converter of Figure 13.8.

In our studies in this section, we have assumed that the fundamental component of the input current is dominant. In fact, we only considered the fundamental component of the current instead of the actual current waveform. However, other harmonics are also present in the waveform of the current. By assuming low ripples for the output current of the rectifier, the current of the constant power load is shown in Figure 13.10.

Fourier expansion of this non-sinusoidal current is given as

$$i_{in}(t) = \frac{2X_{max.}}{\pi} Sin(\omega t) + \frac{2X_{max.}}{3\pi} Sin(3\omega t)$$
$$+ \frac{2X_{max.}}{5\pi} Sin(5\omega t) + \ldots \quad (13.15)$$

In general, for a controlled rectifier with the firing angle φ, sinusoidal voltage and non-sinusoidal current can be written as

$$v(t) = V_{max.} Cos(\omega t)$$
$$i(t) = I_{max.} Cos(\omega t - \varphi) + \frac{I_{max.}}{3} Cos(3(\omega t - \varphi))$$
$$+ \frac{I_{max.}}{5} Cos(5(\omega t - \varphi)) + \cdots \quad (13.16)$$

Average power can be expressed as

$$P = \frac{1}{2}V_{max}.I_{max}.Cos(\varphi) = V_{rms}I_{rms}Cos(\varphi) \tag{13.17}$$

This is the same formula as (13.8). In fact, only the fundamental component of the current is participating for producing of real power. Therefore, our modeling approach is valid for non-sinusoidal currents, assuming sinusoidal voltages. However, in section 13.4, we present the generalized state space averaging method to study non-sinusoidal currents and voltages.

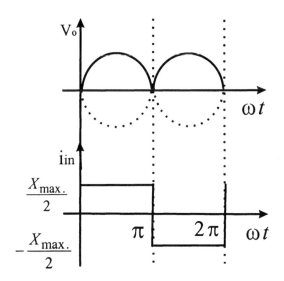

Figure 13.10 Square wave current of the constant power load.

13.2.2 AC Constant Power Loads with Controlled Rectifier at the Front Stage

We consider the continuous conduction mode of operation for the converter. As is discussed in the previous section, the fundamental component of the current is considered. Therefore, voltage and current of the load can be written as

$$v(t) = V_{max}.Cos(\omega t)$$
$$i(t) = I_{max}.Cos(\omega t - \alpha) \tag{13.18}$$

where α is the firing angle of the rectifier. Average power can be expressed as

Effects of Constant Power Loads in AC Vehicular Systems

$$P = \frac{1}{2} V_{max} . I_{max} . Cos(\alpha) = V_{rms} I_{rms} Cos(\alpha) \tag{13.19}$$

For small-signal modeling of the load, small-signal voltage and current perturbations are considered.

$$v(t) = \left(V_{max.} + \tilde{V}_{max.} \right) Cos(\omega t)$$
$$i(t) = \left(I_{max.} + \tilde{I}_{max.} \right) Cos(\omega t - \alpha) \tag{13.20}$$

Since the average power is constant for the load, we can write

$$P = \frac{1}{2}\left(V_{max.} + \tilde{V}_{max.}\right)\left(I_{max.} + \tilde{I}_{max.}\right) Cos(\alpha)$$
$$= \frac{1}{2} V_{max.} I_{max.} Cos(\alpha) \tag{13.21}$$

By neglecting the second order term, we can conclude

$$V_{max.} \tilde{I}_{max.} + I_{max.} \tilde{V}_{max.} = 0 \tag{13.22}$$

This relation, which specifies the small-signal behavior of the load, can be rewritten as

$$\frac{\tilde{V}_{max.}}{\tilde{I}_{max.}} = -\frac{V_{max.}}{I_{max.}} = -R_{CPL} \tag{13.23}$$

R_{CPL} is defined as

$$R_{CPL} = \frac{V_{max.}}{I_{max.}} = \frac{V_{rms}}{I_{rms}} = \frac{V_{rms}^2}{P} Cos(\alpha) \tag{13.24}$$

Therefore, the AC constant power load, considering small-signal variations, behaves as a negative impedance. The absolute value of this impedance is equal to the impedance of the constant power load at its operating point. Furthermore, there is a phase difference between the current and the voltage.

$$\tilde{v}(t) = \tilde{V}_{max.} Cos(\omega t)$$
$$\tilde{i}(t) = \tilde{I}_{max.} Cos(\omega t - \alpha) \tag{13.25}$$

Small-signal impedance of the constant power load can be expressed as

$$\tilde{Z}_{CPL} = -R_{CPL}\exp(j\alpha) = -R_{CPL}Cos(\alpha) - jR_{CPL}Sin(\alpha) \qquad (13.26)$$

Therefore, small-signal resistance and inductance of the load can be defined as

$$\tilde{r}_{CPL} = -R_{CPL}Cos(\alpha)$$
$$\tilde{L}_{CPL} = -\frac{R_{CPL}}{\omega}Sin(\alpha) \qquad (13.27)$$

Figure 13.11 shows the small-signal model of the load.

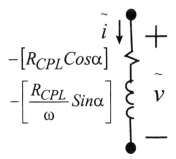

Figure 13.11 Small-signal model of AC constant power loads with controlled rectifier at the front stage.

13.3 Negative Impedance Instability Conditions

In this section, necessary and sufficient conditions for small-signal stability of AC distribution systems containing different constant power and constant voltage loads are presented. Based of the stability conditions, recommendations for the design of stable AC distribution systems are provided.

13.3.1 Constant Power Loads with Diode Rectifier at the Front Stage

Figure 13.12 shows the small-signal equivalent circuit of the AC distribution system of Figure 13.1. In fact, it is the small-signal equivalent circuit at a typical bus driving several constant power and constant voltage loads. *P* and *R* are representing the equivalent constant power and constant voltage loads, respectively. Constant voltage loads have the same large-signal characteristic. However, the impedance of constant power loads is the negative of their operating voltage impedance. In Figure 13.12, the source subsystem includes the equivalent circuits of generators, transmission lines, and cables as well as filters and distribution system at the input side of the bus.

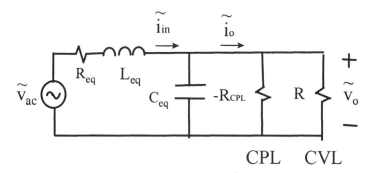

Figure 13.12 Small-signal equivalent circuit of system of Figure 13.1.

The small-signal transfer function of the circuit of Figure 13.12 is expressed by

$$\frac{\tilde{v}_o}{\tilde{v}_{ac}} = \frac{\frac{1}{L_{eq.}C_{eq.}}}{s^2 + \left[\frac{L_{eq.} + \frac{R_{eq.}C_{eq.}RR_{CPL}}{R_{CPL}-R}}{L_{eq.}C_{eq.}\frac{RR_{CPL}}{R_{CPL}-R}}\right]s + \left[\frac{R_{eq.} + \frac{RR_{CPL}}{R_{CPL}-R}}{L_{eq.}C_{eq.}\frac{RR_{CPL}}{R_{CPL}-R}}\right]} \quad (13.28)$$

By considering the condition for which the poles of the system have negative real parts, necessary and sufficient condition for small-signal stability is determined by

$$R_{CPL} > \left(R \parallel R_{eq.}\right) \quad (13.29)$$

and

$$P_{CPL} < P_{CVL} + \frac{R_{eq.}C_{eq.}}{L_{eq.}}V_{o,rms}^2 \quad (13.30)$$

Since $R_{eq.}$ is generally very small, constraint (13.29) is satisfied. Therefore, for stability of the system, satisfying constraint (13.30) is important. The effect of the source subsystem and its distribution system is shown in the second term at the right hand side of (13.30). It is related to the equivalent circuit parameters as well as the output nominal RMS voltage. Therefore, in order to ensure the small-signal stability for the distribution system, the power of constant power loads must be less than the power of constant voltage loads plus $\frac{R_{eq.}C_{eq.}}{L_{eq.}}V_{o,rms}^2$. If this second term is greater than the power of constant power loads itself, the system is stable even without any resistive load. This is an improved stability condition. For example, for the parameters given in Table 13.1 for a typical distribution system, the second term in the right-hand side of relation (13.30) is 13.225 kW. This means that, without any constant voltage load, the system is stable if the power of constant power loads is less than 13.225 kW. However, this power may be much less than the power of loads in a stable system since the main constraint for stability is coming from the difference between the power of constant power loads and constant voltage loads. This difference must be less than 13.225 kW in order to ensure the stability of the system.

Table 13.1 An example of the parameters of the source subsystem.

$R_{eq.}$	$C_{eq.}$	$L_{eq.}$	$V_{o,rms}$
100 mΩ	10 mF	1 mH	115 V

Similar to DC distribution systems, in order to stabilize the system, based on relation (13.30), there are four different possible solutions. The first one is increasing $R_{eq.}$, which is not practical due to the power loss. The second one is decreasing $L_{eq.}$, which is not feasible also. However, in the design stage of the system, it should be considered a priority to have as low an inductance as possible. The third method is increasing $C_{eq.}$, which is easily possible by adding a filter. The last method is increasing $V_{o,rms}$. In most cases, there is no control over the nominal voltages of the system. However, systems with higher base voltages have better negative impedance stability than systems with lower base voltages.

13.3.1.1 Practical Power Electronic Loads

As was explained, the constant power sink behavior of a power electronic converter occurs when the output of the converter is tightly regulated. In fact, fast-response, low-output ripple converters have negative impedance characteristics. Power electronic converters with an open-loop controller or with a poorly performing closed-loop controller do not show the constant power sink behavior.

Figure 13.13 shows a DC/DC Buck converter driving a separately excited DC motor with a mechanical load having a linear torque-speed characteristic. The duty of the closed loop control is to maintain the speed fixed. Assuming a high performance, fast response, and tightly regulated controller, the speed is fixed even if there are perturbations at the input side of the DC/DC converter. This is the constant power sink behavior. However, in most of the practical cases, these are poor controllers for the drives or voltage regulators since high

Effects of Constant Power Loads in AC Vehicular Systems

quality converters are expensive. Open-loop controllers as well as poor closed-loop controllers do not guarantee fixed speed. In fact, the speed changes if there is a perturbation at the input side of the converter. Therefore, torque changes accordingly, and so does power. As a result, the converter does not behave as a constant power load. This is the main reason that most of the practical systems, such as uninterruptable power supplies (UPS) [2], don't have negative impedance instability problem.

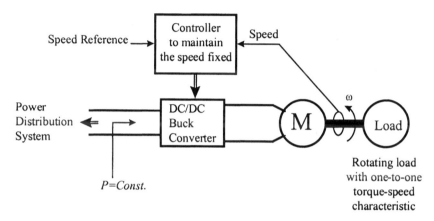

Figure 13.13 A tightly regulated DC/DC Buck converter that sinks constant power from the distribution system.

For example, consider a separately excited DC motor connected to a DC distribution system. Figure 13.14 shows the equivalent circuit.

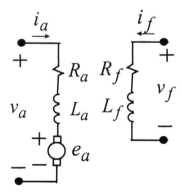

Figure 13.14 Equivalent circuit of a separately excited DC machine.

Voltage and torque equations of this DC machine are as follows:

$$v_a = R_a i_a + L_a \frac{di_a}{dt} + e_a$$

$$T = T_L + J \frac{d\omega_m}{dt}$$

$$e_a = (K\phi)\omega_m \quad (13.31)$$

$$T = (K\phi)i_a$$

$$\phi = \text{constant}$$

The torque-speed characteristic of the load is assumed as

$$T_L = K'\omega_m \quad (13.32)$$

We have simulated the DC motor using the parameters given in Table 13.2. Figure 13.15 shows the simulation results considering an input voltage step change from 220 V to 240 V at 40 sec.

Table 13.2 Parameters of the DC machine and its load.

Power	Voltage	Current	Speed	R_a	L_a	J	K'
35 kW	220 V	175 amp	1000 rpm	0.08 Ω	0.12 H	8 kg-m²	2 Nm/(rad/sec.)

As is depicted in Figure 13.15, a step change in input voltage produces a transient and, during this transient, input power is not constant. In fact an increase in input voltage produces an increase in input current. Therefore, the load has a positive incremental impedance characteristic. This is not a destabilizing effect like the negative incremental impedance characteristic. However, a tightly regulated speed controller maintains a fixed speed even in the presence of voltage changes at the input side. Therefore, the power is fixed. As a result, an increase in input voltage produces a decrease in input current, that is, negative incremental characteristic.

13.3.2 Constant Power Loads with Controlled Rectifier at the Front Stage

Figure 13.16 shows the small-signal equivalent circuit of the AC distribution system of Figure 13.1. We have assumed that the constant power loads in Figure 13.1 have controlled rectifiers at their front stages.

The small-signal transfer function of the circuit of Figure 13.12 is expressed by

Effects of Constant Power Loads in AC Vehicular Systems

$$\frac{\tilde{v}_O}{\tilde{v}_{ac}} = \frac{RR_{CPL}\left(Cos(\alpha) + s\frac{Sin(\alpha)}{\omega}\right)}{As^3 + Bs^2 + Cs + D} \tag{13.33}$$

where

$$A = \frac{L_{eq.}C_{eq.}RR_{CPL}Sin(\alpha)}{\omega}$$

$$B = \frac{L_{eq.} + C_{eq.}RR_{eq.}}{\omega}R_{CPL}Sin(\alpha) + L_{eq.}C_{eq.}RR_{CPL}Cos(\alpha)$$

$$C = R_{eq.}\left(\frac{R_{CPL}Sin(\alpha)}{\omega} + C_{eq.}RR_{CPL}Cos(\alpha)\right) \tag{13.34}$$

$$+ L_{eq.}(R_{CPL}Cos(\alpha) - R) + \frac{RR_{CPL}Sin(\alpha)}{\omega}$$

$$D = R_{eq.}(R_{CPL}Cos(\alpha) - R) + RR_{CPL}Cos(\alpha)$$

By considering the condition for which the poles of the system have negative real parts, using the Routh-Hurwitz method [4], necessary and sufficient condition for small-signal stability is determined by

$$R_{CPL}Cos(\alpha) > (R \| R_{eq.}) \tag{13.35}$$

and

$$P_{CPL} < \lambda\left(P_{CVL} + \frac{R_{eq.}C_{eq.}}{L_{eq.}}V_{o,rms}^2\right) \tag{13.36}$$

where

$$\lambda = \frac{Cos(\alpha)}{\left(P_{CVL}\frac{Sin(\alpha)}{\omega} + C_{eq.}Cos(\alpha)V_{o,rms}^2\right)}\left[V_{o,rms}^2 C_{eq.}Cos^2(\alpha) + \frac{Sin(\alpha)}{\omega} \times \right.$$

$$\left.\left[\frac{R_{eq.}}{\omega L_{eq.}}\left(P_{CVL} + \frac{V_{o,rms}^2}{R_{eq.}}\right) + Cos(\alpha)\left(P_{CVL} + \frac{R_{eq.}C_{eq.}}{L_{eq.}}V_{o,rms}^2\right)\right]\right] \tag{13.37}$$

where α is the firing angle of the controlled rectifier. Considering zero firing angle, from (13.35) and (13.36), we get (13.29) and (13.30) which are for uncontrolled diode rectifiers. From these relations, it can be concluded that a system with diode rectifiers is more stable than a system with controlled rectifiers. Therefore, an approach to stabilize the system, in the case that the system goes to instability, may be decreasing the firing angle. By decreasing the firing angle, λ is increasing and, as a result, the right-hand side of relation (13.36) is increasing, which, in turn, gives a larger margin for stability.

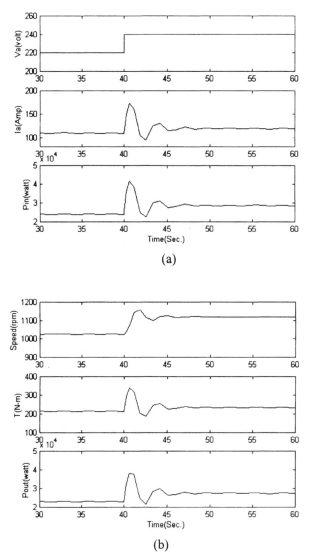

Figure 13.15 Dynamic response of the DC motor to 20V step change in input voltage: (a) input voltage, current, and power; (b) speed, torque, and output power.

Effects of Constant Power Loads in AC Vehicular Systems

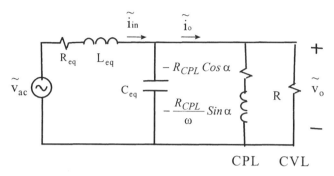

Figure 13.16 Small-signal equivalent circuit of system of Figure 13.1 with controlled rectifiers.

13.4 Hybrid (DC and AC) Vehicular Systems with Constant Power Loads

In this section, we consider DC and AC distribution systems connecting together via a controlled rectifier, as is depicted in Figure 13.17. In the AC side, the equivalent circuit of the source subsystem at the bus driving the power electronic load is considered. At the DC side, after the rectifier, there is a filter to reduce the ripples for the output DC bus. A constant power load and a constant voltage load representing the equivalent loads at the DC bus are connected to the output DC bus. In this section, we study the stability of the system using the generalized state space averaging method.

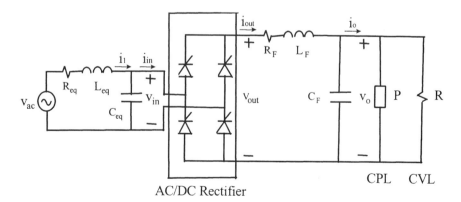

Figure 13.17 Interconnecting constant power and constant voltage loads in a hybrid DC and AC power system.

13.4.1 Modeling of Hybrid Systems Using Generalized State Space Averaging Method

As is explained in Chapter 7, the generalized state space averaging method is a large-signal approach for modeling of power electronic circuits. In order to use this method, we consider an equivalent circuit of the controlled rectifier shown in Figure 13.18.

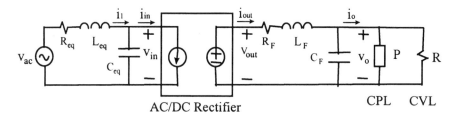

Figure 13.18 Equivalent circuit of the rectifier in the hybrid distribution system.

Input and output of the rectifier circuit in Figure 13.18 are defined as

$$i_{in} = u(t).i_{out}$$
$$v_{out} = u(t).v_{in}$$
(13.38)

where $u(t)$ is the commutation function, which is defined in Figure 13.19.

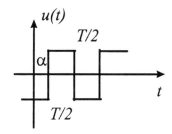

Figure 13.19 Commutation function.

Voltage and current equations of the circuit of Figure 13.18 can be written as

Effects of Constant Power Loads in AC Vehicular Systems

$$\begin{cases} v_{ac} = R_{eq}.i_1 + L_{eq}.\dfrac{di_1}{dt} + v_{in} \\[6pt] C_{eq}.\dfrac{dv_{in}}{dt} + i_{in} = i_1 \\[6pt] v_{out} = R_F i_{out} + L_F \dfrac{di_{out}}{dt} + v_o \\[6pt] C_F \dfrac{dv_o}{dt} + \dfrac{P}{v_o} + \dfrac{v_o}{R} = i_{out} \end{cases} \quad (13.39)$$

Equations (13.38), which show the interconnection between the DC and AC subsystems, along with the set of differential equations of (13.39) describe the behavior of the hybrid system.

13.4.1.1 First Harmonic Approximation

In the set of equations of the generalized state space averaged model, the actual state space variables are the Fourier coefficients of the circuit state variables, which are, in this case, i_1, v_{in}, i_{out}, and v_o. Using the first-order approximation to obtain these circuit state variables, we have twelve real state variables in the model, as follows [3]:

$$\begin{aligned} &<i_1>_0 = x_1 \\ &<v_{in}>_0 = x_2 \\ &<i_{out}>_0 = x_3 \\ &<v_o>_0 = x_4 \\ &<i_1>_1 = x_5 + x_6 \\ &<v_{in}>_1 = x_7 + x_8 \\ &<i_{out}>_1 = x_9 + x_{10} \\ &<v_o>_1 = x_{11} + x_{12} \end{aligned} \quad (13.40)$$

Assuming a sinusoidal voltage source, zero and first harmonics of this voltage source can be expressed as

$$\begin{aligned} v_{ac} &= V_m Sin(\omega t) \\ <v_{ac}>_0 &= 0 \\ <v_{ac}>_1 &= -j\dfrac{V_m}{2} \end{aligned} \quad (13.41)$$

Assuming low voltage ripples compared to the average value at the output DC bus of the hybrid system, zero and first harmonics of the nonlinear term $(1/v_o)$ can be written as

$$< \frac{1}{v_o} >_0 = \frac{1}{x_4}$$

$$< \frac{1}{v_o} >_1 = -\frac{x_{11}}{x_4^2} - j\frac{x_{12}}{x_4^2}$$

(13.42)

By applying the time derivative and convolution properties of the Fourier coefficients in (13.39), considering (13.38), and further substituting the Fourier coefficients of the source voltage and the commutation function $u(t)$, the generalized state space averaged model of the system for the zero firing angle of the rectifier can be written as

$$\dot{X} = \begin{bmatrix} \frac{-R_{eq}}{L_{eq}} & \frac{-1}{L_{eq}} & 0 & 0 & 0 & 0 & 0 & 0 & 0 & 0 & 0 & 0 \\ \frac{1}{C_{eq}} & 0 & 0 & 0 & 0 & 0 & 0 & 0 & \frac{4}{\pi C_{eq}} & 0 & 0 & 0 \\ 0 & 0 & \frac{-R_F}{L_F} & \frac{-1}{L_F} & 0 & 0 & 0 & \frac{-4}{\pi L_F} & 0 & 0 & 0 & 0 \\ 0 & 0 & \frac{1}{C_F} & \frac{-1}{RC_F} & 0 & 0 & 0 & 0 & 0 & 0 & 0 & 0 \\ 0 & 0 & 0 & 0 & \frac{-R_{eq}}{L_{eq}} & \omega & \frac{-1}{L_{eq}} & 0 & 0 & 0 & 0 & 0 \\ 0 & 0 & 0 & 0 & -\omega & \frac{-R_{eq}}{L_{eq}} & 0 & \frac{-1}{L_{eq}} & 0 & 0 & 0 & 0 \\ 0 & 0 & 0 & 0 & \frac{1}{C_{eq}} & 0 & 0 & \omega & 0 & 0 & 0 & 0 \\ 0 & 0 & \frac{2}{\pi C_{eq}} & 0 & 0 & \frac{1}{C_{eq}} & -\omega & 0 & 0 & 0 & 0 & 0 \\ 0 & 0 & 0 & 0 & 0 & 0 & 0 & 0 & \frac{-R_F}{L_F} & \omega & \frac{-1}{L_F} & 0 \\ 0 & \frac{-2}{\pi L_F} & 0 & 0 & 0 & 0 & 0 & 0 & -\omega & \frac{-R_F}{L_F} & 0 & \frac{-1}{L_F} \\ 0 & 0 & 0 & 0 & 0 & 0 & 0 & 0 & \frac{1}{C_F} & 0 & \frac{-1}{RC_F} & \omega \\ 0 & 0 & 0 & 0 & 0 & 0 & 0 & 0 & 0 & \frac{1}{C_F} & -\omega & \frac{-1}{RC_F} \end{bmatrix} X + \begin{bmatrix} 0 \\ 0 \\ 0 \\ -\frac{P}{C_F}\frac{1}{x_4} \\ 0 \\ -\frac{V_m}{2L_{eq}} \\ 0 \\ 0 \\ 0 \\ 0 \\ \frac{P}{C_F}\frac{x_{11}}{x_4^2} \\ \frac{P}{C_F}\frac{x_{12}}{x_4^2} \end{bmatrix}$$

(13.43)

where X is the vector of twelve real state variables as are defined in (13.40). Due to the nonlinear behavior of constant power loads, the model of the system is nonlinear. However, after linearizing the equations (13.43) around the operating point, a linearized model to study the small-signal stability of the system can be expressed as

Effects of Constant Power Loads in AC Vehicular Systems

$$\dot{\tilde{X}} = \begin{bmatrix}
\frac{-R_{eq}}{L_{eq}} & \frac{-1}{L_{eq}} & 0 & 0 & 0 & 0 & 0 & 0 & 0 & 0 & 0 \\
\frac{1}{C_{eq}} & 0 & 0 & 0 & 0 & 0 & 0 & 0 & \frac{4}{\pi C_{eq}} & 0 & 0 \\
0 & 0 & \frac{-R_F}{L_F} & \frac{-1}{L_F} & 0 & 0 & 0 & \frac{-4}{\pi L_F} & 0 & 0 & 0 \\
0 & 0 & \frac{1}{C_F} & \frac{-1}{RC_F}+\frac{P}{C_F V_o^2} & 0 & 0 & 0 & 0 & 0 & 0 & 0 \\
0 & 0 & 0 & 0 & \frac{-R_{eq}}{L_{eq}} & \omega & \frac{-1}{L_{eq}} & 0 & 0 & 0 & 0 \\
0 & 0 & 0 & 0 & -\omega & \frac{-R_{eq}}{L_{eq}} & 0 & \frac{-1}{L_{eq}} & 0 & 0 & 0 \\
0 & 0 & 0 & 0 & \frac{1}{C_{eq}} & 0 & 0 & \omega & 0 & 0 & 0 \\
0 & 0 & \frac{2}{\pi C_{eq}} & 0 & 0 & \frac{1}{C_{eq}} & -\omega & 0 & 0 & 0 & 0 \\
0 & 0 & 0 & 0 & 0 & 0 & 0 & \frac{-R_F}{L_F} & \omega & \frac{-1}{L_F} & 0 \\
0 & \frac{-2}{\pi L_F} & 0 & 0 & 0 & 0 & 0 & -\omega & \frac{-R_F}{L_F} & 0 & \frac{-1}{L_F} \\
0 & 0 & 0 & \frac{-2PX_{11}}{C_F V_o^3} & 0 & 0 & 0 & \frac{1}{C_F} & 0 & \frac{-1}{RC_F}+\frac{P}{C_F V_o^2} & \omega \\
0 & 0 & 0 & \frac{-2PX_{12}}{C_F V_o^3} & 0 & 0 & 0 & 0 & \frac{1}{C_F} & -\omega & \frac{-1}{RC_F}+\frac{P}{C_F V_o^2}
\end{bmatrix} \tilde{X} + \begin{bmatrix} 0 \\ 0 \\ 0 \\ 0 \\ 0 \\ \frac{-1}{2L_{eq}} \tilde{v}_m \\ 0 \\ 0 \\ 0 \\ 0 \\ 0 \\ 0 \end{bmatrix}$$

(13.44)

The small-signal stability of the system can be studied using the set of linear differential equations of (13.44). As an example, we have considered a system with the parameters given in Table 13.3.

Table 13.3 Parameters of the hybrid DC and AC power system.

Power	Voltage	Current	Speed	R_a	L_a	J	K'
35 kW	220 V	175 amp	1000 rpm	0.08 Ω	0.12 H	8 kg-m²	2 Nm/(rad/sec.)

We have simulated the hybrid system using the set of differential equations of (13.44). For different values of constant power and constant voltage loads, the stability of the system has been studied. Necessary and sufficient condition for small-signal stability is determined by considering the condition for which the eigenvalues of the system matrix in (13.44) have negative real parts. Figure 13.20 shows the stable and unstable regions. Necessary and sufficient condition for stability is held only in the stable region.

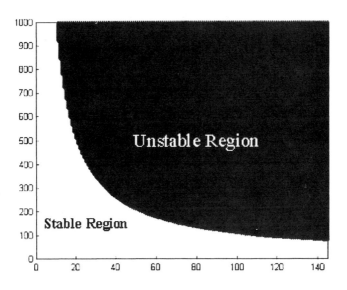

Figure 13.20 Stable and unstable regions of the system of Figure 13.17 in the $P(kW)$-$R(\Omega)$ plane based on the first-order model.

13.4.1.2 Neglecting DC Values at the AC Side and Ripples at the DC Side

In this section, we neglect the ripples and consider only the average values for the DC subsystem. In addition, for the AC subsystem, we only consider the fundamental components of the signals and neglect the DC components as well as harmonics. Therefore, for the system of Figure 13.17, there are six real state variables in the model, as follows:

$$<i_{out}>_0 = x_3$$
$$<v_o>_0 = x_4$$
$$<i_1>_1 = x_5 + x_6 \quad (13.45)$$
$$<v_{in}>_1 = x_7 + x_8$$

Again, by applying the time derivative and convolution properties of the Fourier coefficients in (13.39) considering (13.38), and further substituting the Fourier coefficients of the source voltage and the commutation function $u(t)$, the generalized state space averaged model of the system for the zero firing angle of the rectifier can be written as

Effects of Constant Power Loads in AC Vehicular Systems

$$\dot{\tilde{X}} = A\tilde{X} + B\tilde{v}_m$$

$$\tilde{X} = \begin{bmatrix} x_3 \\ x_4 \\ x_5 \\ x_6 \\ x_7 \\ x_8 \end{bmatrix}, \quad A = \begin{bmatrix} -\dfrac{R_F}{L_F} & -\dfrac{1}{L_F} & 0 & 0 & 0 & -\dfrac{4}{\pi L_F} \\ \dfrac{1}{C_F} & -\dfrac{1}{RC_F} + \dfrac{P}{C_F}\dfrac{1}{V_o^2} & 0 & 0 & 0 & 0 \\ 0 & 0 & -\dfrac{R_{eq.}}{L_{eq.}} & \omega & -\dfrac{1}{L_{eq.}} & 0 \\ 0 & 0 & -\omega & -\dfrac{R_{eq.}}{L_{eq.}} & 0 & -\dfrac{1}{L_{eq.}} \\ 0 & 0 & \dfrac{1}{C_{eq.}} & 0 & 0 & \omega \\ \dfrac{2}{\pi C_{eq.}} & 0 & 0 & \dfrac{1}{C_{eq.}} & -\omega & 0 \end{bmatrix}, \quad B = \begin{bmatrix} 0 \\ 0 \\ 0 \\ -\dfrac{1}{2L_{eq.}} \\ 0 \\ 0 \end{bmatrix}$$

(13.46)

Stability of the system can also be studied using the set of linear differential equations (13.46). We have simulated the hybrid system considering the parameters given in Table 13.3. For different values of constant power and constant voltage loads, the stability of the system has been studied. Figure 13.21 shows the stable and unstable regions.

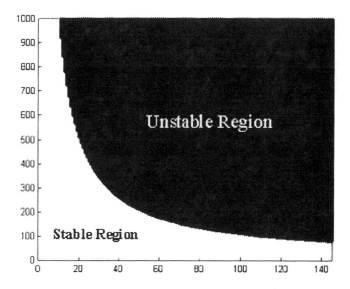

Figure 13.21 Stable and unstable regions of the system of Figure 13.17 in the $P(kW)$-$R(\Omega)$ plane based on the reduced model.

The stable region in Figure 13.21 is a bit larger than the stable region in Figure 13.20. This means that the reduced model cannot guarantee the stability of the system for the operating points very close to the margin of the stable region. However, in order to ensure the stability, we have to make sure that the combination of constant power and constant voltage loads never goes to the unstable region of Figure 13.20.

A simpler approach to get an estimate of the stability condition is considering the stability of AC and DC subsystems individually. However, this method is not precise and cannot ensure the system stability for the combinations of constant power and constant voltage loads close to the margin. By assuming a constant voltage at the output of the rectifier, since $V_o^2 R_F C_F / L_F = 200\,\text{W}$, the DC subsystem is stable provided the power of constant power loads is less than the power of constant voltage loads plus 200 W. This is with the assumption that the AC subsystem is stable. For the AC subsystem, if we neglect R_F and consider a constant zero firing angle for the rectifier, since $V_{in,rms}^2 R_{eq} C_{eq} / L_{eq} = 13.225\,\text{kW}$, it is stable if the power of the constant power loads is less than the power of constant voltage loads plus 13.225 kW. Therefore, the AC subsystem has a better stability condition than the DC subsystem. As a result, the stability condition of the system is the stability condition of the DC subsystem. In fact, the DC subsystem is going to the negative impedance unstable condition sooner than the AC subsystem. It should be mentioned again that this is an estimate of the stability condition. The precise condition is determined by the generalized state space averaging method as is depicted in Figure 13.20. By considering more harmonics, the precision of the method increases.

13.5 Conclusion

In this chapter, we have shown that some of the loads in AC vehicular distribution power systems can destabilize the system. These loads are tightly regulated power electronic converters and motor drives. Most of these loads have a controlled or uncontrolled rectifier at the front end. The concept of negative impedance characteristics of these AC constant power loads has been described. Furthermore, different AC constant power loads have been investigated and modeled. In addition, effects of these loads in conventional AC distribution systems were studied and necessary and sufficient conditions for small-signal stability have been provided. Based on this study, to ensure the small-signal stability, the power of constant power loads must be less than a fraction, which is a function of system parameters, of the power of constant voltage loads plus an additional term which depends on the system parameters. Furthermore, the generalized state space averaging method has been proposed to investigate the negative impedance instability in hybrid AC and DC vehicular distribution systems. The stable region for a typical system has been presented.

Finally, recommendations were given for designing AC power systems to avoid negative impedance instability.

13.6 References

[1] A. Emadi, "Modeling, analysis, and stability assessment of multi-converter power electronic systems," *Ph.D. Dissertation*, Texas A&M University, College Station, TX, Aug. 2000.

[2] S. B. Bekiarov and A. Emadi, "Uninterruptible power supplies: classification, operation, dynamics, and control," in *Proc. 17th Annual IEEE Applied Power Electronics Conference,* Dallas, TX, March 2002.

[3] A. Emadi, "Modeling of power electronic loads in AC distribution systems using the generalized state space averaging method," in *Proc. IEEE 27th Industrial Electronics Conference,* Denver, Colorado, Nov./Dec. 2001.

[4] J. E. Slotine and W. Li, *Applied Nonlinear Control*, Prentice-Hall, Upper Saddle River, NJ, 1991.

Index

A

AC/DC rectifiers, 16-24
  single-phase, full-wave, controlled rectifiers, 22
  single-phase, full-wave, uncontrolled rectifiers, 20
  single-phase, half-wave, controlled rectifiers, 18
  single-phase, half-wave, uncontrolled rectifiers, 16
Aerospace
  modeling, 257-272
  power generation, 234
  stability, 237, 282-289
  state estimation, 272-282
Alternator, 78, 84
Automotive
  communication networks, 138-183
  distribution system, 70
  semiconductors, 132-138
  steering systems, 120-131
  wireless techniques, 138

B

Back-up power supply, 373
Battery, 386
  ideal model, 386
  linear model, 387
  Thevenin model, 387
Bluetooth, 164-182
Brushless DC motor drive
  analysis, 404
  control, 404
  interior-mounted, 400
  machine structure, 399
  sensorless techniques, 407
  sinusoidal shape back-EMF, 404
  surface-mounted, 399
  trapezoidal shape back-EMF, 401

C

Centralized digital controller, 300
Cogeneration, 365
Constant power loads, 428
Control strategies, 198
Control systems, 8
Controller area network, 140145
  CAN error detection, 142
  CAN message arbitration, 141
  CAN message format, 141
  time triggered CAN, 145

Cost analysis, 112
Cyclo-converter, 320

D

DC/AC inverters, 39-47, 266
  current source, 315
  load commutated, 318
  pulse-width modulated inverters, 43, 319
  square-wave, full-bridge inverters, 41
  voltage source, 315
DC/DC converters, 25-38
  Boost converter, 30, 264
  Buck converter, 26, 261
  Buck-Boost converter, 34, 265
DC electrical power systems, 13
DC machines, 55
  compound, 59
  separately excited, 55
  series, 57
  shunt, 56
  slotless armature, 311
Digital systems, 7
Distributed digital controller, 301
Distribution systems
  AC vehicular systems, 465
  advanced automotive systems, 77
  aircraft, 235-237
  automotive, 70
  hybrid, 483
  hybrid electric vehicles, 195
  sea and undersea vehicles, 304
  zonal, 305-307

E

Earth orbiting system, 253
Electric circuits, 1
  RC circuit, 3
  RL circuit, 3
  RLC circuit, 3
Electric dragsters, 219-224
Electric machines, 49-65

Electric vehicles, 189
Electrical loads
  advanced aircraft loads, 232
  advanced automotive loads, 72
  aircraft, 232
  automotive, 72
  sea and undersea, 307
Electrical power systems, 1, 11-13
  14V electrical system, 70
  42V electrical system, 73
  42V electrical system for traction, 208-211
  AC zonal electrical systems, 305
  aircraft, 231-238
  automotive, 67
  DC zonal electrical systems, 306
  integrated, 296
  sea and undersea vehicles, 295-309
  segregated, 295
  space, 241-251
  spacecraft, 251-257
Electromagnetism, 53
Electromechanical systems, 49

F

Feedback control, 9
First harmonic approximation, 485
Flywheel, 249
Fourier series, 7
Frequency domain analysis, 10
Fuel cell, 335
  aerospace applications, 362
  alkaline, 343-345
  direct methanol, 342
  distributed generation, 365
  electrical characteristics, 352
  heavy-duty vehicles, 356
  light-duty vehicles, 353
  modeling, 388
  molten carbonate, 340-341
  operation, 335
  phosphoric acid, 337-338
  portable applications, 369
  power processing system, 346

Index

properties, 335
proton exchange membrane, 338-340
solid-oxide, 341-342
structures, 335
telecommunication applications, 376
utility applications, 364-369
zinc air, 345

G

Gauss-Seidel load flow, 12
Generalized state space averaging, 259

H

Heavy-duty vehicles
  conventional, 212
  fuel cell, 216, 356
  hybrid electric, 214
Honda Insight, 204
Hybrid control strategies
  charge depleting, 201-202
  charge sustaining, 200, 202
  parallel, 200-202
  practical models, 204-206
  series, 202-204
Hybrid electric vehicles, 189
  drivetrain architectures, 194
  drivetrain principles, 190-194
  electrical distribution systems, 195
  more electric hybrid vehicles, 197
Hybridization, 206-208

I

IEEE 1394 protocol, 153-156
Induction machines, 59, 91, 99
Integrated power modules, 136
Integrated starter/alternator, 87-90
  coupling configurations, 115-119
  direct coupling, 117
  offset coupling, 115
International space station, 242

K

Kirchhoff's law, 1

L

Laplace transform, 4
Lightly hybridized vehicles, 209-211
Local interconnect network, 161

M

Matrix converter, 321-325
Media oriented systems transport, 156
Modeling
  AC constant power loads, 468
  aerospace power systems, 257-272
  automotive power systems, 224-229
  energy storage devices, 385-397
  multi-converter, 275
Motor drives
  adjustable speed, 314
  current source, 315
  cyclo-converters, 320
  DC, 331
  drives in sea and undersea vehicles, 309
  multi-level, 325
  variable frequency, 315
  voltage source, 315
Multi-converter systems, 260, 269
Multi-level inverter, 325-331
  cascaded single-phase H-bridge, 327
  diode clamped, 325
  flying capacitor, 326
  hybrid H-bridge, 328
  modulation technique, 330

N

Negative impedance instability, 430, 435, 453, 476

Index

O

Ohm's law, 1

P

Permanent magnet machines, 91-95, 100, 312
  interior-mounted, 400
  surface-mounted, 399
Per-unit quantity, 11
Phasors, 6
Photovoltaic (PV) cells, 390
Power electronics, 15-47, 132, 367
  power electronics building blocks, 298-300
Power management center, 303
Power network models, 11
Power steering, 123-126
  electric power steering, 126-127

R

Routh-Hurwitz criterion, 10
Russian ISS segment, 248

S

SAE J1850, 150-153
Satellite, 254
Sea vehicles, 295
Semiconductor device technology, 134-136
Sensorless operation, 407
Signal-flow graphs, 8
Sinusoidal excitation, 6
Sliding-mode control, 440
Spacecraft, 251
Stability analysis, 10, 237, 282-289, 444, 456
Stabilizing controller, 440
Starter, 78, 81
State estimation, 272-282
State space averaging, 259, 440, 484
State space description, 11
Steer-by-wire, 127-132

Storage systems
  battery, 386
  energy storage devices, 385-397
  low-voltage, 211
  ultracapacitor, 393
Super conducting, 312
Switched reluctance machines, 95-99, 107, 413
  basic operation, 413
  control techniques, 417
  power electronic driver, 419
  torque-speed characteristics, 417
Synchronous machines, 63

T

TERRA spacecraft, 253
Time domain analysis, 10
Torque-speed characteristics, 61, 64
Toyota Prius, 204

U

Ultracapacitors, 393-397
  battery/ultracapacitor, 394
  double layer model, 393
Undersea vehicles, 295
US ISS segment, 248

V

Vehicles
  air, 231
  dynamics and control, 425-462
  fuel cell, 353-356
  land, 67, 189
  sea, 295
  space, 241
  undersea, 295

W

Weighted least squares, 279

X

X-by-wire, 159